SIGHTHILL LIBRARY

AF836329

Small Manufacturing Enterprises

WITHDRAWN

A World Bank Research Publication

NAPIER POLYTECHNIC LIBRARY

3 8042 00122 4375

Small Manufacturing Enterprises

A Comparative Study of India and Other Economies

Ian M. D. Little
Dipak Mazumdar
John M. Page, Jr.

Published for The World Bank
Oxford University Press

NAPIER POLYTECHNIC LIBRARY
AC / / ON 0063
LOC S SHMK 338.6420954 LIT SUP TCF
CON PR 17-50

Oxford University Press

NEW YORK OXFORD LONDON GLASGOW
TORONTO MELBOURNE WELLINGTON HONG KONG
TOKYO KUALA LUMPUR SINGAPORE JAKARTA
DELHI BOMBAY CALCUTTA MADRAS KARACHI
NAIROBI DAR ES SALAAM CAPE TOWN

© 1987 The International Bank for
Reconstruction and Development / THE WORLD BANK
1818 H Street, N.W., Washington, D.C. 20433, U.S.A.

All rights reserved. No part of this publication
may be reproduced, stored in a retrieval system,
or transmitted in any form or by any means,
electronic, mechanical, photocopying, recording,
or otherwise, without the proper permission
of Oxford University Press.

Manufactured in the United States of America
Originally published October 1987
First paperback edition November 1988
Second paperback printing April 1989

The findings, interpretations, and conclusions
expressed in this study are the results of research supported
by the World Bank, but they are entirely those of the
authors and should not be attributed in any manner to
the World Bank, to its affiliated organizations,
or to members of its Board of Executive Directors
or the countries they represent.

Library of Congress Cataloging-in-Publication Data

Little, Ian Malcolm David.

Small manufacturing enterprises.
"Published for the World Bank."
Bibliography: p.
Includes index.
1. India—Manufactures. 2. Small business—India.
3. Developing countries—Manufactures. 4. Small business
—Developing countries. I. Mazumdar, Dipak, 1932– .
II. Page, John M., 1949– . III. International Bank
for Reconstruction and Development. IV. Title.
HD9736.I52L57 1987 338.6'42'0954 87-12365
ISBN 0-19-520619-3
ISBN 0-19-520779-3 (pbk.)

Contents

Preface

Although the development of small-scale enterprises has received increasing attention as an element of industrial policy in many developing countries, India is unique in both the extent and duration of its efforts to promote and protect small-scale firms. This promotion was, however, grafted onto an industrial policy designed primarily to foster the development of large-scale, capital-intensive industry. India therefore provides important lessons for other developing countries on how specific efforts to promote small enterprises affect the efficient growth of small manufacturing industry, and how industrial policies in general and the growth of large-scale manufacturing also affect small-enterprise development.

The rhetoric of industrial policy toward small-scale firms in India, as in other countries, tends to stress their role in the deconcentration of economic power as well as the creation of employment. In attempting to draw lessons from the Indian experience on the desirability of supply-side measures in favor of small firms, we sought in the present volume primarily to evaluate whether smaller enterprises used human and other resources more efficiently than larger ones. The best use of a country's resources involves more than the efficient allocation of a given stock of capital and labor. Savings and the training of entrepreneurs and workers may also influence and be influenced by the size distribution of firms. Thus we have attempted to measure both how well enterprises of varying sizes utilized productive resources and how their use of resources changed over time. From this perspective we hoped to determine whether small firms were more likely than large firms to contribute simultaneously to the growth of both output and employment, and if so why.

This book is one product of a World Bank research project on small-scale enterprise development. Other principal products of the project are Anderson and Khambata (1981); Anderson (1982a, 1982b); Cortes, Berry, and Ishaq (1987); Ho (1980); Mazumdar (1984a); and Page (1979, 1984). The project was initiated by Ian M. D. Little and managed first by Dennis

Anderson and then by Dipak Mazumdar. Armando Pinell-Siles contributed to the Colombian component of the research.

A major feature of the Bank's research was the design and execution of special surveys of industrial firms across all size ranges within well-specified industrial sectors. In India these sectors included shoe manufacturing, printing, metal casting, laundry soap manufacturing, machine tool making, and textile weaving. The main surveys in India were organized and managed by K. B. Suri, who together with J. N. Sharma conducted the soap industry survey. The metal casting survey was carried out by Manoj Majumdar. The Sri Ram Centre for Industrial Relations, under the direction of the late S. M. Pandey, was responsible for the surveys of the printing, shoe, and machine tool industries. John M. Page, Jr., also participated in the machine tool and shoe survey work. The Giri Institute for Development Studies was responsible for the textile surveys (with Dipak Mazumdar participating). Research assistance in the handling and analysis of the data was provided by Deborah Bateman, Theodore Black, Heechol Chang, Eric Manes, and Anne Meyendorff.

We are indebted to our present and former colleagues, Dennis Anderson, Mark Leiserson, Mieko Nishimizu, William Steel, and Larry Westphal, for their support, useful discussion, and constructive criticism at many stages in the research and writing. The manuscript has been through the capable hands of many typists, but we are particularly grateful to Adrienne Guerrero for her continuing administrative and secretarial support.

1

Introduction

This book is primarily about claims made for small-scale enterprises (SSEs) in developing countries. Although the country range is not formally restricted, the book is rather more about SSEs in India than those in other countries. India was the biggest component in the World Bank research project on which the work is largely based. Furthermore, the non-Indian results of the research have been separately published or are otherwise available.[1] Detailed information derived from a number of enterprise surveys in India is presented here for the first time, as are more aggregated results from other Indian sources. However, the book is also intended as a general survey. It therefore draws not only on other components of the research project but also, where appropriate, on other recent research work on SSEs in both developed and developing countries. The Indian results are thus set in a world context, although that context is unavoidably sketchy.

Along with its Indian focus, the book centers on manufacturing enterprises. This is both because nonmanufacturing information is very scarce and because our own research surveys were confined to manufacturing on the ground that small nonmanufacturing enterprises in most developing countries are as yet little threatened by large ones. But this may well be a poor excuse, and the service and construction sectors could usefully be high on the agendas of surveying agencies.

With the exception of India, developing-country governments showed little concern for SSEs until at least the 1970s. They promoted industrialization in three main ways: by regulating trade; by using such investment incentives as interest rate manipulation and credit controls; and by undertaking public sector investments, often with foreign aid. All these discriminated against SSEs. The controlled-foreign-trade regime always discriminated against small firms, since large firms were better placed to obtain import

1. For a full list of the reports, see World Bank (1984). The more important reports are referred to in the preface.

permits for capital equipment, components, and raw materials and were also better able to obtain tariff rebates intended to alleviate some of the harmful effects of high protection. Investment incentive laws sometimes actively discriminated against small firms by restricting tax concessions to firms of some minimum (generally large) size. If there was no such overt discrimination, small firms nevertheless often were ignorant of available concessions or unable to handle the paperwork for the relevant public office. Selective credit controls combined with low controlled real interest rates prevented banks from compensating for the higher cost of small loans by charging more. The limited credit was therefore allocated to large clients. Finally, governments that directly promoted industrial investments favored the large project even when it would be owned and operated by private enterprise, whether foreign, local, or a joint venture.[2]

Bias can also arise because the problems of dealing with the tax authorities and with governmental regulations weigh more heavily on small firms, even in the absence of trade controls and industrial promotion.[3] But in this connection it is important to define "small" more precisely. It seems likely that the hand of the law and the tax collector may often fail to reach the very small firms (say, fewer than 10 workers) in most developing countries. When this is so, nonenforcement of taxes and of some regulations such as those governing minimum wages or working conditions constitutes a bias in favor of the very small (though this in turn may sometimes be outweighed by harassment of the small under zoning regulations in urban areas). Thus taxes and regulations are likely to weigh more heavily on the small modern factory employing 10–50 workers than on either the large factory or household and workshop activities. There is some supporting evidence for this. Small entrepreneurs or entrepreneurial families tend to multiply their firms.[4] The many-firm family may be the counterpart in developing countries of the multiplant firm in developed countries. One can even find two firms to one establishment—for example, the observed case of a son with a small enterprise who supplies uppers to another shoemaking enterprise owned by his father, both in the same shed.[5] A pilot survey in Bombay turned up an improbably large number of enterprises just below the limiting size for registration as a factory (11 employees).[6]

If there had been much concern for small-scale enterprises, ways might have been found to offset the disadvantages imposed upon them, even

2. A large number of reports testify to the biases described in this paragraph, including almost every one mentioned in this book; the biases seem to be nowhere contested. Consequently, we do not feel it is necessary to give references.

3. See, for instance, Lim, Lee, and Thye (1981, pp. 284–85).

4. See Page (1979, p. 38); also see Timberg (1978).

5. Staley and Morse (1965, p. 338) report a similar case.

6. Eighteen out of the twenty-eight enterprises with fewer than 11 employees had 8–10 workers. But this could be at least partly a sampling freak. (Timberg 1978, table 1).

without the major change to a rather free-market economy which would have removed the bias caused by the trade and industrialization regimes actually adopted. Indeed, in most countries, some such measures have been taken. Some credit institution has usually been set up to lend to small and medium-size enterprises (the medium-size ones being favored for the reasons already given). Some industrial extension schemes, management training institutes, and industrial estates (the latter not always mainly for the small) have been started in many countries, often supported by the United Nations Development Programme (UNDP), International Labour Organisation (ILO), or a bilateral aid agency. Many of these schemes have been expensive failures. The reach of all has invariably been extremely limited. Many surveys of SSEs in developing countries show that only a tiny proportion of those interviewed has ever been aided by such means or has even heard of the program intended for them. This applies also to the provision of institutional credit. Small firms (fewer than 50 employees) get very little of the credit reserved for medium-size and small firms; the very small (fewer than 10) almost never get any medium- or long-term loans. Except for India, which we examine at length in later chapters, and where some exceptional measures protect traditional industries, it is safe to conclude that the targeted measures in favor of small enterprises have negligible positive effect compared with the negative effect of the general economic policies of governments.

Limited concern for SSEs is readily understandable. Countries that wanted to acquire a lot of industry very fast and that started from a very small base with a very small (and in Africa a virtually nonexistent) entrepreneurial class had to create large units that were often in the public sector and also often owned or managed by foreigners. They did not want to wait for small men and their small firms to blossom and grow, although they might be a welcome adjunct to government-promoted industry. In any case, most of the small, locally owned industrial enterprises that did exist in Asia (except India) and Africa were of the craft style. As many observers have noted, the craftsperson is not usually able to develop into an entrepreneur manager of even a medium-size modern enterprise. It could well be that the seedbed for entrepreneurs could best be initially formed from persons trained in larger enterprises. This is not to say that countries should not be concerned to create favorable conditions for small and medium-size factories once the first round of industrialization is achieved, though it must also be remembered that the more socialistic developing countries may not want to encourage the formation of an independent private entrepreneurial class. This issue of the role of SSEs as both a training ground and a screening device for potential larger-scale entrepreneurs is addressed at various points in the book.

As will be seen in chapter 2, it has been stated rather dubiously that small enterprises are particularly innovative in the industrial countries. This argument cannot be pressed for the developing countries where, to a much greater extent than in the OECD (Organisation for Economic Co-operation and Development) countries, innovation consists, and should consist, of learning about, choosing, importing, absorbing, and quite often modifying foreign technology. Larger firms have considerable advantages in these respects, and the very small are often able to acquire modern technology only via larger enterprises.

Although the dynamic issues raised above concerning the role of SSEs are important, the central claim made for SSEs in developing countries is that they are good for employment. Any inquiry into the efficiency with which resources are used in pursuit of a country's ultimate aims is by definition an exercise in applied welfare economics. Thus both our investigation of the claim that SSEs have a special role in the promotion of efficient use of the labor force—an investigation that constitutes the core of this book—and much of the rest of our analysis are cast in this framework.

It was not until the late 1960s that international institutions and some private development propagandists stated that they had discovered massive and rising unemployment in developing countries. The ILO's World Employment Program was launched. The concern for employment was soon linked to advocacy of SSEs, which were believed to offer much more employment for any given investment. The ILO's Kenya mission took up and popularized the concept of the "informal" sector. Discussion of how it should be defined and its role has filled thousands of pages. But whatever it exactly was, it was labor-intensive, it provided goods and services by the poor for the poor, and it should be encouraged and not regarded as an unproductive slough which development would and should eliminate. We have not found the informal sector to be a concept of either analytic or operational value, and we do not use it.

Apart from the informal sector, there was much conceptual and analytic confusion as it came to be recognized that unemployment was not closely related to inequality or poverty and that the evidence of any general increase in unemployment was weak and even false. We need not follow the changes in stress from unemployment, to inequality, to poverty in international forums. It suffices to say that a consensus formed within the international development establishment in favor of a pattern of development that would be more demanding of unskilled labor and that the case for this pattern does not depend on evidence of either unemployment or underemployment.[7] Essentially the case is that while a minority of people work with a lot of capital (or land), a large majority work with very little indeed and thus can earn very little. If new investment called for more participating labor, then either more work would be done or real wages would rise. Either way, the unskilled (or initially unskilled) workers benefit.[8] It is by no means clear that

7. Anderson and Khambata (1981) argue strongly that in the Philippines insufficient demand for unskilled labor showed up in a fall in real wages, not in unemployment.

8. See, for example, Little (1982a, p. 214).

most developing-country governments are convinced of the merits of this case, although some countries (such as India and the Philippines among those we have studied) have in the 1970s strengthened their small-industry programs with the aim of relieving poverty and reducing inequality.

The main link between encouraging SSE development and reduced poverty and inequality is the increased demand for unskilled labor that is supposed to follow. But it is obvious that output cannot be neglected. For a good case to be made, it also has to be shown that output would at least not fall, for this could offset the hoped-for redistributive effect. Even better, it may be that SSEs employ labor and capital more efficiently, given their real social costs, than do large-scale enterprises (LSEs), so national output would rise as well as being distributed more equally.

Thus it is not enough to show that SSEs use more labor and no more capital per unit of output. If that were all, one could as well employ more men to do nothing in larger enterprises. Labor almost always and everywhere has some opportunity cost—that is, if not employed wherever it is, it would somehow earn something elsewhere. There is no escaping the fact that the static case for promoting any particular type of enterprise is that it uses factors more efficiently than do other types, given their social costs. The so-called employment case for SSEs rests on this—so there has to be at least a reasonable presumption of greater efficiency. However, there is a strong case *against* SSEs if they are shown to use both more capital and more labor: if such is the case, then the promotion of SSEs in order to increase the demand for labor implies that there is no way of increasing labor demand without reducing output—and this is clearly an unacceptable proposition.

Efficiency of course must be defined and assessed in social terms—taking account of the true opportunity costs of factors of production, not the market prices. It is often alleged that SSEs face an unjustifiably high supply price of capital in the informal credit market. This is offset to some extent by alleged distortions in the labor market, with large firms having to pay institutionally determined high wages. The coexistence of small and large firms in the same industry may be partly caused by widely different factor price ratios facing producers—with the small sector consequently opting for the more labor-intensive techniques. The removal, if possible, of distortions in factor markets is the first-best solution. If this is not feasible, there may be a case for special encouragement of SSEs, particularly if it is felt that capital is the binding constraint reducing their ability to take advantage of cheap labor. There may also be a distributional argument in favor of SSEs. The relatively very high wages that prevail in large firms in many countries imply that income distribution can be improved by encouraging small firms. But if this were all, the better policy would be to reduce real wages in the large-scale sectors. The issues raised in this paragraph are discussed at length, though rarely settled, in later chapters.

Suppose now that it has indeed been shown that SSEs both use more labor and are more socially efficient than large-scale enterprises (LSEs). The question may still be asked whether they should be promoted (since promotion may itself have some economic or political costs). Much now depends on the extent to which SSEs and LSEs are competitive. Consider two extremes. In the first, SSEs and LSEs make different things which are not good substitutes for each other. Both also are far from being able to compete in the other's market. In this case, reducing costs for SSEs could result in only a very small increase in their output and their demand for labor, at least in the domestic market. Exports might or might not come to the rescue. But in the home market any considerable increase in the demand for labor by SSEs would depend on twisting the pattern of demand in favor of their products. Redistribution might or might not accomplish this and would tend to do so only to the extent that SSEs produce things favored by the poor. Of course, in the opposite case where SSEs and LSEs are close competitors, producing highly substitutable products, a small change in the relative costs of production could eventually produce a large change in the size distribution of plants and enterprises and hence in the demand for labor. Thus to know whether there is much to be gained by a policy of favoring SSEs (or ceasing to disfavor them) requires disaggregated inquiry into the nature of markets because the degree of competitiveness between SSEs and LSEs will vary a great deal with the production sector (just as indeed the relative efficiency of small and large is also likely to vary widely). It was for these reasons that the World Bank's research focused on narrowly defined sectors of production. One might add that many of the failures of targeted approaches to aiding small enterprises, including the training of handicraft workers, seem to be attributable to a failure to ask whether an increased output—of products or of skilled workers—would find a market.

It might be objected that the above arguments have been based too exclusively on the demand for labor. Would not an increase in the productivity of SSEs be beneficial, even without any rise in labor demand? The answer is normally yes, if costlessly achieved; but then this is true also of LSEs. One should, however, also consider who is likely to get the benefits. If productivity is raised for a group of competing SSEs that have a very inelastic demand for their product, the beneficiaries will not be the owners of these enterprises, or their employees, or potential employees, but the consumers. In such cases, there is a special argument for SSEs only if SSEs have relatively more poor customers than LSEs because SSEs produce goods of lower quality and price. We believe that this is a subsidiary argument. No one has established its general validity, though in certain products it is doubtless true (and also true that some handicraft workers mainly serve the wealthy). Our research was not closely directed to the point, but where our surveys threw any light (necessarily very limited) on it we record the fact, and the relation of size and product quality is examined in chapter 13. But the main thrust of our inquiries was directed to the relative efficiency of small and large enterprises and to the problem of adequate labor demand.

We conclude this introductory survey of the issues with an outline of the book. Chapter 2 gives background information on the relative decline of small enterprises in both developing and industrial countries. It also discusses reasons for being concerned about the evolution of the industrial size structure and arguments for intervention. Chapters 3 and 4 are exclusively concerned with India. Indian industrial policies are described in chapter 3, and chapter 4 discusses their effects on the textile, sugar, and light engineering industries. Chapter 5 describes the size structure of Indian manufacturing in general and makes comparisons with other countries. Chapter 6 is concerned with changes in the Indian industrial structure. Chapter 7 opens with a brief statement of the relations between factor proportions and productivities that the theory leads one to expect and goes on to survey some estimates of these relations obtained from industrial censuses for such developing economies as India, the Republic of Korea, and Taiwan. Chapters 8 through 12 are about the surveys made in India that were the core of the research. A major reason for conducting the surveys was to obtain more and better information about inputs and outputs for narrowly defined industries than is obtainable from censuses in order to investigate in depth the relationship between size and the efficient employment of factors of production. The growth and profitability of firms were also investigated, and an attempt was made to explain the variations. Chapter 8 gives a general account of the surveys. Chapter 9 is about production theory and the econometric estimation of production relations (it can be skipped by any reader prepared to take this on trust). Chapter 10 gives the results for the variability of factor proportions and productivities with size and estimates the degree of substitutability between capital, skilled labor, and unskilled labor. Chapter 11 explores the relationship between technical efficiency and size. Chapter 12 relates growth and profitability to such explanatory variables as the age of the firm and various attributes of the entrepreneur.

With chapter 13 the tune changes. This and the following two chapters are about product and factor markets. Chapter 13 mainly probes two issues: first, the extent to which factor proportions may be partly explained by differences in the quality of the product rather than factor prices; and second, the extent to which markets have become segregated, with SSEs producing mainly for the poorer segment of the market. Chapter 14 is about labor markets, mainly in India but including some comparisons with other developing countries. Chapter 15 deals similarly with capital markets and the experience of lending to small borrowers. The rationale for including these chapters is that distortions in factor markets are often considered to be the main reasons, especially in developing countries, for special measures in favor of SSEs. Chapter 16 summarizes the book and gives some tentative conclusions.

2

Changes in Industrial Size Structure in Developing and Developed Countries

Statistics on changes over time in the size structure of industry in both developing and developed countries are hard to come by. But the share of very small establishments (fewer than 10 workers) in manufacturing, while remaining very high in very poor countries, is falling; and it is falling fast where the growth of manufacturing output is high.[1] In Europe and North America such small-scale manufacturing is almost extinct. The medium-size firm (50–99 workers) has suffered a much smaller relative decline in the developed countries, but no generalization is possible for the developing countries. Unfortunately, however, there is no way of measuring the extent to which these relative declines are caused by the biases discussed in the previous chapter; undoubtedly, the dominant cause is natural economic evolution induced by the growth of the market, scale economies, falling transport costs, and changes in factor prices.[2]

The Experience of Developing Economies

Before we turn to a survey of available statistics, let us be more precise about the meaning of "small," "medium-size," and "large." For developing economies, "small" firms will generally mean those with fewer than 50 workers, and the "very small" those with fewer than 10 workers. The term "cottage shop" will be used for enterprises with fewer than 5 workers that manufacture at home (household manufacturing) or in small workshops. "Medium-size" firms will usually mean those with 50–99 workers. The data preclude any rigid adherence to these numbers, but variations will be noted. For

1. In the case of developing countries, we shall use the terms "establishment," "plant," "enterprise," and "firm" synonymously. The vast majority of establishments are independent enterprises.

2. A much fuller discussion of this evolution is in Cortes, Berry, and Ishaq (1987, chap. 1).

India, small firms are often defined in terms of capital employed, and details of this definition will be recorded where appropriate. We now turn to the fate of SSEs in those economies for which studies were commissioned by the research project—Taiwan, Korea, the Philippines, Colombia, and India.

In Taiwan, there is evidence that employment in very small (nonfactory) manufacturing establishments with fewer than 5 workers has been falling since 1915. It halved between then and 1940 and fell from three-quarters to one-quarter as a proportion of total manufacturing employment (Ho 1980, table 2-2). In the postwar period, small factory manufacturing employment (4–49 workers) fell from 45 to 26 percent of the total between 1954 and 1971; almost all the decline took place in the period of extremely rapid labor-intensive growth from 1961 to 1971. Medium-size establishments (50–99 workers) held their own, while firms with 100–499 workers increased their share dramatically from 17 percent in 1961 to 29 percent in 1971 (Ho 1980, table 3-1). Firms with 100 or more workers accounted for 73 percent of the total increase in manufacturing employment of nearly 800,000 workers. Firms of all size groups except the very smallest (1–3 workers) showed some absolute increase (Ho 1980, table 2-1).

We have no information for Korean manufacturing employment before 1958 or for establishments with fewer than 5 workers. But the firm size group 5–49 workers reduced its share of manufacturing employment between 1958 and 1975 from 54 to 17 percent. This was a still more rapid decline than in Taiwan; and in Korea, unlike Taiwan, the size group 50–99 also lost ground, while the biggest gainer (from 12 to 44 percent of employment) was in the largest size group with more than 500 workers to an establishment. But Korea is similar to Taiwan in that the smaller the establishment the larger the relative decline and in that these declines were most pronounced in the later period of most rapid growth (1963–75). Korea seems to have become big-establishment oriented even further and faster than Taiwan. From 1963 to 1975, there was an increase in manufacturing employment of more than 1 million persons, of whom the size group 100 or more workers accounted for 86 percent and the 500 or more group for 52 percent (Ho 1980, table 6-3; see also figures 6-1–6-4 below). The smallest size group (5–9 workers) showed no change in employment.

Outside manufacturing, there is very little information on the size structure of enterprises, although in most economies services are much greater employers than manufacturing, even excluding government and military service. For Korea, we have no information on the size structure of nonmanufacturing enterprises.[3] For Taiwan, there is some information. This confirms that service industries are the home of small enterprises to a much

3. Information exists for construction and trade in Korea, but the data were not available for the Ho study.

Table 2-1. *Employment Growth in Some Nonmanufacturing Sectors and in Manufacturing, Taiwan, 1961–71*
(thousands of workers)

Plant size	*Construction*		*Trade*		*Other*[a]		*Total nonmanufacturing*		*Manufacturing*[b]	
(number of workers)	*Percent*	*Workers*	*Percent*	*Workers*	*Percent*	*Workers*	*Percent*	*Workers*	*Percent*	*Workers*
1–3	−8	−0	62	88	60	9	52	97	−12	−8
4–9	35	2	205	71	97	29	144	102		
10–19	42	4	384	27	160	26	170	58	70	72
20–49	76	16	252	13	166	44	139	74	125	72
50–99	130	21	363	8	245	35	195	64	252	80
100–499	16	8	213	9	242	42	87	58	454	279
500 and more	−7	−3		4	131	77	82	78	230	302
Total	35	48	112	220	125	262	97	530	175	796
Percent a year	3.1		7.8		8.5		7.0		10.7	

Note: The recorded increase in employment of 1.326 million in this table accounts for about 85 percent of the total increase during the period (compared with Galenson (ed.) 1979, tables 6-2, 6-3). Electricity, gas, water, and mining, which are almost exclusively the preserve of large enterprises, have been excluded from the table. They also recorded a slight reduction in employment, a fall in mining offsetting a rise in electricity, gas, and water.

a. Includes hotels, restaurants, transport and storage, banks and other financial institutions, commercial services, recreation services, personal services (excluding repair shops and tailors), and other unclassifiable industries.

b. Includes repair shops and tailors.

Sources: Ho (1980, tables 2-3, D-3); Galenson (1979, tables 6-2, 6-3).

greater extent than manufacturing. Thus enterprises with fewer than 100 workers in the nonmanufacturing industries, whose employment *growth* is detailed in table 2-1, employed 773,000 in 1971 as opposed to 475,000 for manufacturing. For still smaller enterprises, the contrast is greater. SSEs with fewer than 20 workers employed 550,000 in nonmanufacturing industries in 1971 as opposed to 235,000 in manufacturing.

But in Taiwan, although these nonmanufacturing industries account for much more SSE employment, they no longer account for more employment in total (though they did in 1961). Manufacturing, especially in large enterprises, accounted for far more of the *growth* of employment in 1961–71. This is shown in table 2-1. Since figures for nonmanufacturing industries are rarely available, this table deserves close study. The remarkable annual growth rate of manufacturing employment of 10.7 percent resulted in an absolute increase of 796,000 workers, of whom 581,000 were in enterprises with 100 workers or more. Employment in minienterprises (1–3 workers) in both manufacturing and construction fell absolutely. In trade and other service industries, employment in minienterprises fell relatively while growing at the modest rate (for Taiwan) of about 4.9 percent per annum. In construction, where employment grew surprisingly slowly, the size group 10–99 grew at above average rates. In trade, all except the smallest group were above average. In other service industries, the fastest growth was in the range of 10 workers and above, with some tendency for those of size 50–499 to increase employment fastest. Looking at the total increases in nonmanufacturing employment, it is shown that the size group 1–10 still provided the most new jobs in absolute terms, but this was very largely caused by trade—for which, however, the *percentage* increases in employment, which were far greater in large enterprises than in small ones, may be a clue to the future.

In the Philippines, employment in household manufacturing has declined relatively while still showing a slow absolute increase (Anderson and Khambata 1981, para. 6-7). Figures for employment in manufacturing establishments of different sizes are confusing,[4] and there is no clear evidence of relative success by either large or small, though employment in all groups has risen substantially. We have no information concerning nonmanufacturing industries.

In Colombian manufacturing, the category cottage shop includes both household activities and those of workshops with 1–4 workers. Cottage shop employment appears to have fallen from 77 percent of all manufacturing employment in 1939–41 to 44 percent in 1978 while nevertheless continuing

4. For instance, table 6-1 in Anderson and Khambata (1981) shows employment in establishments of 1–19 workers growing faster than larger ones from 1967 to 1975 (census data). Table 2-20 in Bautista (1981) shows employment in establishments of 5–19 workers growing much slower than in larger ones from 1968 to 1974 (survey data). It would be extremely rash to conclude that those in the size class 1–4 had grown exceptionally fast.

to rise absolutely at least since 1951 (Cortes, Berry, and Ishaq 1983, table 2-20). Within the factory sector, small enterprises (5–49 workers) grew more slowly than medium-size (50–99) or large (100 or more) from 1953 to 1973. But a remarkable reversal seems to have occurred around 1973: small factory employment began to grow faster than medium-size; since then, medium-size has grown faster than large (Cortes, Berry, and Ishaq 1983, table 2-2).[5]

We shall deal at length with India in chapter 6. At this point, a murky sketch must suffice—murky because that is the way the statistics are. From a comparison of the 1961 and 1971 population censuses, it is clear that household manufacturing employment fell relatively and may have fallen absolutely, though a change of definition makes it impossible to say this for sure (Mazumdar 1983, pp. 26–28, citing work by Banerjee). This finding is not inconsistent with what is known of the production of some traditional industries. First, during roughly the same period, there was only a very slight increase in *khadi* (cloth made from hand-spun yarn), handloom, and raw silk production.[6] Second, during the same period, the number of establishments grew most rapidly in the range 10–49 workers (5.9 percent a year), while larger ones grew faster than those with fewer than 10 workers (5.3 as opposed to 3.7 percent a year). This is consistent with national income statistics, which show some fall in the share of output contributed by unregistered establishments.[7] Third, in the short period 1974–77, establishments in the range 10–99 rapidly increased their share of employment at the expense of the large (500 or more), both overall and in most two-digit (according to the International Standard Industrial Classification) industries (Mazumdar 1983).[8] Fourth, Mazumdar has selected twenty-five industries which seem to be comparable between the 1974–77 period (when a size breakdown of Annual Survey of Industry [ASI] data was available), and the Census of Manufacturing Industry for the period 1953–58. In almost all, the share of employment in the size groups 10–19, 20–49, and 50–99 has risen.

The Indian data can be summed up in a few propositions:

5. The discussion of this turnaround in chapter 7 dates it from the late 1960s. The recrudescence of small enterprises coincided with an acceleration of the overall growth rate. Explanation is still tentative, and in contrast, as shown above, booming conditions in Korea and Taiwan coincided with a more rapid growth of large firms and actual decline of the equivalent of cottage shop enterprises.

6. See fourth five-year plan, para. 13-7, and Sandesara (1981, app. D, table 3). But it should be added that during the 1970s, output of these traditional industries grew somewhat faster (see Sandesara 1981 and sixth five-year plan, table 12-1). For handloom production, see also chapter 4. There are substantial divergences between the Planning Commission figures and those of chapter 4. There is no direct information on handloom production, and estimates can vary widely.

7. The share fell from 45 percent in 1960–61 to 35 percent in 1965–66 but subsequently stabilized at this later figure. See Sandesara (1981, app. D, table 1). However, there appears to be no sound basis for this particular component of national income statistics.

8. One can say nothing about very small establishments from this ASI source.

- Household manufacturing has fallen relatively. It may have fallen absolutely in the 1960s, but this is less probable in the 1970s.
- Almost nothing statistical is known of the very small nonhousehold manufacturing establishments (fewer than 10 workers). They have probably suffered a relative decline, but this is far from sure.
- In the 1960s, employment in small manufacturing establishments in the size range 10–99 workers rose relative to the large, especially those with 500 or more workers.

There are two further instances of rapid industrialization in Asia, Malaysia and Singapore, for which some historical figures are available, although these countries were not part of World Bank research.[9] In Malaysia, informal sector manufacturing employment (no paid employees) has not significantly increased since the 1950s. But employment in the formal sector has increased very fast, over 11 percent a year from 1957 to 1973.[10] Within the formal sector, all size groups increased employment, but the largest the fastest. The share of establishments with fewer than 100 workers fell from 71 to 38 percent (Amjad 1981, table 4-9). In Singapore, we have no information on very small establishments or household manufacturing. But while establishments with 10 or more workers increased their employment by almost 13 percent a year, the *share* of those in the range of 10–99 workers fell from 55 to 29 percent (Fong and Tan 1981, table 4-9).

Secure generalizations clearly cannot be made from the limited information presented, but some tentative suggestions are in order. Everywhere there is a relative fall in employment in household manufacturing. It is probably falling absolutely wherever total manufacturing employment is growing very fast. The same may well be true of minifactories or workshops with 1–3 workers. Likewise, where manufacturing employment has grown very fast (Korea, Malaysia, Singapore, and Taiwan), factories with 10–99 workers have increased employment more slowly than have larger ones. Where manufacturing has been more sluggish, firms of this size have probably held their own or been in the lead. Colombia in the 1970s seems to be an exception in that accelerated manufacturing growth coincided with a reversal of the relative decline of small establishments.

The Experience of Developed Countries

Supporters of SSE development in developing countries often point to the fact that small-scale businesses (or firms) and establishments (or plants)[11]

9. The other country studies included in Amjad (1981), Indonesia, the Philippines, and Thailand, yield no evidence of the pattern of historical change. It also appears from Page (1979) that Africa yields no evidence. We have not searched the literature on other countries.

10. Calculated from Lim, Lee, and Thye (1981, table 6-12).

11. In this section, the terms "business" and "firm" are reserved for independent enterprises. "Plants" and "establishments" may be subsidiaries.

survive in large numbers and still employ many people in the industrial countries. While this is true, it is also true that the average size of a manufacturing plant and firm is much larger in developed countries, both overall and by industry (the definition of small is often fewer than 200 workers; of medium, 200–499 workers). The main reasons for this are clear: the size of the market is larger, and national income thus is higher; the smaller developed countries export a high proportion of their manufactures; and the domestic and international transport network is more fully developed and costs are lower. These factors permit economies of scale and marketing to be more fully exploited. In addition, there may be an independent tendency in expanding economies for the size of firms and plants to grow with their age.

Excluding agriculture, small firms with 1–100 employees accounted for about 36 percent of U.S. private employment in 1978.[12] Establishments with fewer than 100 employees accounted for 49 percent, but many of these belong to multiplant firms. Firms with 1–20 employees accounted for about 19 percent (U.S. Small Business Administration 1984). In Japan, firms with fewer than 20 employees (in manufacturing) or fewer than 5 (in commerce and services) employed 31 percent of the private nonprimary work force in 1975 (Kaneda 1980, p. 86). Such overall figures are hard to come by. More is generally known about the manufacturing sector, although small firms are more typical in other sectors. For instance, only 21 percent of manufacturing sales in the United States in 1982 were reckoned to come from businesses with fewer than 500 employees, compared with 80 percent in construction and 60 percent in retail trade. Corresponding figures for employment were 27, 81, and 62 percent (U.S. Small Business Administration 1984, table A-1-36). In the United Kingdom in 1963, only 20 percent of manufacturing employment was in firms with fewer than 200 employees, while 33 percent of construction workers were in firms employing 25 or fewer workers (Bolton Committee 1971, table 6-1).

Establishment data are more available than enterprise data. The Bolton Committee (United Kingdom) compared the proportion of manufacturing employment in establishments with 200 or fewer workers for thirteen industrial countries for various years from 1961 to 1968. The range was wide—from 31 to 66 percent, with the United Kingdom and Italy occupying the extreme positions. Japan occupied a middle position, together with Belgium, France, the Netherlands, and Sweden (Bolton Committee 1971, table 6-1). No relation of these figures to per capita income was apparent. The committee also found that 20 percent of manufacturing employment was in small

12. U.S. Small Business Administration, *The State of Small Business: A Report to the President* (March 1984, table A-1-33) gives a figure of 34.1 percent for 1982. The data base for industrial size distributions in the United States and all other countries is very imperfect. Intercountry comparisons are especially difficult. All figures given in the text must be regarded as imprecise orders of magnitude. The figures of 36 and 49 percent are from Armington and Odle (1982, table 1).

enterprises (fewer than 200 workers) in the United Kingdom compared with 30 percent in the United States (1963).[13] It also concluded that for the economy as a whole the small firm sector was more important in the United States than in the United Kingdom. Some more recent figures are available for manufacturing in European countries. Enterprises with fewer than 200 employees account for more than 40 percent of total employment in most of the countries for which there are figures. The United Kingdom was the least small-firm oriented, with 22.6 percent of employment in such enterprises (Storey 1982, table 2-1).

The present situation has been reached after a long period of relative decline of small firms and establishments. Until World War II, the decline was quite small[14] and probably affected mainly very small plants. More figures are available for the 1950s and 1960s, on the basis of which the Bolton Committee found a decline in all countries examined. The decline is not confined to manufacturing. In the United States, the relative output of small businesses (fewer than 500 employees) fell in all sectors between 1958 and 1977 (U.S. Small Business Administration 1984, tables A-2-1 and A-2-18). In the Federal Republic of Germany in the 1960s, the employment share of all firms with fewer than 100 workers fell from 47.6 to 44.0 percent, but it rose in the size range 100–499 (Bannock 1976, table 16).

The relative decline of small firms is greater the smaller the size. In the United States from 1958 to 1977, the percentage sales contribution of the size group with fewer than 20 employees fell from 28.2 to 20.2, while that of the size range 20–499 was virtually unchanged (U.S. Small Business Administration 1984, table 2-3). Employment in small firms was better maintained, the group with fewer than 20 employees declining only from 23.3 to 19.8 (U.S. Small Business Administration 1984, table 2-13). Similarly in Germany, the relative fall was greatest in the size group 1–9 workers (Bannock 1976).

We have written above of relative decline. In at least one country and sector, U.K. manufacturing, the decline was absolute. From 1935 to 1968, the number of small manufacturing firms (fewer than 200 employees) fell from 136,000 to 58,000 (Bannock 1976, table 3), with no significant change in the number of large ones. In Germany overall, the absolute number of firms in the size range 1–9 fell during the 1960s but rose in other ranges.

13. However, it seems that the committee may have exaggerated the role of small enterprise in U.S. manufacturing. A recent estimate is that only 26 percent of gross output came from manufacturing enterprises with fewer than 500 workers (see White 1981). The U.S. Small Business Administration (1984, table 2-13) gives a figure of 30 percent for enterprises with fewer than 250 employees in 1958.

14. Some figures for plants for Australia, Canada, Germany, the United Kingdom, and the United States are given in Jewkes (1952). Other sources are International Labour Organisation (1961), Organisation for Economic Co-operation and Development (1971), and Kaneda (1980). Except for the last named, the figures are collected in Sandesara (1981).

Table 2-2. *Germany's Relative Number of Manufacturing Plants, Average 1970–73*

Size	*U.K. = 100*	*U.S. = 100*
1–4	488	148
5–9	210	100
10–19	145	69
20–49	132	53
50–99	127	48
100–199	116	44
200–499	104	45
500–999	100	48
1,000 or more	100	52
Total plants	241	94
Total employment	124	53

Source: Prais (1981, table 23). It is known that German definitions include under manufacturing some small units which would count as services or trade elsewhere. But Prais has excluded trades from the statistics given where such incomparability was serious.

German manufacturing remains relatively very slanted toward small plant sizes. The contrast with the United Kingdom as shown in table 2-2 is striking. The table also compares Germany with the United States. Here the distributions are very similar, except that Germany has relatively many more plants with fewer than 20 workers. Germany seems to be the shining example for those who argue that hundreds of thousands of very small plants are compatible with a very rich and progressive society. Very large numbers of small plants also exist in Japan,[15] but the fact that, even after improvements in recent years, these have relatively low productivity and pay low wages makes the Japanese example less shining.

In the 1970s, however, the decline of the small and medium-size manufacturing firm (fewer than 200 employees) seems to have been arrested (and even reversed in some countries). Between 1968 and 1976, the percentage of employment in small firms in the United Kingdom rose from 20.8 to 22.6 percent (Storey 1982, table 2-3). During about the same period in the United States, small firms gained and larger firms lost workers (Storey 1982, table 2-6). The reasons for this reversal of a long-term trend have not been fully analyzed, but high unemployment, structural change in manufacturing, and bad industrial relations in large firms are almost certainly contributory factors.

In the United Kingdom and other European countries in the late 1970s and 1980s, there has also been a marked change of sentiment. In the first

15. By a new definition, "small" in Japan means fewer than 21 workers in manufacturing and fewer than 6 in commerce and services. In manufacturing, the percentage of small establishments actually rose from 86.4 in 1960 to 87.2 in 1975, although in the aggregate of nonprimary sectors it fell from 84.4 to 80.7 percent (Kaneda 1980, table 18). Absolute numbers rose from about 3 million to 4.3 million overall and from 478,000 to 709,000 in manufacturing.

twenty-five years after World War II, governments favored large size and encouraged industrial concentration. Now quite suddenly small has become beautiful, and there has been a wave of legislation favoring the small (Storey 1982, chaps. 2, 11; *Financial Times Survey* 1984).

The new concern for SSEs in Europe has a goal different from that in developing countries where, as we have seen, the main objective is increased demand for labor. Political concern is distinguished from economic concern. Strong political or social preference may be felt by some for a structure of production which, for as many trades as possible, contains a large number of small enterprises that are not dominated by or in any way subordinate to a few large companies. The class of active independent but relatively powerless entrepreneurs is to be admired for its own sake, rather like the yeoman farmer or homesteader of the past. This vision is clearly linked to those other utopias where the independent peasant proprietor was to be the backbone of society. Capitalism is easier to defend in such a guise. Fear of a concentration of production in a few large companies is obviously related. But these two concerns are not exactly opposite sides of the same coin. Any substantial trade or subindustry may have low concentration without any large number of small enterprises (say, those with fewer than 100 employees). Similarly, a large number of very small enterprises is consistent with three or four firms being responsible for, say, 80 percent of the output.

The Lessons of History

The history of and the present state of affairs in the industrial countries may provide a glimpse of what is in store for developing countries—at least for those that will ever attain such high levels of income. Similarly, the more advanced developing countries may also provide a glimpse of the future for the less advanced.

In manufacturing, the very small establishments (fewer than 10 workers) are destined to near extinction unless they become a protected species. In this respect, as we shall see, India is trying to swim against the tide of history. In many other developing countries, relative or even absolute decline is in evidence. In services, very small establishments also suffer a relative decline but are nevertheless likely to survive in large numbers. Manufacturing establishments with 10–50 workers will decline relatively and probably also in absolute numbers; but the future of the middle ground in manufacturing, say 50–500 workers, is more obscure. There are wide variations among developed countries that are unrelated to gross domestic product (GDP) per head, and there is some evidence of a reversal of the trend toward giant plants and enterprises. It is possible that changes in technology and in industrial relations, as well as in the composition of manufacturing output, will reinforce this trend in the future. And the relative decline of manufacturing in total output and employment that takes place at very high levels of national income implies an increase in the proportion accounted for by small establishments.

But is there any need to worry about what the future has in store for the size distribution of plants or firms? The next section presents a brief survey of the economic arguments that raise concern and may support intervention in the industrial countries.

Arguments for Intervention in Favor of Small Enterprises in Industrial Countries

Economic concerns may be about static or dynamic efficiency, or both. Static efficiency is taken to mean the maximization of national output given the products and means of production. As chapter 1 showed, the main argument advanced for favoring SSEs in developing countries is that their allocative efficiency is supposedly higher from a social point of view because they face lower wage and higher capital costs than larger enterprises and because this better reflects the social cost of labor and capital.

In the industrial countries there is no generalized presumption of a pervasive distortion of relative factor prices, with the possible exception that small enterprises may face either unduly high interest rates or experience undue difficulty in raising loans or equity finance. We shall not discuss the putative problem of imperfect capital markets in industrial countries, but many Western governments have from time to time tried to ensure that small or medium-size firms are not seriously disadvantaged in this respect. (Capital market imperfections are discussed in a developing country context in chapter 15.)

Dynamic arguments, which include a time dimension, have recently been more stressed in the OECD countries than the static arguments outlined above. The two most favored are, first, that small firms are more innovative, and second, that they constitute a seedbed or nursery which is an essential part of the forest of firms whose component trees are decaying as well as growing (see Marshall 1920, book 4, chap. 13).

There is evidence that the contribution to innovation of small firms (fewer than 200 employees) is less than their contribution to output or employment and that this is reversed for only a very few industries (scientific instruments, some kinds of machinery, and paper and board). In many U.K. industries, especially those that are capital-intensive, small firms have apparently made no innovations. Since small firms accounted for only 4 percent of total manufacturing expenditure on research and development, however, it could be concluded that their innovative (and inventive) performance per dollar of research and development expenditure was high (Freeman 1974, chap. 6; Freeman n.d.). For the United States it has been found that small firms are two to three times as innovative per employee as large firms. (U.S. Small Business Administration 1984, p. 59). The argument is likely to remain open, since researchers would be the first to admit that it is very hard to measure contributions to innovation and invention. We have already remarked in chapter 1 that we do not believe that the innovation argument is very relevant for developing countries, though there are some anecdotes of appropriate adaptation arising in SSEs.

The seedbed argument has been well put by the Bolton Committee in the following terms:

> We believe that the health of the economy requires the birth of new enterprises in substantial number and the growth of some to a position from which they are able to challenge and supplant the existing leaders of industry. We fear that an economy totally dominated by large firms could not for long avoid ossification and decay, nor can we think of any long-term alternative to the maintenance of a thriving small firm sector as a safeguard against this. This "seed-bed" function, therefore, appears to be a vital contribution of the small firm sector to the long-run health of the economy. We cannot assume that the ordinary working of market forces will necessarily preserve a small firm sector large enough to perform this function in the future.

In most developing countries, there is certainly no lack of a seedbed. It is mainly in some African countries that the problem arises. There the relative lack of small-scale entrepreneurship and its confinement to a few trades may require at least making sure that the formation of enterprises is not discouraged by overall macroeconomic and industrial policies.

Very few small firms are ever going to grow even to medium size, let alone to a challenging size (see, for example, Storey 1982, table 2-8). Most small businessmen lack the drive, imagination, and managerial ability, as well as the ambition and luck, required to set a small business on a strong growth path.[16] This does not invalidate the argument, for only a few need grow. But it does raise the question of how large the seedbed and how small the seedlings need be. It is often said that all large businesses started small. This is clearly untrue when businesses are promoted in the public sector, which is to some extent the case in almost all developing countries. But we later present some evidence that it is largely true at least for medium-scale private sector enterprises.

16. An interesting typology of businesspersons has been proposed in Smith (1967). He distinguishes two polar types, craft entrepreneurs and opportunist entrepreneurs. The former have a narrow, mainly technical education and low social awareness and involvement. They are bad at delegation, hire on a personal basis, and have limited horizons in the realm of finance and marketing. They have no long-range plans which might involve a change in the character of the business. Opportunist entrepreneurs are of course just the reverse. They build more adaptable firms, and success stories mostly concern such entrepreneurs and their firms.

We have singled out two of the most common dynamic arguments made by SSE proponents. Of course, small enterprises enjoy many advantages and suffer many disadvantages. But few of these constitute arguments for policy intervention for the reason that most of the advantages rightly favor survival and growth and most of the disadvantages, decline and death.[17] Apart from the political argument mentioned above, there is an argument for special measures for (or against) small business only in two cases: first, if the market left to itself would produce a bias, which it would do if a particular size of business produced costs or benefits for the rest of society that were not properly reflected in its purchases or sales; or second, if government economic measures, whether fiscal or regulatory, unintentionally and for no good reason distort the size structure.

The seedbed argument was an example of the first case. The innovation argument, if a valid ground for favoring SSEs, would also have to fall under the first case—that is, it would need to be shown that innovations were inadequately rewarded. Another reason for special measures would be, and has been, those alleged imperfections in the capital markets which have been referred to above and which are discussed in chapter 15.

As for the second case, it is likely that taxation in industrial countries discourages small enterprises more than large ones unless special relief is given. First, founding a business is very risky, and taxation reduces the potential reward that compensates for the risk. Second, taxation increases the likelihood of sale, usually to a larger firm, at the retirement or death of the entrepreneur (and may induce early retirement). Third, both direct and indirect taxation and also social security provisions are often very complex, and the need to deal with them constitutes to some extent an indivisible overhead.[18] This latter aspect of taxation is very similar to regulations. Regulations concerning accounts, employment, training, health and safety, the supply of statistics, planning permission, pollution, and so on have become very complex and demanding for firms too small to employ specialist staff economically.

These aspects of taxes and government economic measures and regulations apply in both developed and developing countries, though with differences.

17. An important advantage is that strikes are almost unknown in small firms, which is partly a reflection of the fact that many people prefer to work in small groups and that access to management is much easier. This preference should be reflected both in higher productivity (other things being equal, of course) and in lower wages (provided that minimum-wage legislation does not prevent this), which would rightly allow some small enterprises to survive and grow that otherwise could not. The very strong relationship between plant size and strikes (stronger in the United Kingdom than elsewhere) has been investigated by Prais (1981).

18. This highly abbreviated treatment is all that seems justified in a work primarily concerned with developing countries. A more extended and balanced discussion may be found in the *Bolton Committee Report* (1971, chap. 13). Other references include Boswell (1972) and Hay and Morris (1984).

Very small firms in developing countries may be able to escape taxes and regulations altogether. But most of these economies are run in such a way that some materials and other inputs are allocated. The consequence of this is that small firms may be able to obtain supplies only on a black market. Many inquiries and reports cite shortages or excessive prices of inputs as one of the prime problems of SSEs in these countries. This is not a complaint that one often finds in the industrial countries.[19] Again, as chapter 1 showed, in developing countries special investment incentives often favor large firms and may even be formally denied to small ones.

Although the economic principles of market intervention are universal, the need for and the nature of intervention may differ greatly with differing economic conditions and institutions. This is indeed the case, as the above paragraphs and chapter 1 have already indicated in a preliminary way. The rest of this book is exclusively about developing countries, mainly India. It throws some light both on the effects of those overall economic and industrial policies that create distortion discouraging to SSEs and on the effects of particular interventions in their favor, which are sometimes very different from what was expected.

19. Our own surveys in India do not suggest that material shortages were more than rarely a problem in the industries studied.

3

Indian Industrial Policies and Measures to Promote Small Enterprises

This chapter reviews the policy environment for small-scale enterprises in India. The first section of the chapter places small enterprise development policies within the context of India's overall industrial policy since independence. Specific measures for small-industry promotion are considered in the second section.

Industrial Policy and Establishment Size

Indian governments and leaders have for a long time shown concern for the size structure of industry. This dates back to the early years of the century. M. K. Gandhi wrote, "I do not remember to have seen a handloom or a spinning wheel when in 1908 I described it in *Hind Swaraj* as the panacea for the growing pauperism of India" (1940, p. 360). He subsequently had considerable difficulty in finding anyone who could teach him and his followers to spin. The making of khadi was a resuscitation of an almost vanished craft, and it became both a political and cultural symbol. No other country has revived and then greatly expanded during a half century both very primitive methods of making cloth, sugar, soap, and other products, and also equally primitive methods of processing cereals and vegetable oils. At the same time, India became notable in the 1950s for its insistence on the value of capital-intensive heavy industry. This reflects Gandhi's and Nehru's opposing visions of India's future. Although Nehru's views prevailed for the vast bulk of industrial investment, Gandhian thinking has continued to be influential to this day. The Mahalanobis model, which inspired the second five-year plan and set the stage for India's heavy industrial scene, reserved a place for the Gandhians in that small and household industries were supposed to be able to supply the increased demand for consumer goods with very little investment but greatly increased employment. Factories were to make only investment goods and intermediates, and the traditional household industries were to be protected against them (Mahalanobis 1963, pp. 22–25). In some measure, this policy has continued ever since, even if less starkly and with imperfect implementation.

The Mahalanobis model was able to give a theoretical justification for the continued existence of the small-scale sector: it would provide an elastic supply of consumer goods needed to support workers in the large-scale sector of heavy industries. But even at this simple theoretical level, two crucial points seem to have been slurred over. First, the mechanism in the model depended on the small-scale (consumer goods) sector being able to provide enough surplus or savings for the development of the large-scale (capital goods) sector. Was the former, with its low productivity, capable of doing so? Second, the ultimate purpose of the development of the large-scale capital goods sector would be the production of machinery for producing consumer goods in a capital-intensive way. The model therefore implicitly assumed that the small-scale sector would gradually yield ground to modern technology. But the question of when and how this transition was to take place has never been properly addressed in policy discussions.

It is important to be clear that the Gandhians wanted to expand traditional, mainly rural manufacture and not create modern small urban factories. But Mahalanobis refers to "small and household," and it must be supposed that he included nontraditional enterprises if they were not factories (which by definition were units with 10 or more employees with power or 20 or more without), and even the small labor-intensive factories which came under the newly established Small Scale Industries Board (SSIB). A series of boards had been established in about 1950 to deal with traditional industries. It was only later (1954), after the visit and report of a Ford Foundation team, that the SSIB was set up to encourage those industrial enterprises that were too small to be regulated under the Industrial Development Act and did not fall under the traditional industry boards. The SSIB thus was concerned largely with small units producing modern products, usually though not always by power-using methods. The initial definition of "small," later frequently amended, excluded enterprises with more than Rs500,000 of fixed assets or more than 50 employees (100 if without power).

The main recommendations of the Ford Foundation team were accepted. The recommendations included setting up small-industry service units (for technical assistance and industrial extension) under the Central Small Industries Organization, and a Small Industries Corporation (for marketing, including government purchase, and supplying machinery on hire purchase). Thus by the beginning of the second five-year plan (1956), the main administrative framework for government encouragement of both traditional and modern small enterprises was in place. The same is true of the policies which have been continued ever since.

These policies included the protection of traditional industries in several ways, by taxing and banning the expansion of medium- and large-scale enterprises that competed with them, by direct subsidies, by preference in government purchase, and by reserving certain products for exclusive production by traditional and small industry. Initially, except in the case of textiles, which is examined in chapter 4, these measures mostly benefited traditional rather than modern small-scale enterprises. But the industrial estates that were being planned, and the activities of the small-industry service units, would mostly help small modern enterprises.

The Village and Small Scale Industries (Karve) Committee of 1955 was evidently influential in securing an increase in the size of the programs for these industries from the tiny amount of Rs42 million in the first five-year plan to Rs187 million in the second plan. The name of this committee, which is also the title of a chapter in every plan from the first to sixth, invites confusion and the false idea that small-scale industries can somehow be assimilated to village industries. The Karve Committee had the essentially Gandhian vision of a very decentralized economy with an industrial structure like a pyramid whose base would be dispersed small-scale and village industries. To this day (for example, in the sixth plan), small-scale and village industries are frequently referred to as the decentralized sector. This is wishful thinking. Of course, village industries are dispersed, though more in some states than others. But the important handloom sector, while certainly small and traditional, is by no means mainly a village industry and is quite highly concentrated in certain states. Nor is modern small-scale industry decentralized; furthermore, it is a poor candidate for decentralization since, as several critics have pointed out, it needs the economies of agglomeration more than large-scale factories which to a much greater extent can supply themselves with components and services or afford to buy them from a distance. The effort was nevertheless made, and continues to be made, to disperse small-scale modern industry by building small rural industrial estates. Virtually all studies of industrial estates in India have concluded that they have failed to promote small entrepreneurs and labor-intensive small-scale industry in a cost-effective manner, but at least the urban estates have been more successful than the rural. "Small industry is a follower rather than a pioneer."[1] Despite this, successive plans have emphasized rural estates.

This book is not, however, primarily concerned with industrial decentralization or the means by which it can be achieved. The other main reason advanced by the Karve Committee (and repeated in the Industrial Policy Resolution of 1956) for encouraging and protecting village and small industries was of course employment. This is more closely our concern. There may well have been the same confusion here, as the research of Dhar and Lydall

1. Dhar and Lydall (1961, p. 61); chapter 3 of that work provides an early critical account of the Industrial Estates Programme. Sandesara (1981) also reviews both a number of studies of industrial estates and the reasons why small manufacturing units are not particularly good candidates for industrial dispersion.

(1961) was soon to suggest. Important positions were certainly taken up without any investigation of the extent to which different industries and techniques would produce output with more labor and less capital or of the relation between size and technique.

It is important not to exaggerate the weight given to village and small-industry programs, and therefore implicitly the weight given to the concerns of decentralization and employment which these programs were specially intended to serve. In the first five five-year plans, public sector outlay on these programs was 2.1, 4.0, 2.8, 1.5, and 1.3 percent of the total, as compared with 2.8, 20.1, 20.1, 18.2, and 17.4 for "organized" industry and minerals (Sandesara 1981, app. B, table 1). The traditional or village industries got the lion's share, generally about 60 to 70 percent, of these relatively small amounts; except in the case of textiles, they probably also benefited most from the protective measures that did not involve public money. There is no doubt that these measures were effective in that some traditional industry which could not compete with factories was preserved and even expanded (see, for example, sixth five-year plan, 1976, table 12-1, and earlier plans). Finally, however, we may note, with the Planning Commission, that "the precise impact of the various policy measures and development programs cannot be assessed for want of adequate data, particularly in respect of the traditional village and household industries" (sixth five-year plan, 1976, para. 12-3). The small-scale textile industry, which is both traditional and modern, is a special case and is examined in chapter 4.

Various committees appointed by the Indian government have followed the basic reasoning of the Industrial Policy Resolution of 1956 and the five-year plans. The most influential of these was the 1973 committee under the chairmanship of H. S. Bhat to formulate specific recommendations to accelerate the growth of small and medium-size entrepreneurs. The committee recommended expanded government action in support of small enterprises in the following areas:

- Development of infrastructure for small-scale enterprises on a priority basis
- Promotion of regional industrial development centers
- Implementation of effective measures to ensure the supply of raw materials
- Implementation of small-enterprise promotion measures at the state level
- Reinforcement of the tax concessions available to new entrepreneurs.

The Industrial Policy Statement of December 1977 went even further than the Bhat Committee in defining the role of small and medium-size enterprises in India's industrialization strategy. The main emphasis of the Industrial Policy Statement was on the effective promotion of cottage and small industries, and the government committed itself to the concept that any product which could be produced by cottage and small industry should be exclusively produced within the small-enterprise sector.

To implement the new policy, the government expanded the "reservation" of product lines for exclusive production in the small-scale sector. The list of industries exclusively reserved for small-scale production was expanded from about 180 items in 1976–77 to 500 in 1978 and to more than 800 in 1979.

Apart from textiles and traditional industries, a whole range of "modern" consumer goods industries (for example, electronics) and even capital goods industries (for example, machine tools) have been encouraged through the policy of reserving particular products for the small-scale sector (this will be shown quantitatively below). The policy of reservation seems in each case to have been based on the physical capability of small-scale units to produce the commodities rather than on any economic investigation of relative cost-benefit ratios.

Policy Measures to Promote Small-Scale Enterprises

For policy purposes, small-scale enterprises are defined by the value of their fixed assets, excluding land and buildings. The Budget Speech of 1985 defined a small-scale unit as one employing less than Rs3.5 million in fixed assets and an ancillary unit as one with less than Rs4.5 million in machinery and equipment (these limits were raised from Rs2 million and Rs2.5 million, respectively).

The policy instruments designed to encourage the development of firms below these investment limits consist of fiscal incentives; preferential pricing policies; quantitative restrictions on the output of large-scale firms (reservation); and infrastructure, marketing, and industrial extension services. Both the central and state governments provide incentives in each of these areas. Central government incentives are uniform across states but vary with the level of development of the region within which the enterprise is located, while state government incentives vary both according to the individual state and with the level of development of the region.

Fiscal Incentives

The primary fiscal incentive provided by the Indian government to small-scale enterprises is exemption from the excise tax. Initially, two categories of small-scale firms were defined: first, firms producing goods on a list of seventy-two product groups with very low levels of output (Rs500,000) were exempted provided that total output was below the maximum threshold; and second, the remainder of firms producing less than Rs3 million were exempted from the excise tax on the first Rs1.5 million of output and were required to pay duty at a reduced rate on the remaining output.

The Budget Statement of 1985 introduced a revised formula for the computation of the excise tax exemption. The new law increased the tax exempt limit to Rs2.5 million and provided for gradual reductions in the level of excise tax exemption up to total sales of Rs7.5 million. This represented a substantial increase in the volume of sales covered by excise tax exemptions.

The implicit subsidy to small-scale producers resulting from excise exemption varies according to the level of tax payable. Excise tax rates vary from 100 to 10 percent, with the higher rates payable on luxury and consumer goods and the lower rates applicable mainly to intermediate and capital goods.

In addition to the excise tax exemption allowed by the central government, states grant exemption from state taxes, primarily sales taxes and turnover taxes on purchases of inputs. These exemptions are frequently limited to specific products or are limited in duration; in general, they provide a lower level of implicit subsidy than the excise tax exemption.

Other incentives include preferential pricing policies and various subsidies. Small-scale units, defined by their level of investment, are eligible for a 15 percent price preference over medium- and large-scale units in public sector purchases. An additional price preference at varying rates is also given by different states for purchases by the state government. The marketing of small industries' products for public procurement is done through the National Small Industries Corporation.

The central government provides a capital subsidy at the rate of 15 percent of the value of fixed investment for new industrial projects and for rehabilitation and expansion projects undertaken in specified backward areas. In addition, state governments provide capital subsidies at varying rates to new and existing firms located in economically backward districts not covered by the central government scheme.

State and local governments also provide financial subsidies to small-scale firms, both within and outside defined backward areas. These subsidies consist of interest rate and capital subsidies, water and electricity subsidies and exemptions from electricity tariffs, and subsidies for the acquisition of land.

Table 3-1 presents a breakdown of the major financial incentives available to small enterprises for selected products and states. Four main findings emerge from the data in the table:

- Excise tax exemption tends to dominate all other incentives and introduces the greatest variability into the level of subsidy because of wide variations in excise tax rates.
- Within industries, interstate variations in subsidies and incentives are small.
- Within states, interindustry variations in levels of incentives are substantial, primarily because of the incidence of the excise tax exemption.
- Additional incentives for location in disadvantaged areas are small.

Table 3-1. *Financial Incentives as a Percentage of the Value of Output, Selected Products and States, India, 1980*
(percentage of ex-factory value of output)

Products and states	*Excise tax exemption*	*Price preference*	*Sales tax exemption*	*Power subsidy*	*Electricity tax exemption*	*Total*
Cosmetics and toilet preparations						
Uttar Pradesh	25.1	20.0	n.a.	0.4	n.a.	45.5
Andhra Pradesh	25.1	17.1	3.1	0.4	n.a.	45.6
Punjab	25.1	15.0	1.4	0.2	0.1	41.8
Rajasthan	25.1	15.0	n.a.	0.6	0.4	41.1
Glass and glass products						
Uttar Pradesh	10.4	20.0	n.a.	4.0	0.3	34.8
Andhra Pradesh	10.4	17.0	2.6	2.5	n.a.	32.5
Punjab	10.4	15.0	1.2	2.0	1.3	30.0
Rajasthan	10.4	15.0	0.1	6.4	0.4	32.0
Rubber products						
Uttar Pradesh	7.7	20.0	n.a.	0.7	0.1	28.5
Andhra Pradesh	7.7	17.0	2.7	0.4	n.a.	28.0
Punjab	7.7	15.0	1.2	0.4	0.2	24.5
Rajasthan	7.7	15.0	n.a.	1.1	0.1	23.9
Paints and varnishes						
Uttar Pradesh	6.2	20.0	n.a.	0.6	0.1	26.9
Andhra Pradesh	6.2	17.0	2.7	0.4	n.a.	26.3
Punjab	6.2	15.0	1.3	0.3	0.2	23.0
Rajasthan	6.2	15.0	n.a.	1.0	0.1	22.3
Hand tools						
Uttar Pradesh	4.8	20.0	n.a.	0.8	0.1	25.6
Andhra Pradesh	4.8	17.0	2.7	0.5	n.a.	25.0
Punjab	4.8	15.0	1.3	0.4	0.7	22.1
Rajasthan	4.8	15.0	n.a.	1.3	0.1	21.1

n.a. Not applicable.

Note: Based on a value of output of Rs1.5 million. Interest rate subsidies are negligible and are excluded from the value of the benefit stream.

Source: Tulsi (1980).

Before the introduction of the new excise tax scheme in 1985, the disincentives to growth implied in the excise tax exemption were substantial. Table 3-2 shows the decline in the incidence of excise tax exemptions with firm size as well as their overall importance in the total level of fiscal incentives to small firms. The sharp reduction in the proportion of the excise tax exemption in total incentives is apparent from the table.

Quantitative Restrictions on the Output of Large-Scale Firms

A major implicit subsidy to small-scale enterprises originates in the reservation of more than 800 product groups for production by firms that have fixed investment in machinery and equipment below the stipulated level

Table 3-2. *Percentage of Excise Tax Exemption in Total Incentives, Selected Products, India, 1980*

Product group	Output level Rs500,000	Rs1 million	Rs1.5 million
Cosmetics and toilet preparations	71.1	62.1	55.1
Glass and glass products	52.2	38.1	30.0
Rubber products	49.2	34.8	27.0
Paints and varnishes	44.7	30.6	23.2
Hand tools	38.5	25.0	18.6
Industrial gasses	28.3	17.0	11.9

Source: Tulsi (1980).

defining small firms. Table 3-3 provides a sectoral breakdown of the proportions of the 805 product groups currently subject to reservation. More than half of all product lines reserved for small-scale production are in the light engineering sectors, followed in number by chemical products, rubber products, wood products, and textiles. The relative frequency of products reserved does not, however, provide a guide to the importance of reservation for the structure of production in each industry (see, for example, the case study of textiles in the next chapter).

With the introduction of reservation in a product line, firms with levels of investment exceeding the SSI limit are restricted to production at levels equal to their installed capacity at the time reservation was imposed. Thus all expansion in the product line is limited to small-scale enterprises. Small firms

Table 3-3. *Percentage Distribution of Restricted Commodities, India, 1980*

Product group	Percent
Food and food products	2.0
Textile products	2.5
Paper products	3.7
Wood and wood products	1.5
Leather products and footwear	2.5
Rubber products	2.9
Plastic products	3.7
Chemicals and chemical products	21.1
Electrical apparatuses	6.1
Electrical equipment and components	2.7
Glass and ceramics	2.9
Machinery and parts	24.5
Transport and equipment	0.4
Auto parts components and ancillaries	7.6
Bicycles and parts	5.5
Stationary items	1.6
Other	8.8
Total	100.0

Source: Indian Development Commissioner for Small Scale Industries.

are not required to obtain a license for production in reserved industries but may register to obtain other incentives.

The implicit subsidy to small-scale producers conferred by capacity restrictions on the output of large-scale enterprises depends upon the cost structure of each individual industry. In industries characterized by economies of scale or substantial productivity differentials between small and large firms, the effect of the capacity restriction is to displace the industry supply curve upward at the level of approved capacity of large-scale firms and thus to increase the market clearing price by the cost differential between the marginal small-scale and large-scale enterprise. In industries with competitive cost structures among enterprises of varying size and in those activities exposed to external competition, the extent of the displacement of the supply curve would be less and the implicit subsidy to small-scale producers correspondingly lower.

Infrastructure, Marketing, and Industrial Extension Services

In addition to the financial incentives offered to small-scale enterprises, both the central and state governments offer a substantial range of services to small firms. In broad terms, these consist of the provision of infrastructure, primarily in the form of industrial estates; marketing services, including export marketing; and other industrial services, including technology development, extension, and training.

Infrastructural activities are undertaken by the industrial area development authorities, which are also empowered to provide power, interest, and capital subsidies and to register small firms for benefits. The industrial area development authorities are limited to specific geographical areas in which the industrial estates are located. Outside the industrial areas, district industry centers fulfill a similar function by providing a full range of services to small-scale firms, including project identification and preparation, arrangements for the supply of machinery and equipment, provision for the supply of raw materials, credit facilities, and marketing and extension services.

Registration with industrial area development authorities or with district industry centers is entirely voluntary, but unless it is done no government assistance can be claimed. Moreover, any small firm which encounters financial difficulties can apply to the district industry center to register as a "sick unit" and thereby gain access to additional benefits and incentives.

The institutional framework described briefly above is both elaborate and costly, but its actual assistance to small-scale units is of uncertain efficacy. Reaching potential entrepreneurs is a problem because in many instances small firms are expected to come to the organization that provides technical and financial assistance rather than the organization being expected to identify and seek out new small firms. Small-enterprise support organizations thus favor existing small-scale entrepreneurs.

Coordination is lacking between the promotional agencies and the financial intermediaries entrusted with the administration of small-enterprise lending schemes. State finance corporations and commercial banks frequently refuse to provide credit to projects which have been approved and forwarded by the relevant promotional agency. Generally this is because the bank's technical appraisal of the proposed investment finds it to be of doubtful viability.

Finally, there is a perception that the relevance of the technical advice offered by promotional agencies is limited by their staff's lack of technical expertise and experience. Thus, while the staffs of small-enterprise promotional agencies help small businesses choose products and technologies, they themselves are not experienced in technology development, and the information they provide may be secondhand, nonspecific, and out of date.

Summary

The Indian government's policies of protecting and subsidizing small-scale enterprises have a long history. Gandhian thought emphasized the desirable social and employment consequences of promoting village and small-scale enterprise. The Mahalanobis model formalized the thinking of Nehru, in which the small-enterprise sector was viewed primarily as the source of an elastic supply of consumer goods to support the development of heavy industry.

Although small and village development programs have never been allocated large public resources in Indian plans, several high-level committees have recommended a wide range of incentives to small-scale enterprises, including excise and other tax exemptions, provision of physical infrastructure, and (perhaps most important) reservation of many lines of industrial production for small-scale firms.

Reservation means that licenses for the commodity in question are in general not granted for large-scale units. However, commodities with the same name may vary widely in quality. Typically, higher-quality products of the same name cannot be produced by small units. Thus in many cases the effect of the policy has been to encourage large units to specialize in the upper range of the quality spectrum rather than to shut them out completely from the production of the item in question. (This aspect of the problem is discussed further in chapter 13.)

A second point to be noted in this connection is that the policy favoring the extension of the small-scale sector to even wider areas of manufacturing has resulted in fairly regular increases in the ceiling that defines the upper boundary of the small-scale sector. This upward revision constitutes the only opportunity for the continued expansion of individual small units. The growth of firms beyond the limit of the ceiling is economically almost prohibitive. Not only would small firms have to cope with a much more difficult licensing policy, but they would also have to contend with higher labor costs (including wages and fringe benefits as laid down by labor laws) and substantially higher excise duties. The result has been that although the potential for the growth of units within the small-scale sector has been fairly high, the opportunity for such firms to grow organically into successful large firms has been severely limited.

4

Small-Industry Policy in Practice: Case Studies of Three Industries

This chapter reviews the effects of the Indian policy of protection of the small-scale sector in three specific industries—textiles, sugar, and light engineering. These industries have been chosen for special attention because, apart from their importance on the Indian economic scene, substantial research has been done on the effects of small-enterprise promotion policies on their structure and performance.[1] They also represent widely varying technologies; we thus can examine the consequences of the policy environment in a variety of settings.

Textiles: A Case of Unplanned Growth of Intermediate Technology

The textile industry is by far the most important of those industries that have been profoundly affected by the government's policies to protect and promote traditional and other small-scale industries.

In the early 1950s, the handloom and spinning wheel continued to be for Gandhians symbols of independence and hallowed instruments of the most desirable way of life. It was envisaged that handlooms would produce mainly khadi. Khadi still absorbs much of the budget devoted to village industries but is quantitatively negligible because handloom operators are supplied almost entirely by mill yarn (as has been the case ever since the turn of the century). There was also in the 1950s much disagreement over the role of powerlooms. Before World War II, a few handloom workers had shifted to small-scale operation of powerlooms bought secondhand from the mills, there being little or no technical difference between powerlooms and the

1. The three major studies on which the findings are based are Mazumdar (1984a) for textiles, Haan (1985) for sugar, and Berry (1983) for light engineering. The reader is referred to the works cited for further results.

great bulk of nonautomatic looms used by the mills. There had been considerable powerloom development during the war, but there were still far fewer of them than handlooms. Some (for example, the Kanungo Committee of 1954) saw the transformation of handloom weavers into powerloom operators as a natural development to be encouraged. The more Gandhian Karve Committee wanted protection of handlooms not only from the mills (this had been initiated by reservation of some products for hand weaving in 1950) but also from powerlooms. The latter won the day as far as *planning* was concerned. The government banned the installation of new looms by the mills (except for replacement or for export production) in 1956, and the second five-year plan envisaged that powerlooms would account for only 10 percent of the increased cloth production.

The actual turn of events was very different. Mill cotton cloth production fell from its peak of 4.9 billion meters in 1956 to 3.1 in 1981, while total cloth production rose from 6.5 to 8.1 billion meters (billion is 1,000 million). But powerlooms accounted for 74 percent of the increase of the small-sector output of 3.3 billion meters (Mazumdar 1984a, tables 1-3 and 1-7, which cite original sources).[2] There were several reasons for the remarkable growth of powerloom output in these years. First, weaving units with fewer than five looms suffered no excise duty and could make reserved kinds of cloth until 1974. Second, for larger powerloom units, prohibited lines of production made a much shorter list than for the mills. Third and probably most important, unlicensed units which evade excise duty and the restrictive regulations have mushroomed.

Small powerloom units using essentially the same technology as the mills thus paid less tax and could make most or all of the products banned for production in mills. They were also free from the restrictions on expansion which applied to the mills. Not only this, they paid wages less than half those of the mills (Mazumdar 1984a, p. 45). Their expansion would probably have been still greater if the government had not during the 1970s taken over and continued to operate about a third of the mills (with about a sixth of the spindles); these had gone bankrupt partly because of the protection given to the small sector but also because they were forced to produce a certain proportion of their output as coarse cloth to be sold at a controlled price. The nationalized mills have continued to produce this subsidized coarse utility cloth, mainly in competition with handlooms, which are the predominant producers of coarse cloth (also subsidized) within the "decentralized" sector.

The powerloom sector was also encouraged by the government's policy toward synthetic fibers, a policy which (as will be shown) had other important effects. As a result of import controls and the licensing of domestic

2. It should be noted that the division of output between powerlooms and handlooms is more of an estimate than a recorded fact.

production, synthetic fiber and yarn—especially polyester—came late to India. Rapid growth came only during the 1970s, despite the fact that polyester was burdened by extremely high excise and import duties (Mazumdar 1984a, p. 30). The main reason for this policy was the desire to protect cotton growing. Fiscal discrimination fell heaviest on polyester because viscose fiber and yarn production came earlier and had succeeded in building up its own protective lobby. The mills are forbidden to make purely synthetic cloth. As a result, the small-scale sector accounted for 69 percent of synthetic and blended cloth in 1981, which in turn accounted for 27 percent of total cloth production (Mazumdar 1984a, table 1-7). (In the world at large, more than half of the cloth is synthetic or blended.) Handlooms, which are less suitable for weaving synthetic yarn, accounted for only 7 percent of the total synthetic production.

An Overview of the Economic Effects of Textile Policy

The upshot of governmental production policy in the twenty-five years from 1956 to 1981 was a slow average annual fall in mill production (−0.7 percent a year), a somewhat faster rise in handloom output (+2.2 percent a year), and a rapid rise in powerloom output (8.2 percent a year).[3] We turn to the wider implication of the changes in production caused by policy. One effect has been an increase in specialization by the three sectors. The mills in particular have turned more to higher-quality cloth, the finer counts, and closer weave. The powerlooms are relatively specialized for synthetic fibers. Cotton production is fairly evenly spread over all counts but the coarsest. The handlooms continue with their traditional relative specialization on the coarsest cotton cloth, though they also produce some of the finest. No doubt it was believed that, by encouraging handlooms, the restrictions on the mills would result in greater production of the cloth used by the poorest people. It is not, however, clear that the increase in handloom output was caused by the policies, for handloom production has steadily increased throughout this century. Moreover, the very high taxation of synthetics has meant that the poorer if not the poorest sections of society have been largely denied the use of the more durable and at first view "appropriate" synthetic cloth. These quality problems are further discussed in chapter 12.

Perhaps more important are the effect on employment and the question whether any increase in the demand for labor has been bought too dearly. There is no doubt that textile policies have increased employment in weaving, given the level of wages in the mills. The powerlooms use about twice as much labor per loom compared with the mill sector and about three times as much labor per unit of output. Wages are much lower, however, and this

3. Calculated from Mazumdar (1984a, tables 1-3, 1-7) on the assumption that the handloom share of synthetic and blended cloth was constant at 7 percent.

can be seen as a means of keeping in production old looms which otherwise could not be retained in the mill sector. The powerloom sector also has new looms that use basically the same technology as the nonautomatic mill looms but incorporate fewer labor-saving devices and are cheaper. Although the powerloom sector apparently makes less demand on power capacity, the mills supplement the public supply by generating their own power in order to be able to work more shifts without interruption.[4] It is thus possible that the powerloom sector uses less capital per unit of output, despite the fact that it works fewer shifts (about one and a half compared with two and a half for the mills). If this were not so, the restrictions on the mills could clearly be condemned from a social point of view, since it cannot be right to use both more capital and more labor per unit of output and since the powerloom sector has been overwhelmingly the beneficiary of government policy.

Before summarizing the cost-benefit analysis by which we attempt to quantify the above arguments and also compare powerlooms and handlooms, two further points must be stressed. First, if wages had not been so high in the mill sector, the mills would probably have economized in capital, especially by continuing to use the older and cheaper looms now used by the powerloom sector. This raises the question, discussed at length in chapter 13, as to whether the high mill wages are socially efficient (which they could be if in fact they are least-cost wages despite being high) or whether they are a distortion induced by institutional features of the Indian scene. The upshot of the discussion of chapter 13 is that although mill wages were relatively high before such institutional factors were significantly in evidence, nevertheless the differential has widened in the postwar period as a result of government policies. Therefore, part (though by no means all) of the differences in wages between the mills and powerlooms (and also the handlooms) must be regarded as a distortion. To this extent, one arm of government policy has been twisting the other.

The second point is that the direct increase in employment in weaving resulting from the relative shift to powerlooms may have been more than offset by the effect of India's textile policies on exports of textiles and especially on the export of clothing, which is highly labor-intensive. From 1960 to 1979 the share of Indian textile exports in world trade fell from 8.8 to 4.1 percent, while that of Hong Kong, Korea, Pakistan, and Taiwan rose from 11.3 to 24.9 percent. India's clothing exports were negligible until after 1970 and have remained very small compared with those of the East Asian countries—for example, 11 percent of those of Hong Kong and 16 percent of those of Korea (Wolf 1982, p. 35). We cannot undertake here a full analysis of the incentives and disincentives to export which underlay the Indian failure, but it seems clear that a principal impediment was the very high protection

4. However, in the cost-benefit calculations given below the unit of output is value added, so differences in power inputs are part of the product.

Table 4-1. *Inputs Required to Produce Value Added of Rs10,000 a Year in the Three Sectors of the Weaving Industry*

Input	*Handlooms*	*Powerlooms*	*Mills*
Number of workers (adult male equivalents)	3.36	1.76	0.51
Fixed capital costs (rupees)			
Machinery	534	3,426	
Land and building	5,155	4,480	
Total	5,689	7,906	
Working capital	3,024	2,600	
Total capital	8,713	10,506	19,000

Source: Mazumdar (1984a, table 2-1).

and inefficient production of synthetic fiber and yarn. Indian cotton also is of low quality, and imports have been restricted. However, these policies were essentially independent of that of discriminating against the mills in favor of the small-scale sectors. But it cannot be ruled out that this discrimination itself reduced the ability of the mills to compete in export markets.[5]

Textiles: Private and Social Cost-Benefit Analysis

The analysis of textiles that follows is somewhat limited. It is largely based on surveys of forty handloom and thirty-seven powerloom units in the textile town of Mau in Uttar Pradesh. While we have some confidence that reasonably reliable figures resulted from these intensive surveys, we are in no position to reject the possibility that Mau's units are not typical. For this reason we make some comparisons with the Reserve Bank of India (RBI) survey (1979), referred to elsewhere in this book, the samples for which were drawn from all of India. We have no personal knowledge of the quality of the RBI data, but it has to be noted that there are some marked differences.

While we are reasonably confident that the comparison we make between handlooms and powerlooms is broadly valid at least for Mau, we cannot have as much confidence in the comparison made between powerlooms and the mills. Mill inputs and costs are all taken from official and unofficial secondary sources (described in the appendix to this chapter). There are marked divergences between these estimates, and we have used either an average or those that appeared to be the more reliable. It is also unclear for what range of mills the estimates were made.

The details of the samples; of the technology and organization of the two sectors; and of the estimations of inputs, factor costs, and value added are all given in the appendix to this chapter. Here we present and discuss only the results. These are presented in table 4-1 and summarized in figure 4-1, which

5. For further discussion of India's lack of competitiveness, see Wolf (1982) and Nayyar (1976).

Figure 4-1. *Unit Isoquant for Cotton Weaving*

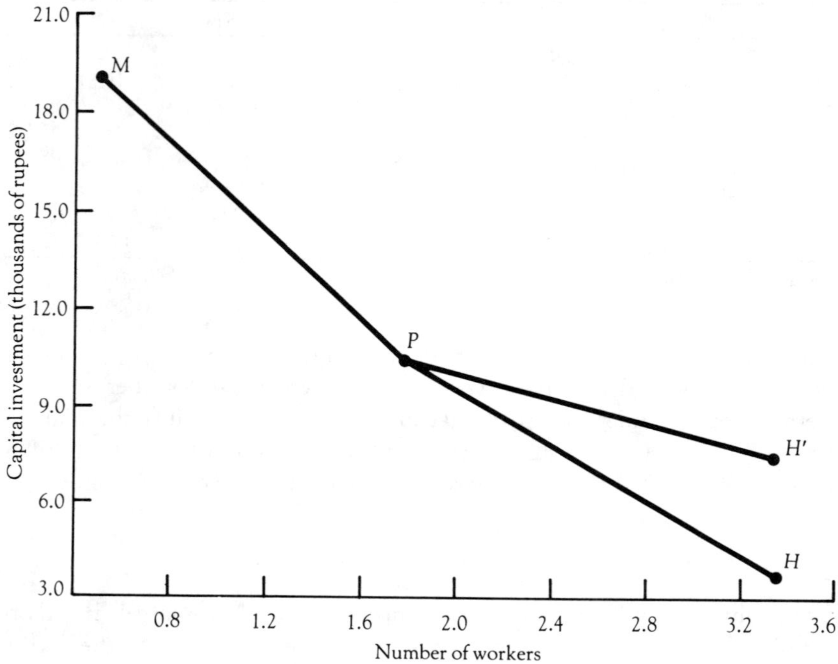

Note: The isoquant unit is one thousand square meters. M = mills; P = powerlooms; H and H' = handlooms.

discloses a well-behaved isoquant with the variation in capital intensity being in the generally expected direction. Note that two alternative points are plotted for handlooms: point H represents only the cost of machinery plus working capital, while point H' adds 75 percent of the value of land and building to capital costs. As explained more fully in the appendix, the former is relevant for calculation of private profitability. The master weaver in the handloom sector typically advances raw material and maintains the loom in the weaver worker's residence. He pays the weaver a piece-rated wage based on the amount of cloth produced. Thus for the master weaver's calculation of private profits, the capital cost of the handloom and the working capital used are relevant—not the value of land and building, for which no rent is paid. However, for social profitability calculations some of the latter have to be added to the cost of capital.

Given the isoquant, we need to know the wage cost per adult male worker in the three sectors in order to find the range of interest rates over which the different sectors are profitable from the point of view of a private entrepreneur. In Mau, the earnings of an adult male worker were very similar in the powerloom and handloom sectors—about Rs2,100 a year. Hired male workers worked by themselves on powerlooms, while on handlooms adult males were assisted by members of their families to varying degrees. The participation of family members—and how to value it—is important for the evaluation of shadow wages to be used in social profitability calculations. From the point of view of the private profitability of an entrepreneur, however, the earnings of each adult male worker are his labor cost. Wages in mills are far higher—our best estimate is Rs4,999 (see appendix to this chapter).

Private Profitability

Production by powerlooms requires more capital for each unit of value added but less labor cost than that by handlooms. Thus whether or not a switch to powerlooms will be profitable for a master weaver depends on the rate of interest entering his calculation.

In the case of handlooms, master weavers (as discussed earlier) bear the cost of the machinery and of working capital amounting to Rs3,558 (per Rs10,000 of value added), but not of the land and buildings of the weaver's residence in which the looms are located. For powerlooms, they bear the whole capital cost of Rs10,506 shown in table 4-1. Thus for the same amount of value added, a switch to powerlooms implies an additional capital cost of Rs6,948. The saving in the wage bill is about 1.6 times Rs2,000 a year. If the expected life of the capital stock is assumed to be twenty-five years, the internal rate of return which makes master weavers indifferent between the two processes is 46 percent. At any lower figure, they have an incentive to shift to powerlooms. This is a very high figure, and one would expect handlooms to be declining if it is valid, especially as powerlooms are very profitable in absolute terms (our sample estimates for the mean profitability were more than 50 percent; see chapter 11). But this is not the case. So we may have underestimated the capital required for powerlooms.

We next compare the mill sector with powerloom production. Table 4-1 shows the saving in capital cost of producing value added of Rs10,000 in powerlooms rather than in mills to be Rs8,494. Although wage rates are lower in powerlooms, the number of workers needed to produce the given value added of Rs10,000 is sufficiently larger to result in an increase in the wage bill of Rs1,120. The rate of interest that equates the present value of the increase in wage cost to the saving in capital cost is about 13 percent. At lower rates of interest, mill production is preferred. Thus, for a very wide range of interest rates (13–46 percent), powerlooms appear to be the most profitable.

Very different results emerge from using the RBI sample. For a move from handlooms to powerlooms, the saving in labor is less, and the wage is (unlike that in Mau) about 40 percent higher, with the result that the wage bill is higher. Since capital costs are also higher, powerlooms are dominated by handlooms. The same survey found that gross profit as a percentage of total net assets was only 14 percent for powerlooms as opposed to 31 percent for handlooms. In view of the very rapid growth of powerloom units, our results seem more plausible, even if we have somewhat overestimated powerloom profitability.

Social Profitability

The shadow wage of a millworker appropriate to the evaluation of social costs may differ from the market wage. There is probably no difference in the case of powerlooms, since the labor market is highly competitive. But institutional influences have (at least in the postwar period) raised mill wages, and the shadow wage is clearly below the actual wage in this sector. In the case of handloom workers, it can be argued that their shadow wage is somewhat less than the market wage for powerloom workers, since they have the advantage of working at home, their hours are flexible, and they can be assisted by other family members. Finally, part of the value of the land and buildings of handloom workers should be included in the social costs of production.

THE POWERLOOM-MILL COMPARISON. At one extreme, if distributional considerations are ignored, the shadow wage for the mills could be equated to that of the powerlooms. Mill production would then represent a shadow wage cost saving of Rs2,520 for a capital cost of Rs8,494, which would give a social rate of return of 30 percent (again supposing a twenty-five-year life of all equipment). This compares with the 13 percent private return. The other extreme is to suppose that the real cost of mill labor is the actual cost (this would make some allowance for the distributional disadvantage of paying mill labor so much more than powerloom labor), which reduces the social rate of return to equality with the private—that is, 13 percent. Given that the target social rate of return in India is reckoned to be 12 percent, even this lower extreme yields the result that mill operation is socially more desirable than powerloom operation. If we based our analysis on the RBI sample, this would be even truer, since that sample has the powerlooms using more labor and more capital than they do in our sample.

THE HANDLOOM-POWERLOOM COMPARISON. If half the cost of land and buildings is added and the rather extreme assumption is made that the shadow wage in handlooms is half the market wage for powerlooms, the result is that a switch from handlooms to powerlooms has a social rate of return of 35 percent. If the RBI sample is used, the very different result is that powerlooms have a higher social wage bill for each unit of output and also use more capital.

CONCLUSION FOR SOCIAL PROFITABILITY. From the Mau sample and the adopted estimates for the mills, the switch from handlooms to powerlooms and the switch from powerlooms to mills are both socially profitable. This remains true even on the extreme assumption that shadow and market wages are equal for the mills if the target social rate of 12 percent is used. In any event, the range of interest rates (13–46 percent) within which powerlooms are the profitable sector is no longer so wide when shadow wages are substituted for market wages. In the RBI sample, powerlooms are socially inferior to both mills and handlooms. The comparison between mills and handlooms—if we use the RBI sample for handlooms and our adopted estimates for the mills—is too shaky to give a figure, and therefore it cannot be reasonably demonstrated that handlooms are socially less profitable than mills. In short, a preliminary case has been made that powerlooms are socially unprofitable, but much more detailed work on costs is clearly needed.

A principal conclusion that can be drawn from the above cost-benefit analysis is the doubtful social profitability of intermediate technology in the textile-weaving industry. This sector has grown at a spectacular rate because of the restrictions on mills in the product market and the large wage advantage accruing to powerlooms. When the shadow wage is adjusted downward for both mills and handlooms, the range of interest rates within which powerlooms are profitable is narrowed significantly.

Sugar: A Weak Case for Intermediate Technology?

A somewhat parallel story emerges in the case of another intermediate technology promoted in India. In the sugar industry, the open plan sulfitation (OPS) process of producing khandsari in small-scale operations has been hailed by many as a technology that is preferable to processing sugar in large mills using the vacuum plan sulfitation process (VPS). The cases of the sugar and weaving industries are not exactly similar; powerlooms grew as an unintended consequence of official policy, while khandsari has been deliberately supported in preference to mill-produced sugar. But they are somewhat similar: the sugar produced in the household sector, called gur, was never discussed in the khandsari mills controversy, and therefore the detrimental effect of the expansion of khandsari on the gur industry has been an unintended result of policies which had not been properly evaluated.

Apart from the omission of gur in the discussion of alternative technologies in the sugar industry, policy formulation in this area was not undertaken in an atmosphere of ignorance or silence. In fact, a long tradition of debate exists on the relative merits of the OPS and VPS processes. Unfortunately, two diametrically opposite views prevailed in interested circles on the OPS method, and there is no authoritative evaluation of the reports and studies on which these views were based. The first current of thought was represented by the staff of the National Sugar Institute at Kanpur. Its reports maintained that the VPS process was superior, both technically and economically, but that since the shortage of investment funds prevented crushing all cane by this method, less efficient techniques should be accepted to meet the demand for sugar in the medium to long run. This opinion has been challenged by representatives of the Appropriate Technology Development Association and of the Planning Research and Action Division, both located at Lucknow.

While they have recognized the technical superiority of the VPS process, they have maintained that the lower recovery rate of the OPS process could be more than compensated for by the lower capital costs and the higher degree of employment generation. They also argued that OPS provided better opportunities for dispersal of economic activity and linkages at the local level. Occasionally they referred to the possibility of improving the economic performance of OPS units through technical innovations; this view received a boost from a study by the ILO which concluded that the present value of the costs of both processes would be equal for a crushing season of 150 days but that the costs of OPS would be lower for seasons of shorter duration. Haan (1980, p. 3) has pointed out serious deficiencies in these studies, both in their assumptions and in methodological errors in their cost-benefit calculations. At the time of the commencement of Haan's careful study in Uttar Pradesh, "the evidence . . . did not permit one to draw unambiguous conclusions." The number of studies in favor of VPS was remarkably small, but this only reflected the greater zeal displayed by partisans of intermediate technology.

The Planning Commission in the draft five-year plan for 1978–83 proposed to ban the licensing of new sugar mills. As shown by Haan, the linear-programming exercise undertaken by some of the staff of the commission to provide the "scientific" basis for such a far-reaching policy was unsatisfactory (Haan 1980, section 4, pp. 22–27).

When a new government came to power in 1980, a new plan was prepared for the period 1980–85 which dropped the earlier proposal for a complete ban on new mills. This was hardly a surprise, for India was undergoing a severe sugar crisis and sugar prices were rising to record levels.

Evidently, the rapid growth of khandsari units in the past twenty years has not been able to make up for the loss in utilization of sugar-producing capacity in the mills. Apart from direct restriction, sugar mills, like textile mills, bore a much higher burden of taxation. Mills also suffered from a shortage of the supply of cane. Sugar mills have been subjected to the so-called dual price system, under which the price of their output as well as the price of cane inputs have been controlled. The official policy allowed khandsari units to be set up only far from the large mills so that the two types of establishments would not compete for sugarcane from the same fields. But during periods of bumper crops of sugarcane, the government in practice allowed khandsari units to multiply in the same areas as the mills. In later years of lean crops, the problem of raw material supply became acute for the mills. Thus the strange coexistence of shortage of sugar and low-capacity utilization in the mills has been a common phenomenon in the industry.

The cost-benefit analysis undertaken by Haan showed that correcting only for the excise taxes made "the NPV (net present value) of VPS (mill sugar) exceed that of the other technologies so much that it would be profitable to have existing OPS (khandsari) units replaced by sugar mills since the value added forgone by the entrepreneurs and the workers of OPS could be easily financed from the extra benefits obtained by VPS" (Haan 1980, pp. 15–16). Even when simulated free market prices were used by Haan, his sensitivity analysis showed that the NPV of VPS was much larger than that of OPS except when the recovery rate of sugar from cane under the OPS process *as well as* the price ratio between khandsari and mill sugar were set at unreasonably high levels (Haan 1980, pp. 16–18, table 3-4, p. A-3-5).

Haan also examined the distributional aspects of the government preference for khandsari. Paradoxically, since the mill sector is the one that pays income tax, the principal consequence of official policy is a redistribution of income from the government to owners of khandsari units. Sugar production differs from the textile industry in that relative labor costs are not of great significance. "Since the wage costs are a small proportion of total costs, even a social price of zero for semi-unskilled labor will hardly affect the appraisal" (Haan 1980, p. 33). Employment of unskilled labor—and the associated distributional effects—would probably be of more importance in gur, but government policy did not address the question of gur as a possible alternative to khandsari.

According to Haan, the main beneficiaries of the development of khandsari have been rich trader families and large landholders. The khandsari industry is a very profitable outlet for the rural surplus—and also adds to the profits of farmers by providing an alternative market for the lucrative cane crop. The official backing for this sector—and its low burden of taxation, in particular—is connected with the need to get political support from the rich rural lobby. Although we did not examine the political economy of the development of powerlooms in any detail, we would not be surprised if similar factors were found to be responsible for the spectacular (although unofficial) growth of this sector.

Light Engineering Industries: A Case of Absent Linkages between the Small and the Large Sectors

Engineering activities, which primarily involve the casting, forming, or shaping of metal and manufacture of electrical apparatus, have been subject to increasing levels of intervention by the government to restrict the capacity of large-scale units and to promote small-enterprise development. The principal incentive to small-scale enterprises in the light engineering sector has been the reservation of incremental capacity in a wide range of product groups. With the introduction of product reservations, firms with assets exceeding the SSIB limit were confined to production at levels equal to their installed capacity at the time reservation was imposed. Thus, all incremental expansion in those product groups was confined exclusively to small enterprises.

More than half of all activities reserved for small-scale enterprises are in light engineering sectors; approximately one-quarter are in mechanical engineering. Among mechanical engineering product lines of particular importance are castings, machine tools, diesel engines, and hand tools.

Light engineering products reserved for small-scale production are also eligible for excise duty exemptions and preferential purchasing arrangements. For the most part, light engineering firms benefit from the excise tax regulations applicable to activities with output below Rs3 million, although a limited number are on the schedule of the product groups with lower output ceilings for excise exemption. In addition, a 15 percent price preference is granted to small firms by the central government. The combined effect of excise exemption and preferential purchasing arrangements is equal to approximately 30 percent of the ex-factory value of output.

Small-enterprise promotion policies have significantly influenced the structure of India's engineering industries. Engineering industries in most countries are characterized by relatively large-scale manufacturing and assembly operations which are supplied by a diversified base of component manufacturers and ancillary firms. Present incentives and reservations in India, however, encourage SSEs to manufacture complex finished products, such as diesel engines, consumer durables, and bicycles, rather than to specialize in the production of individual components. For such final-stage manufacturing, large-scale producers report difficulty in obtaining an adequate supply of acceptable quality parts from existing SSEs. Many have had to resort to manufacturing more components in-house than they would have preferred in order to meet minimum acceptable standards of quality.

Surveys of small-scale engineering enterprises have revealed that component manufacturers often use badly worn machine tools without basic jigs and fixtures. These deficiencies make it difficult for SSEs to meet quality and standardization requirements, despite impressive labor force skills. In some activities, moreover, lower quality is not compensated for by lower prices. In bicycle manufacturing, for example, the captive ancillaries of the principal manufacturers produce superior components at lower production costs. Small-scale diesel engine component suppliers, however, have lower unit costs than the component manufacturing operations of larger firms. But their low quality standards are an incentive for continued vertical integration by large units.

As a result of the high degree to which large firms are vertically integrated, there is little competition between small and large enterprises in the markets for individual components. Rather, specialization has tended to occur within the small-enterprise sector with some firms manufacturing individual parts for other small-scale assemblers. The principal technical need of SSEs in the engineering sector is to upgrade their general purpose machine tools and to use suitable jigs and fixtures to assure improved machining tolerances. The current structure of small-enterprise protection provides few incentives for such upgrading of the capital stock since enterprises can sell all of their existing output regardless of quality defects.

The structure of the engineering sector has given rise to significant spatial concentration of light engineering activities. Small-scale firms tend to group in manufacturing centers in order to be close to downstream users and to benefit from agglomeration economies. Thus, the structure of the sector tends to work against the objective of the decentralization of industrial activity, which was one of the motivations for small-enterprise promotion.

In the following sections, we explore in greater detail the effects of SSIB promotion policies on the structure and performance of some main engineering subsectors.

Diesel Engines

India produced about 475,000 diesel engines in 1978, of which 94 percent were 1–20 horsepower. The two largest manufacturers, Kirloskar, and Ruston and Hornsby, produced about 16 percent of total output, while small enterprises centered in Rajkot (Gujarat) produced 42 percent of the total.

The small-enterprise manufacturing networks in the three principal producing centers, Rajkot, Kolhapur, and Punjab, have a unique mode of production based on an extensive breakdown of production among specialized component suppliers and assemblers. In Rajkot, for example, about 2,500 units manufacture components for approximately 150 engine manufacturing firms. Of these manufacturers, however, the three largest account for 30 percent of total output of completed engines, and the 200 largest firms produce the majority of completed engines, parts, and components.

Low-horsepower Indian diesel engines are primarily of slow-speed, long-stroke design. These large and heavy engines consist of cast iron blocks; crankcases; cylinder heads; and heavy forged-steel crankshafts, camrods, and camshafts. The design makes the engines tolerate wide dimensional discrepancies during manufacture, rough handling, and limited maintenance. These attributes, combined with the engines' tolerance for low-quality fuels, have made them particularly suited to slow-speed stationary applications in agricultural irrigation, but—in contrast to modern high-speed diesels—at the sacrifice of fuel economy, versatility, and reliability.

SSIB reservation has resulted in limited linkages between small and large manufacturing enterprises in the diesel engine industry. For the most part, small component-manufacturing firms supply small assemblers with parts which have become standardized during the fifty-year history of the development of slow-speed engines. Larger firms have had difficulty obtaining enough components of sufficient quality and tolerances from small independent producers and have therefore resorted to more in-house production than is common in the industry internationally. Nevertheless, large firms purchase about 50 percent of their components from captive ancillaries and subcontractors.

Table 4-2 contrasts the production cost structure of a vertically integrated large-scale engine manufacturer with the production costs of a typical small manufacturing network. For engines of comparable quality, there is no appreciable difference in unit costs between large and small manufacturing processes. Indeed, in the absence of capacity restrictions, large-scale production arrangements would probably lower unit costs to the level of small-scale units or below because of greater vertical integration.

The principal effect of SSIB reservation has been to limit the ability of the industry to innovate and upgrade technology in the high-speed engine market. Tooling requirements and the length of production runs, coupled with the lower tolerance of the lightweight engines for manufacturing discrepancies, would make it difficult for existing small enterprises to compete successfully with large-scale enterprises and foreign manufacturers in the production of the newer engine design. The larger Indian firms have the capability to manufacture the newer lightweight models, but much tighter linkage between the large-scale manufacturing operations and an expanded network of specialized ancillary and component manufacturing firms would be required for them to do so.

Bicycles and Components

India produced about 4 million bicycles in 1979–80, of which approximately 9 percent were exported. The ten largest manufacturing firms account for

Table 4-2. *Production Cost Structure of Small and Large Diesel Engine Manufacturers*

Cost item	Large-scale manufacturer, export quality	Small-scale manufacturing network: Current quality	Small-scale manufacturing network: Export quality
Materials	45.0	39.0	41.0
Labor	10.0	6.2	7.8
Depreciation	1.6	0.3	5.2
Power and fuel	1.5	1.0	1.5
Financial costs	2.0	1.5	2.0
Overheads	5.0	1.6	5.0
Total ex-factory cost	65.1	49.6	62.5

Note: Production cost given as a percentage of dollar cost of production for developed-country competition.

Source: Barry (1983).

approximately 90 percent of total production, with the largest three producing more than 3 million units. The industry is centered in Ludhiana (Punjab), where two of the three largest firms and the majority of small enterprises are located.

Large-scale manufacturer-assemblers are permitted to manufacture components only for their own requirements, and the larger component manufacturers are limited to their licensed capacity by SSIB reservation. Large firms produce up to 80 percent of their own components in-house or through closely affiliated ancillaries. Many of the bicycle components currently reserved for SSEs are not produced to adequate quality standards. As with diesel engines, SSEs fail to use simple jigs and fixtures which would improve the speed and accuracy of production.

The principal Indian model is a utility bike, capable of transporting heavy loads and of withstanding substantial abuse. The standard light roadster (SLR) design intended primarily for recreational use requires substantially more sophisticated manufacturing methods and is not well suited to the technological levels of small manufacturing firms. It therefore appears likely that further development of the industry will involve segmentation of the market by firm size, with large-scale firms producing the SLR design and small-scale manufacturers concentrating on the utility market.

Hand Tools

India produces a wide variety of hand tools—such as spanners, vices, pipe wrenches, and pliers—which are sold domestically and exported to both developing- and developed-country markets. The industry is split between forged tools, the production of which is dominated by large-scale producers, and cast tools, which are produced primarily by small-scale enterprises. In 1979–80, the forged tool output, valued at approximately Rs328 million, was produced by 12 large and medium-size enterprises. The same year's cast tool production, valued at Rs230 million, was produced by approximately 200 registered small enterprises and more than 1,000 unregistered workshops. About 70 percent of total forged and cast hand tool output was exported.

The dominance of large-scale firms in the forged tool market is mainly caused by indivisibilities in the capital costs associated with the drop forging process. The forged tool manufacturing process is labor-intensive, and general purpose machines with fixtures are the standard capital equipment required. There is wide latitude in the dimensions of the finished product permitted, and there are few critical tolerances involved. For the most part, Indian manufacturers meet international requirements in the do-it-yourself market, and with some upgrading of design and finish they could penetrate the industrial tools market in industrial countries.

Hand or bench tools made from castings are produced primarily for the Indian market. Tolerances are set to the thinnest sections which can be conveniently cast and therefore normally exceed minimum specifications at the expense of material wastage. Major changes in casting methods and quality would be required to meet the minimum standards of the international industrial tools market.

Conclusions

This chapter has illustrated some of the difficulties in administering India's policy of protecting the small-scale sector. In all three industries considered, two points of criticism stand out. First, there was a lack of information about the industry in question available to policymakers at the time when they made their decisions. Second, in part as a consequence of this gap in knowledge, blunt policy instruments were used uniformly across industries.

The restriction on the licensed capacity of the large-scale sector and the much heavier incidence of taxation (both income and excise taxes) on this sector were the major instruments for protecting the small-scale sector. But as the cases of textiles and sugar illustrate, it was not clearly recognized in formulating policies that the small-scale sector itself contained at least two subsectors with widely varying technologies. In textiles, the announced policy of protecting handlooms was defeated by the unintended growth of small (and sometimes not so small) powerloom units whose importance and potential were apparently not recognized. In sugar, in contrast, the concentration of attention on khandsari led to virtual neglect of the traditional small-scale gur industry.

In both industries, the consumer of the products bore a major part of the burden of the industrial policy. In sugar it may be possible to argue that, insofar as consumers of processed sugar were largely urban and gur was a predominantly rural product, the cost of protection of the khandsari industry was borne by relatively richer households. But the consequence of protection in the textile industry is not at all clear as far as the poorest consumers are concerned. Cloth produced by the powerloom sector may have been costlier and less durable than mill-made cloth. The point becomes more important when we take into account the strong barriers raised against the development of the synthetic fiber industry, which has become in most countries the producer of cloth for mass consumption.

The distribution effect on households as producers, rather than as consumers, was probably not so regressive. Wages of workers employed in the units that used intermediate technology were substantially lower than the wages in the large-scale sector. The increase in employment in the small units relative to the large-scale sector therefore shifted the distribution of labor earnings toward low-wage workers. As against this, the significant redistribution of income from the state to the owners of khandsari and powerloom units (who were often rich landowners) must be considered regressive.

The light engineering sectors produce mainly capital goods. Hence the distributional effects of the small-enterprise reservation policy are difficult to trace. The policies have had a dramatic effect, however, on the structure of the engineering industries and on their adoption of technology. SSIB reservation has resulted in reductions in the linkages between small and large enterprises and in the limited development of specialized subcontracting and ancillary firms. The structure of the industries which developed subsequent to the capacity restrictions on large-scale firms has inhibited the adoption of new technologies for both products and processes.

It is obvious that such issues could not be explicitly considered by India's policymakers because cost-benefit analyses of alternative techniques were either nonexistent or misleading. An illustration of the latter is provided by the studies of sugar technology reviewed by Haan. It is difficult to avoid the conclusion that the results of the cost-benefit work on this industry reflected the preconceptions of the analysts. Textile policy, in contrast, seems to have been formulated in an environment of almost total lack of information.

The absence of relevant information was also responsible for a uniform application of reservation policies in the engineering industries. These policies have clearly acted to reduce any dynamic benefits which might have accrued from innovation in small-scale enterprises and to encourage an industrial structure which inhibited technological development and product upgrading in the more technologically complex product groups. Thus, the effect of capacity restrictions on diesel engines and bicycles has reduced the performance and international competitiveness of these industries more than in the case of hand tool manufacturing.

Appendix

The Textile Sample

HANDLOOMS. Our survey included two separate types of handloom establishments: the small handloom unit, which is essentially a family enterprise located in the house of the owner; and the large master weaver's unit, which typically employs weavers as hired workers who, however, work in their own homes. The distribution of the two types of units in our sample is given in table 4-3.

The difference in organization between the two types of units merits further explanation. Weavers do not typically do their own marketing. They sell their output at a piece rate to master weavers, who normally advance the yarn needed for the product. This relationship between the producer and the master weaver is as true for the small unit as for the master weaver's establishment. The difference is that in the case of the small establishments, the weavers themselves own the looms, while in the larger enterprises the

Table 4-3. *Distribution of Handloom Units in the Sample Size*

Number of looms in unit	*Mean employment*	*Mean capacity (meters per day)*	*Number of units*
		Small units	
1	3.4	8.2	14
2	4.7	13.6	9
3	5.7	21.1	7
4	6.0	30.0	1
5–6	6.0	20.0	1
		Large units	
15–20	37.5	310.0	2
30–35	62.5	363.0	4
45	98.0	540.0	1
70	140.0	770.0	1

weavers are set up in their own residences with equipment paid for by the master weaver. This shows in the figures for initial capital required to start the business. They are typically small for the independent units and large for the master weavers' establishments (varying with the number of looms owned). The difference between the two types of weaving units is critical in evaluating the cost of both fixed and working capital as well as value added and earnings per worker.

Powerlooms. The powerloom units in Mau used secondhand as well as new looms. Unlike the case of handlooms, the sheds in which these powerlooms are located are not always part of the owner's dwelling. The cotton and staple yarn is first sent to large-scale sizing plants in the town for preparing beams. These beams are wound onto bobbins, which in their turn are fixed onto the shuttles of the powerloom for weaving the cloth. At the end of the process, the manufactured cloth is again sent for calendering to a larger unit to give a glaze to the product. The size distribution of the powerloom units covered in our sample is given in table 4-4.

Type of Products. The handlooms produced only saris. Most produced only one quality of sari—medium 40s × 40s count.[6] The powerlooms produced dhoties and chaddar as well as saris. It was not possible to get cost data separately for the three types of output. The comparison between handlooms and powerlooms must therefore be in terms of a rupee of value added

6. "The thickness of yarn is called its count. Several measures of count are currently used. In the nonmetric Anglo-American one [used here] a count of 1s indicates that one hank, or length of yarn [amounting to] 840 yards, weighs one pound. A count of 20s . . . thus implies that twenty lengths of 840 yards (16,800 yards) weigh one pound. The higher the count, the finer the yarn, because it weighs less per hank." (Adapted from Howard Pack, *Productivity, Technology, and Industrial Development: A Case Study in Textiles* [New York: Oxford University Press, 1987].)

Table 4-4. *Distribution of Employment and Units by Number of Looms in the Powerloom Sample*

	Number of looms						
Employment and units	*1–5*	*6–10*	*11–15*	*16–20*	*21–30*	*31+*	*Total*
Average number employed							
Hired	8.50	14.75	21.70	27.88	43.00	43.00	19.9
Hired plus family	8.63	15.25	21.70	27.88	43.00	43.00	20.0
Number of units	8	8	10	8	1	2	37

of the cloth produced. The comparison makes sense because the powerlooms, like the handlooms, concentrated on the medium 40s × 40s count range in the cloth woven, even if they did not specialize in saris.

The Economics of Handloom Weaving

THE EARNINGS AND PRODUCTIVITY OF WEAVERS. The basic data on labor and production for the two types of handloom units are given in table 4-5. The most important difference between the two types is the type of labor involved in weaving and the output per loom. In the small units, most of the family is working at least part time. In the larger units, only adult male workers are reported to be working as hired labor. It is conceivable that they are assisted by some family members who were not recorded in the master weavers' statements to the interviewers. There seems to be a real difference, however, between small and large units in the number of adult male workers per establishment. If one looks at the figures on earnings accruing to the workers, the average monthly wage per adult male worker is roughly the same in the two types of units (about Rs160).

There is one loom per adult male worker in the independent units, and a half in the master weaver's units. But productivity *per loom* in the small units is little more than half that in the master weavers' establishments. Thus the larger number of looms per worker in the independent units means that there is much underutilization of loom capacity in such units, which leaves the productivity per worker more or less the same in the two types of units (if only adult male workers are considered). This seems reasonable if the organizational difference noted above between the two types of units is considered. When the looms are owned by the master weavers, they are more likely to be maintained in good working order, while many of the looms owned by the small-scale weavers are old and of low efficiency. For the purposes of this exercise, we therefore take the number of weavers per loom, the productivity per loom shift, and the earnings per worker as given in table 4-5 for the larger units.

THE COST OF FIXED CAPITAL. The low price of looms used in the handloom industry makes the cost of machines a minor part of total fixed capital,

Table 4-5. *Labor, Loom, and Production in Handloom Sector*

Type	Number of looms per unit	Number of workers per unit					Pieces produced per month	Average wage per price	Earnings per month per adult male worker (rupees)	Pieces produced per loom per month	Pieces produced per adult male worker per month
			Family[a]								
		Hired	M	WF	WP	C					
Large	34.9	70.4	—	—	—	—	2,304	—	161.0	66.0	32.7
Small	1.9	—	1.94	0.58	0.54	1.42	58.7	5.258	157.2	30.6	29.9

— Not available.

a. M = adult males; WF = full-time adult women; WP = part-time adult women; C = children.

most of which is accounted for by land and buildings. The figures are given in table 4-6. The low value of land and building for the larger master weaver's firm is misleading. The master weavers gave only the value of the premises used for their trading purposes—the space needed for storing raw materials and cloth received from the handloom weavers in their employment. The value of the premises of the weavers actually devoted to production was not included in the questionnaires filled in by the master weavers. For estimating costs of production in the handloom industry, the figure of the value of land and building given for the small producer is more appropriate. The reported cost of land and buildings in the larger "master-weaver" units is given for the part of the premises used for business purposes only. It therefore represents the cost of buildings for storing the clothing before it reaches the wholesaler. Thus, our figure of fixed capital cost per adult male worker is the value of machines in *larger firms* (to get around the problem of underutilization of looms in small firms) *plus* the value of land and buildings in small firms *plus* the value of land and buildings in large firms. This gives a total figure of Rs1,693 per adult male worker.

THE COST OF WORKING CAPITAL. As has been explained, the master weaver advances raw materials (and perhaps wages as well) to the producers, who get paid at the stipulated rate per sari when they deliver the product. The cost of the working capital involved is borne largely by the master weaver. The amount required is a function of the period of production.

In the accounting data obtained from a typical industrial firm, we do not have to consider the period of production because the statistics on raw materials and semifinished products in stock, as well as cash on hand, together add up to the working capital required for the rate of production actually observed. For the handloom weavers' units surveyed by us, however, no such statistics for working capital can be clearly identified. The small independent units did not provide any data on stocks of raw material or products carried. They did, however, indicate the amount of yarn obtained from the master weaver that was processed in their homes. The lag involved in producing the cloth was stated to be one to two months. The larger master weavers' establishments did have stock data, but it is not clear if the stocks of materials advanced by them to their workers (and presumably retained in the latter's homes) were included in the figures given to the interviewers. We compare estimates of working capital derived from the different approaches. Table 4-7 gives three sets of estimates—two for the large

Table 4-6. *Current Value of Fixed Capital per Adult Male Worker*
(rupees)

Unit size	*Machines*	*Land and building*	*Total*
Small	214	1,138	1,352
Large	159	396	555

Table 4-7. *Alternative Measures of Working Capital*
(rupees)

Firm size	*Material cost*	*Wage cost*	*Total*	*One-sixth of total*[a]	*Working capital per adult male worker*[b]	*Current assets*	*Current assets per adult male worker*[c]
Large	306,000	135,000	441,000	73,500	1,044	53,250	756
Small	6,646	3,660	10,306	1,718	886	n.a.	n.a.

n.a. Not available.

Note: All figures are mean values of each group of firms.

a. Hence, the finance required to support the pipeline of production is one-sixth of the total value of annual output.

b. Assumes that the period of production is two months.

c. As directly reported by the large firms.

units, one based on reported stock data and the other on the basis of recurring costs; a third estimate is for small firms based on materials and wage costs. The three estimates are not too far apart. We have already mentioned that the stock data for larger firms probably underestimate working capital. Thus an acceptable figure for working capital in handloom weaving would seem to be Rs900 per adult male worker.

Value Added per Worker. The large units provided data on value added in the statement of accounts given to the interviewers. From these data, the value added per worker was found to be Rs251, as compared with an average wage of Rs161 per worker. Payments for capital and land and the profits of the master weavers account for the difference between the two. Since we have seen that the productivity per adult male worker (in quantity terms) is the same in the two types of units, and that they produce the same type of fabric, we accept the figure of Rs251 as the value added per adult male worker for the handloom sector as a whole. We are thus able to get the labor and capital cost required to produce value added of Rs10,000 a year given in table 4-1.

Powerloom costs

The cost data available from our survey of powerlooms in Mau needed less doctoring. The relevant statistics are available directly from the balance sheets of the firms collected during the survey. Table 4-8 summarizes the data. (Because of the large interfirm variability, geometric means are used.)

However, the number of workers required for production had to be adjusted. Many of the powerloom units suffered from power shortages. Thus although the number of shifts which the plants were planning to work was generally 1.5—involving twelve hours of work—the actual number of hours

Table 4-8. *Inputs Required for Powerlooms to Produce Value Added of Rs10,000 a Year*
(thousands of rupees)

Adult male workers (person-years)	*Fixed capital*			*Working capital*	*Total capital*	*Annual wage per worker (rupees)*
	Machinery	*Land and building*	*Total*			
1.76	3,426	4,480	7,906	2,600	10,506	2,136

worked was less. The stock of labor employed reported by each firm gave the expected number of person-hours worked if each of the workers were employed for an average of 1.5 shifts. The actual number of person-hours fell short of this amount. Since labor was paid on piece rates based on the actual amount produced, the flow of labor time involved in the production of the actual output observed was adjusted downward from the expected flow. Two sets of wage data available in our survey enabled us to make this adjustment. First, there was the average wage per worker actually paid, based on the wage bill for the year given in the balance sheet of the establishment. Second, there was the monthly wage for the workers employed in the establishment (appropriately weighted by major occupational categories), based on the expected level of output. The ratio of the former to the latter, 0.88, could be used as an index of underemployment. The reported stock of labor per unit was multiplied by this factor to give the effective rate of person-years of labor used by the unit.

Relative Costs in Mills and Powerlooms. We did not attempt to collect cost data from mills. It was decided to make broad comparisons from other recent studies. Three sets of data could be identified:

- A study of relative costs by Padmanabhan for 1974 for weaving coarse medium gray cloth (34s × 34s counts). The cost data were derived from relevant published and unpublished sources available from organizations connected with the textile trade rather than from a sample survey.
- A Planning Commission study made under the direction of Raj Krishna in 1979. Again, the basic data were in the nature of "most informed" estimates gathered by the staff of the project from a variety of sources, which included interviews with selected establishments (India 1978).
- A National Industrial Development Corporation Ltd. study of the relative costs of producing coarse cloth (UNIDO 1978). The data seemed to be mainly derived from the technical experience of textile engineers, particularly those working in the Ahmedabad Textile Industry's Research Association.

The following attempt to quantify costs of production uses all three sources cited above and also refers to the statistics reported in the Annual Survey of Industry (although these refer to registered spinning and weaving mills together). The estimate of relative costs of producing gray cloth of medium count (no finer than 40s × 40s count) will be built up step by step.

CAPITAL COST PER UNIT OF VALUE ADDED. To understand the capital cost per unit of value added, several factors must be considered: the cost of machinery, productivity per loom, nonloom capital costs, and the employment–value added ratio.

The nonautomatic looms used by the mills are essentially the same as those used by the powerloom sector, although there is a difference in the condition of the looms. The powerlooms used in the small sector are older, are often bought secondhand, and are less well maintained. There is thus a large variation in the price of a nonautomatic loom used in the two sectors. Padmanabhan's study gives the ratio of the price of looms in mills to that in the small units as 3:1. The Planning Commission study gives a higher ratio, 5:1. We take the average of the two, 4:1, for our calculations.

Productivity depends on the number of shifts worked per day and on the output per loom shift. The relevant data from the two studies cited are:

	Powerlooms	*Mills*
Padmanabhan		
Number of shifts	1.5	2.25
Output in meters per loom shift	27	34
Total output in meters per day	40.5	76.5
Planning Commission		
Production, annual output in meters	10,000	25,000

The Planning Commission figures include difference in shifts per day. Taking into consideration the fact that the number of shifts in the mill sector has been increasing continuously and that the Planning Commission figures are more recent, we accept them as representing overall differences in physical productivity per loom.

Mill cloth is priced higher than powerloom cloth. The Planning Commission study gives the value added per meter as 36 percent higher for mills. This difference represents intrinsic difference in the quality (for example, durability) of the cloth produced, and not the effect of excise duties imposed on mill-made cloth.

In our survey of powerlooms, loom cost accounted for only 30 percent of fixed capital (including land and building) and 23 percent of total capital (including working capital). It is pertinent to assume that the proportion of fixed capital other than looms would be a larger proportion of the total in the mill sector than in powerlooms (nothing very much can be said about the relative importance of working capital without much more intensive inquiry). Thus the capital–value added ratio would be higher in the mill sector than a ratio derived from looms alone. This is borne out by figures available from two other studies. In our own survey, the capital–value added ratio in powerlooms in Mau was 1.05—taking geometric means. The Annual

Table 4-9. *Machine Productivity in the Three Sectors, India*

Aspect of productivity	*Handloom*	*Powerloom*	*Machinery*
Output per loom shift[a]	11	27	34
Number of shifts	1	1.5	2.5
Number of workers per loom	2.02	1.60	0.8
Capital costs per loom[b]			
Machinery	321	3,358	13,435
Land and building	3,100	4,392	n.a.
Working capital	1,818	2,652	n.a.
Total costs	5,239	10,402	57,000

n.a. Not available.

a. Meters.

b. Thousands of rupees.

Survey of Industries (ASI) for 1977–78 gives the capital–value added ratio in the factory sector as 1.91 (but spinning is included). In the study by the Working Group on Textile Technology, figures are given on capital costs—including machines other than looms. The capital-output ratio for mills in this study comes to 2.05 times the ratio for powerlooms (UNIDO 1978, table 9, p. 36). We adopt the ASI figure and thus estimate the value of overall capital required to produce Rs10,000 in the mills as Rs19,000.

We next need to calculate the employment per unit of value added in the two sectors. This is a function of two terms: value added per loom, as has already been discussed; and the number of workers per loom shift. From Padmanabhan's data, the value added per loom shift ratio between mills and powerlooms is equal to (34/27) x 1.36 = 1.71. The number of looms looked after by a worker in a weaving mill varies between 2 and 4, and we take the average as 3. In powerloom units, according to our survey, the number of looms is 1 to 2, with a mean value of about 1.5. We assume that the ratio of other workers to weavers is the same in the two industries. Thus the ratio of employment per unit of value added in mills to that in powerlooms is 0.5/1.71 = 0.29.

Summary of Machine Productivity

Table 4-1 summarizes labor requirements and capital costs per Rs10,000 of value added produced in the three sectors. Table 4-9 summarizes the data on machine productivity (which lie behind table 4-1) which have been discussed in this appendix.

5

The Changing Size Structure of Indian Industry

The policy environment described and illustrated in the preceding two chapters provides substantial incentives for the growth of small-scale firms. This chapter and the following one assess the consequences of that policy regime for the growth and structure of the Indian industrial sector. In this chapter we describe changes in the size structure of Indian industry from the 1950s to the present. Chapter 6 places the current size structure of Indian industry in international perspective.

"Small" is now defined for policy purposes solely in terms of the original or undiscounted book value of plant and machinery, the criterion of less than fifty employees having been sensibly dropped.[1] The limiting amount has been raised several times over the years and now stands at 3.5–4 million. Bearing in mind that the value of plant and machinery is usually less than half of total capital employed, it is clear that factories that would usually be thought of as medium size are included. However, recently a "tiny" sector has been defined with a limit of Rs200,000. It is unclear what the operational significance of this will be.

Most information about the size structure of industry comes from sources that do not usually include (or at least collate and publish) figures for capital, which thus makes it impossible to marry descriptive categories to those in use for policy measures. The only sources of information covering all shapes and sizes of industrial activity are the population censuses. These permit a distinction of household from other activities.[2] The Annual Survey of Industries—and, earlier, the Census of Manufacturing Industry (CMI)—covers only "factories," these being establishments with ten or more employees with power, or twenty or more without. This permits a division between factories and the rest, so that one has a tripartite classification: household enterprises;

1. At first, the limiting capital size referred to all fixed assets, not just plant and machinery.

2. Household enterprises are those carried on mainly by family members. In urban areas they have to be located in the homes of the workers; in rural areas, anywhere within the same village.

factories; and nonhousehold, nonfactory enterprises (for convenience we shall refer to the third as "workshops"). Almost all household enterprises and workshops are small; but, of course, many factories are also small by the official definition. There may be only a rough correspondence between this classification and such technological distinctions as handloom or powerloom, or more generally, traditional or modern. For instance, probably about three-quarters of handloom enterprises are households, almost all the rest being workshops, but with a few factories. The great majority of powerlooms are workshops, but there are certainly some factories and probably a few household enterprises.

Census Comparisons: Household versus Nonhousehold Manufacturing

We compare the censuses of 1961 and 1971.[3] Unfortunately the definition of a worker was changed in the 1971 census, which resulted in a large fall in recorded participation, especially for women in rural areas. This limits the conclusions that can be drawn, for a fall in apparent employment may be either genuine or the result of redefinition. Nevertheless, it is possible to demonstrate a rise in the share of a sector in total male or female employment in some cases.[4] These cases are collected in table 5-1. It must be remembered in reading the table that it does not follow that there has been no rise where no rise is recorded.

The most important result shown by the table is that nonhousehold manufacturing employment grew relatively faster than in the rest of the economy, more decisively in rural areas than in urban areas. It is particularly remarkable that this sector can be shown to have almost certainly increased its share of total female employment in rural areas when the recorded participation rate for rural women declined from 31 to 13 percent as a result of the change of definition.

Household manufacturing has been much affected by the change in definition, for men as well as women. As the table shows, little can be said. Only in Assam, Jammu and Kashmir, and West Bengal, in either rural or urban areas, can household manufacturing employment be shown to have increased its share of the labor force, and that only for men. In these states it

3. The census of 1981 was not yet available. The methods used and results given in this section are drawn from Banerjee (1977).

4. The method of comparison is as follows. A projected work force is calculated by applying 1961 participation rates to 1971 populations. Projected employment in household and nonhousehold manufacturing is next calculated by assuming that their shares of total employment are the same as in 1961. If the recorded 1971 employment exceeds projected employment, then the rise in the employment share is real (it is reasonably assumed that real participation rates have not risen).

Table 5-1. *Changes in Employment Shares of Sectors in Fifteen States, India, 1961–71*

Nonhousehold enterprises				*Household enterprises*			
Urban		*Rural*		*Urban*		*Rural*	
Male	*Female*	*Male*	*Female*	*Male*	*Female*	*Male*	*Female*
Rise in five states,[a] very probable rise in two others[b]	No statement	Rise in eleven states[c] (and in all India)	Rise in six states,[d] very probable rise in one other[e]	Rise in three states[f]	No statement	Rise in two states[g]	No statement

a. Andhra Pradesh, Tamil Nadu, Karnataka, Punjab and Haryana, and Rajasthan.
b. Gujarat, Maharashtra.
c. The exceptions are Assam, Bihar, Jammu and Kashmir, and West Bengal.
d. Andhra Pradesh, Kerala, Tamil Nadu, Maharashtra, Karnataka, Orissa.
e. Madhya Pradesh.
f. Assam, Jammu and Kashmir, and West Bengal.
g. Assam and Jammu and Kashmir.
Source: B. Banerjee (1977).

Table 5-2. *Male Work Force, India, 1961–71*
(millions)

	Rural		*Urban*		*Total*	
Sector	*1961*	*1971*	*1961*	*1971*	*1961*	*1971*
Mining	0.6	0.5	0.2	0.3	0.7	0.8
Household industry	4.6	3.8	1.2	1.3	5.9	5.0
Nonhousehold industry	2.1	3.2	5.3	7.2	7.4	10.4
Construction	1.0	1.0	0.9	1.0	1.8	2.0
Total secondary	8.4	8.5	7.6	9.8	15.8	18.2
Total primary	86.3	100.8	2.2	3.3	88.6	104.1
Total tertiary	12.1	11.1	12.6	15.6	24.8	26.8
Grand total	106.8	120.4	22.4	28.7	129.2	149.1

Source: Original census figures. Quoted in Haan (1980, part II, table 3).

could not be shown that nonhousehold manufacturing had increased its share, and the probability is that household manufacturing gained relatively as a result of the industrial stagnation that afflicted these states in the 1960s.

While it is clear that relative nonhousehold manufacturing employment has risen, especially in rural areas, it cannot be strictly shown, because of the definitional change, that this is at the relative or absolute expense of household manufacturing employment. However, it has been argued that the definitional changes scarcely affect male workers aged 15–60, and since these account for 85 percent of the male work force the effect of the change in definition should have been small (Krishnamurty 1973). Yet in rural areas the decline in recorded census employment in household manufactures was 19 percent for India as a whole, some absolute decline being recorded in all states except Assam, Jammu and Kashmir, and Rajasthan (Banerjee 1977, table A-3). It seems reasonable to conclude that there was certainly a relative and quite likely an absolute fall, and that the growth in nonhousehold manufacturing male employment was partly at the expense of household manufacturing in rural areas. Recorded male employment rose absolutely in all other sectors, except rural mining, as is shown in table 5-2.

It is not certain that the decline of household manufacturing continued into the 1970s. According to figures calculated by Mazumdar (1984, table I-3), handloom output stagnated. But according to the Planning Commission it grew at 5.5 percent a year from 1973 to 1979, and other traditional industries grew quite fast (sixth five-year plan, table 12-1).

Census Comparisons: Growth of Industrial Establishments in Rural and Urban Areas, 1961–71

The census has tables which permit one to study how the number of establishments of different sizes changed, in both urban and rural areas, and by state and industry. Unfortunately, however, the 1961 census did not record average employment in each employment size class, so that employment estimates cannot be made except on the questionable assumption of no change in these averages. Table 5-3 summarizes the data for all of India. It shows that the number of establishments grew nearly twice as fast in urban areas. Yet in the previous section we found that there was some evidence of a more rapid growth of manufacturing employment in rural areas. The two observations are reconcilable only if the number of larger establishments grew faster in rural areas. This is indeed the case, as the table shows. Small establishments (nonfactories) grew more than twice as fast in towns. Medium-size and large factories (more than fifty workers) grew twice as fast in the country; this growth supports the view that the road to decentralization is not paved with very small establishments.

Table 5-4, which has a different size classification, indicates that the slower growth in rural establishments is due to the very slow growth of one-person establishments. This, however, is an unreliable statistic since the census authorities admit to missing a great many such establishments: if census takers have become more efficient over the years, it could be that one-person establishments have actually declined. In the urban areas the 2–19-employee size class grew the fastest at the rate of 6.4 percent a year. This contrasts with the experience of rural areas, where the units in the larger size classes grew relatively faster. The two tables together indicate that establishments with 50–100 workers were the fastest growers in rural areas. Finally, it is worth mentioning that the tails of the distribution have grown relatively slowly, so that Dhar and Lydall's (1961, pp. 30–31) law of the excluded Indian middle may be in course of repeal. This matter is discussed at greater length in chapter 6.

Census Comparisons for Selected States

We decided to look at data for individual states. However, to make the work manageable we had to confine our attention to five states of North India.

Table 5-3. *Annual Growth Rate of Establishments in Manufacturing, Processing, and Service Industries by Employment Size and Rural or Urban Location, India, 1961–71*

		Number of workers		
Location	*Total*	*1–9*	*10–49*	*50+*
Total	3.763	3.7	5.9	5.3
Rural	2.970	2.8	7.4	7.7
Urban	5.446	5.8	5.7	3.9

Source: Population Census 1971 (establishment tables). Table adapted from Mazumdar (1983, part I, table 11).

Table 5-4. *Number and Growth Rates of Industrial Establishments by Size and Location, India, 1961–71*

Number of workers	*Number of establishments (thousands)* 1961	1971	*Annual growth rate (percent)*
Rural			
1	931.7	1,127.4	1.9
2–19	678.8	1,101.1	5.0
20–99	8.8	19.0	8.0
100+	1.5	3.0	7.2
Urban			
1	238.8	408.9	5.5
2–19	414.4	770.4	6.4
20–99	15.4	23.5	4.3
100+	3.6	5.3	3.9
Total			
1	1,170.5	1,536.3	2.8
2–19	1,093.2	1,871.5	5.5
20–99	24.2	42.5	5.8
100+	5.1	8.3	5.0

They had experienced contrasting industrial development—from stagnation and decline in Madhya Pradesh and Bihar to rapid growth of the number of establishments in the other three.[5] The data are presented in table 5-5. The overall picture for these North Indian states is very similar to that shown for all India in table 5-3. The rate of growth of rural establishments is little more than half that of the urban ones but increases with the size class of units. The highest rate of growth in the urban areas, by contrast, is found in the smallest size class of 1–9 workers. However, despite the overall result, the number of establishments grew faster in rural areas in the three progressive states, Uttar Pradesh, Punjab and Haryana, and West Bengal. The rural pattern of a higher growth rate for manufacturing establishments in the size group of over fifty workers is true for all states with the exception of West Bengal. The urban pattern of highest growth in establishments of the smallest size is true of the three progressive states but not of Bihar or Madhya Pradesh.

The question arises whether the different experience for rural and urban areas in the more rapidly growing states can be traced to the development of individual industries. The establishment tables of the census give information only at the two-digit level of industrial classification, so we can look only at fairly broad groups of industries. The relevant material is presented in tables 5-6–5-9.

5. In 1971 the region studied comprised six states, but for comparison over time Haryana has to be amalgamated with Punjab, from which it was formed, reducing the number to five.

Table 5-5. *Growth Rates of the Number of Industrial Establishments by Size and Location, Selected States, India, 1961–71*
(percentage per year)

State and location	*Total*	*Number of workers* 9 or fewer	10–49	50+
Uttar Pradesh				
Rural	8.8	8.7	11.6	14.9
Urban	8.2	8.2	7.8	5.0
Total	8.6	8.5	9.7	10.3
Punjab and Haryana				
Rural	9.5	9.5	9.1	19.9
Urban	8.7	8.9	5.2	6.8
Total	9.1	9.2	5.9	10.8
Bihar				
Rural	−2.3	−2.3	0.1	1.7
Urban	2.0	2.0	3.7	4.2
Total	−1.4	−1.5	1.8	2.8
West Bengal				
Rural	7.7	7.9	3.3	3.5
Urban	6.1	6.8	1.9	1.2
Total	7.0	7.4	2.3	1.9
Madhya Pradesh				
Rural	−0.4	−0.4	3.3	11.0
Urban	1.9	1.9	3.1	2.9
Total	0.0	0.0	3.2	5.1
All five states				
Rural	3.0	3.0	6.1	9.5
Urban	5.8	5.9	4.2	3.1
Total	3.8	3.7	5.0	5.6

Source: 1971 and 1961 Census of India, all India and state establishment tables. Mazumdar (1983, part I, table 13).

Let us consider the rural economy first. In all the northern states, units of fifty or more workers are dominated by two industries: food, tobacco, and beverages, and nonmetallic mineral products; textiles are a poor third. Between 1961 and 1971 the shares of these two industries changed dramatically in some states. In Punjab, Uttar Pradesh, and West Bengal—the progressive states—the share of nonmetallic mineral products increased spectacularly at the expense of the food group. Tables 5-7–5-9 show that in the rural areas of these states the former was one of the fastest growing industries and that the number of units with fifty or more workers rose much faster than the number with fewer workers. The number of establishments in the food group rose much more slowly, although the number of larger units in this industry had a relatively high growth rate in most states. In all the states the growth in textiles favored the smaller units, particularly those with fewer than ten workers. It is thus clear that the nonmetallic minerals industry is very largely responsible for the rapid growth in the number of medium-size

Table 5-6. *Percentage Shares of Different Industries in the Total Number of Establishments, Five States, India, 1961 and 1971*
(establishments employing fifty or more persons)

	Bihar		*Madhya Pradesh*		*Punjab*		*Uttar Pradesh*		*West Bengal*	
Industry	*1961*	*1971*	*1961*	*1971*	*1961*	*1971*	*1961*	*1971*	*1961*	*1971*
Rural										
Tobacco, beverages, and foodstuffs	26.3	23.0	65.4	58.7	16.9	7.8	61.3	26.0	67.3	52.3
Cotton textiles	6.3	13.8	8.6	6.9	9.2	1.4	19.8	1.1	4.6	2.4
Miscellaneous textiles	1.7	1.0	0.0	2.8	15.4	3.2	0.0	0.1	4.2	2.1
Wood and wood products	1.3	0.6	0.0	0.5	3.0	0.5	0.6	0.5	0.8	0.6
Leather and leather products	0.8	0.6	3.7	2.8	1.5	0.0	1.1	0.4	1.4	0.7
Nonmetallic mineral products	55.0	49.3	12.3	17.0	15.4	74.3	12.0	68.0	12.9	35.2
Chemicals	0.3	1.2	0.0	2.8	1.5	0.0	0.3	1.1	3.2	1.5
Basic metals and metal products	7.1	8.4	8.6	5.5	12.3	5.1	4.7	0.9	3.8	2.6
Machinery and electrical equipment	1.2	2.4	1.2	3.2	24.6	7.8	0.3	1.1	2.0	2.6
Urban										
Tobacco, beverages, and foodstuffs	20.0	24.6	46.4	36.1	7.6	13.2	23.9	16.9	16.5	13.7
Cotton textiles	5.2	2.8	12.5	25.7	12.8	20.5	16.6	7.8	6.3	7.6
Miscellaneous textiles	3.9	2.8	6.5	5.4	22.7	21.5	10.1	4.6	15.9	9.7
Wood and wood products	6.5	2.4	5.7	2.1	4.4	1.0	4.9	1.8	5.8	3.0
Leather and leather products	1.3	1.2	0.4	0.6	0.8	0.3	5.5	5.9	1.3	1.6
Nonmetallic mineral products	37.4	33.7	6.5	4.8	2.8	6.9	4.7	31.6	5.4	14.0
Chemicals	2.6	5.2	1.5	4.8	7.2	2.5	3.1	5.2	9.7	9.9
Basic metals and metal products	20.6	21.4	11.8	9.3	25.1	9.5	7.7	13.8	28.5	22.0
Machinery and electrical equipment	2.6	6.0	8.7	11.3	16.7	24.6	4.4	12.4	10.6	18.4

Source: Mazumdar (1983, part I, table 14).

Table 5-7. *Annual Rate of Growth of Industrial Establishments in Selected Subsectors by Size Groups, Urban and Rural West Bengal, 1961–71*

	Rural				*Urban*					
		Number of workers				*Number of workers*				
Subsector	*Total*	*9 or fewer*	*10–49*	*50+*	*Total*	*9 or fewer*	*10–49*	*50+*	*1961 rural unspecified units*[a]	*1961 urban unspecified units*[a]
Foodstuffs	−4.9	−4.9	−5.7	1.4	6.1	6.6	1.7	−1.5	21,479	1,304
Miscellaneous textiles	6.9	7.0	1.9	11.6	4.6	4.9	1.1	−8.2	1,178	3,408
Wood products	19.9	20.4	3.6	−4.6	2.8	3.6	−2.6	−7.2	197	812
Leather products	−1.5	−1.4	−2.8	2.2	0.4	0.7	−2.1	5.2	132	272
Cotton textiles	6.3	6.5	0.5	−4.2	4.7	5.1	1.9	2.4	2,312	224
Nonmetallic mineral products	7.4	7.4	5.7	12.9	5.8	6.3	1.1	10.0	873	180
Basic metals and mineral products	0.5	0.6	−8.0	−1.9	0.7	1.5	−2.3	−2.6	2,017	751
Chemicals	17.5	18.7	2.2	−4.8	2.4	4.1	−3.1	0.9	36	77
Machinery and electrical equipment	20.3	22.0	15.1	2.3	4.6	4.5	5.0	4.0	21	668

a. Unspecified units were proportionately distributed according to the percentage distribution of specified units.

Sources: 1961 and 1971 Census of India, establishment tables, West Bengal; Mazumdar (1983, appendix).

Table 5-8. *Annual Rate of Growth of Establishments in Selected Industries by Size Groups, Urban and Rural Uttar Pradesh, 1961–71*

	Rural				*Urban*			
	Number of workers				*Number of workers*			
Industry	*Total*	*1–9*	*10–49*	*50+*	*Total*	*1–9*	*10–49*	*50+*
Tobacco, beverages, and foodstuffs	7.0	6.8	13.9	6.3	6.1	6.1	6.9	2.3
Cotton textiles	10.8	10.8	12.9	−13.8	12.1	12.1	18.7	−1.9
Miscellaneous textiles	9.2	9.4	−2.2	—	9.9	9.9	12.2	−2.8
Wood and wood products	15.6	15.6	14.9	11.6	10.6	10.7	7.8	−4.5
Leather and leather products	5.6	5.7	3.2	7.2	4.2	4.3	1.8	6.7
Nonmetallic mineral products	13.1	12.5	17.2	37.4	5.1	4.8	5.4	8.4
Chemicals	7.9	7.5	37.7	32.0	9.8	9.6	12.6	11.3
Basic metals and metal products	11.3	11.4	1.7	4.3	7.5	7.9	3.8	4.5
Machinery and electrical equipment	29.1	29.9	14.7	32.8	22.3	22.4	21.9	17.3

—Growth rate not calculated, 1961 or 1971 units = 0.

Sources: 1961 and 1971 Census of India, establishment tables, Uttar Pradesh; Mazumdar (1983, appendix).

and large units in the rural economy of North India. This presumably is much the same as saying that construction was primarily responsible.

Turning to the urban economy, the larger units (with fifty or more workers) are dispersed much more widely among the different industrial groups. Table 5-6 shows that basic metals and metal products, and machinery and electrical products, account for an important proportion of the large industrial units, in addition to the three industries that were more important in the rural sector. Thus the overall experience of the urban economy between 1961 and 1971 in favoring the growth of small units is less easy to isolate in

Table 5-9. *Annual Rate of Growth of Industrial Establishments in Selected Subsectors by Size Groups for Urban and Rural Punjab and Haryana, 1961–71*

	Rural				*Urban*			
	Number of workers				*Number of workers*			
Industry	*Total*	*1–9*	*10–49*	*50+*	*Total*	*1–9*	*10–49*	*50+*
Tobacco, beverages, and foodstuffs	1.5	1.3	17.2	22.9	14.0	14.2	10.1	11.8
Cotton textiles	22.1	22.5	5.4	−8.0	3.0	2.9	2.3	11.5
Miscellaneous textiles	14.0	14.1	6.8	10.6	7.9	8.0	4.9	8.7
Wood products	8.6	8.6	9.9	11.6	7.4	7.5	6.2	−7.6
Leather and leather products	14.8	14.7	6.9	0.9	3.9	3.9	8.0	9.0
Nonmetallic mineral products	21.2	21.4	10.8	37.8	14.1	14.6	4.7	17.5
Chemicals	14.8	15.3	8.5	27.1	12.6	13.3	8.7	−1.8
Basic metals and metal products	5.4	5.2	15.1	18.0	7.4	7.6	5.2	3.9
Machinery and electrical equipment	20.1	23.5	6.3	11.1	15.2	16.7	9.8	8.8

Note: Haryana 1971 units are added to Punjab 1971 units to allow a 1961–71 comparison of the entire region. This procedure is necessary because the State of Haryana was created from districts which were part of Punjab in 1961.

Sources: 1961 and 1971 Census of India, establishment tables, Punjab and Haryana; Mazumdar (1983, appendix).

terms of the development of individual industries. Tables 5-7–5-9 show that the smallest units multiplied fastest in most industries in the three fast growing states. Indeed, nonmetallic minerals is the only industry in which the medium-size or large units (fifty or more workers) multiplied fastest in all three states.

Changes in the Size Distribution of Factories and Factory Employment

Since 1974 the ASI has given information by size class for factories. Table 5-10 shows a relative rise of employment in factories with fewer than 100 employees at the expense of those with more than 500. In the same period, the average factory size declined from 94 to 82 employees (Shetty 1982). This tends to confirm the suggestion made in the previous section that the lack of medium-size factories is no longer as remarkable as it once was.

But four years is a short period to establish a trend. A longer period comparison is difficult. The CMI gave size distributions of employment for 1953–57, but the industry coverage was much less than that of the ASI, so no aggregate comparison is possible. However, in table 5-11 we have selected those industries that are covered by both the CMI and the ASI and that seem to have the same definition. In twelve of the twenty-two industries, the size distribution changed markedly in favor of small factories (fewer than 100 employees). In one, brewing and distilling, both large and small establishments increased their share of employment. Of the other nine, only three—biscuit making; aluminum, brass, and copper; and tanning—recorded a significant rise in employment in large units (more than 500 employees). A few industries—notably chemicals, and fruit and vegetable processing—increased employment, mainly in middle-size units. However, most of the largest industries in terms of employment—iron and steel, jute, cotton textiles, and sugar—have a high concentration of employment in large units. In two of these—jute and sugar—there was no shift toward smaller factories; in cotton textiles, and iron and steel, although the shift was noticeable, about three-

Table 5-10. *Percentage Distribution of Factory Employment by Size Group, India, 1974–77*

Size group (number of workers)	*1974–75*		*1977–78*		*1978–79*
1–20	3.1	17.8	4.9	24.8	23.4
20–49	6.8		10.9		
50–99	7.9		9.0		
100–499	23.4		23.3		22.6
500+	58.8		51.9		54.0

Source: For 1974–75 and 1977–78, ASI; for 1978–79, Shetty (1982).

Table 5-11. *Percentage Distribution of Factory Employment by Size Groups, Selected Industries, India, 1956 and 1977*

	Number of workers						*Total employed (thousands)*	
	1–99		*100–499*		*500+*			
Industry	*1956*	*1977*	*1956*	*1977*	*1956*	*1977*	*1956*	*1977*
Iron and steel	4.4	12.0	7.0	12.8	88.6	75.3	88.0	181.2
Cement	0.0	8.5	10.8	15.5	89.2	76.0	26.2	43.6
Ceramics	5.5	47.5	36.6	33.3	57.9	19.2	21.6	93.8
Matches	3.1	25.5	26.7	40.4	70.2	34.0	16.5	30.0
Woven textiles	5.5	25.2	26.4	27.4	68.1	47.4	15.4	40.9
Biscuit making	49.2	33.0	50.8	26.4	0.0	40.6	5.5	14.1
Jute	0.0	0.7	0.1	1.7	99.9	97.7	273.7	223.3
Bicycles	18.7	41.6	21.2	26.6	60.2	31.8	10.2	19.2
Distilleries and brewers	33.9	23.6	66.1	56.9	0.0	19.4	5.2	13.0
Cotton textiles	0.5	11.1	3.1	16.9	96.4	72.0	782.4	922.5
Grain mill products	79.5	86.9	20.5	12.2	0.0	1.0	75.0	84.9
Chemicals	16.9	24.0	32.9	34.1	50.2	41.9	57.4	173.7
Vegetable oil	68.9	79.2	29.4	19.0	1.7	1.7	51.0	51.3
Aluminum, brass, and copper	27.2	23.7	28.2	15.3	44.7	61.0	23.9	25.4
Soap	13.3	28.8	17.2	21.4	69.5	49.7	5.9	18.5
Paints and varnishes	27.6	32.0	43.5	38.7	29.0	29.3	6.3	15.4
Tanning	43.5	42.2	51.1	35.1	5.5	22.7	11.0	23.0
Sugar	0.6	0.1	2.4	1.0	97.1	99.0	134.7	253.1
Edible oil	0.5	5.2	70.1	66.5	29.3	28.3	8.1	12.2
Starch	7.6	47.9	65.1	25.4	27.3	26.7	1.9	4.8
Paper and paper products	4.1	20.6	5.6	17.5	90.3	61.9	29.9	80.2
Fruit and vegetable processing	68.2	40.0	31.8	51.8	0.0	7.8	1.5	5.9

Sources: CMI, 1956; ASI, 1977; Mazumdar (1983, part II, table 4).

quarters of employment remained with the large units. One may conclude from this limited twenty-year comparison that there is a definite trend toward smaller factories but that its importance has been limited by the predominance of some of the old large-scale industries.

Summary

Indian industrial policies in the 1950s favored very small nonfactory establishments and very large factories. Later, especially in the 1970s, the reservation of many products for small units (and some of medium size when categorized by employment) has contributed to a rapid growth of small workshop and factory establishments in urban areas. The number of medium-size factories has grown fastest in rural areas, especially in the more progressive states. Much of this latter growth is accounted for by one industry, nonmetallic minerals (mainly building materials), while the small-establishment boom in urban areas is spread over several industries. Urban development may thus be more closely related to the policy of product reservation, and rural development to rapid growth. Overall, the number of medium-size factories has been catching up, with a filling-in of a trough which had earlier been remarkable (see also chapter 6). Finally, household manufacturing has grown relatively slowly, but it would have grown still more slowly or declined without the official policies favoring it.

6

India's Industrial Size Structure in International Perspective

This chapter attempts to provide a statistical picture of the importance of the small-scale sector in Indian manufacturing in the early 1970s and to evaluate the Indian situation in an international setting. The statistical sources on which the Indian size distribution is based are diverse and complicated. A review of the sources and of problems of reconciling different sets of figures is given in the appendix to the chapter. The text distills the best estimates for India and makes comparisons between selected states within India and between India and other economies whenever possible.

Household and Cottage-Shop Manufacturing Establishments in India and Other Developing Economies

Estimates of the proportion of manufacturing employment accounted for by household enterprises can vary greatly according to the definitions used. The range of figures for numbers employed varies with the treatment of the part-time employment that is often a feature of household enterprises. In some surveys or censuses, a worker may have done as little as one hour's work in a week, and in others part-time workers are excluded altogether. The treatment of children, the borderline between a household and a workshop or a minifactory, and the definition of manufacturing all vary. The zeal, efficiency, and resources of the surveyors also vary. Comparisons therefore are at best very rough. Only broad impressions not subject to scientific analysis are possible.

The above is well illustrated by figures for the Republic of Korea in 1975. It is possible to produce an estimate of the proportion employed in household manufacturing ranging from 7 to 29 percent depending on whether temporary workers (even for an hour a week) are excluded or included. Nonhousehold workshop (1–4 workers) employment is firmer, and the proportion lay between 7 and 9 percent (producing approximately 3 percent of the value added by factories) (Ho 1980, table 2-1).

In Taiwan, the industrial census of 1971 showed workshops with 1–3 workers as having only 3 percent of manufacturing employment (Ho 1980). No statistically based estimate for employment in household manufacturing was possible, and the figure of 3 percent also excluded some kinds of establishment that were counted as manufacturing in Korea (tailoring and repairs). Ho finally guesses that the proportion employed in the nonfactory sector in Taiwan (less than 4 workers) was little different from Korea (less than 5 workers) (Ho 1980, p. 6). What is certain is that nonfactory manufacturing is far less important in both Taiwan and Korea than in the great majority of developing economies.

In Colombia, Cortes, Berry, and Ishaq have estimated cottage shop manufacturing employment (1–4 workers) as about 45 percent of the total in 1978 (1987, table 2-2). About two-thirds of this nonfactory employment was in households (in Population Census Year 1973). For the Philippines, Anderson and Khambata have estimated employment in household manufacturing to be 53 percent of the total in 1975 (1981, table 6-1). Bautista gives a figure of 63 percent for household and workshop employment (less than 5 workers) in 1974 (1981, table 2-20).

For Indonesia, 68 percent of manufacturing employment was estimated to be in establishments with less than 5 workers in 1974–75 (Poot 1981, table 3-7). A similar estimate of 3 percent for Malaysia is obviously incomparable (Lim, Lee, and Thye 1981, table 6-1): by including only paid full-time employment, almost all household manufacturing is excluded. Figures for African countries are even more shaky and less comparable, but it is likely that household and workshop (less than 5 employees) constituted over half of manufacturing employment in Ghana, Kenya, Nigeria, and Tanzania (Page 1979), and this was probably true of almost all African countries.

In most developed countries the proportion of manufacturing employment in firms with less than 5 workers is very small. For instance, it is estimated at 1 percent for the United States (but 6.2 percent for all sectors) (U.S. Small Business Administration 1982, table 1-28). We do not have figures for this very small size category for other developed countries. But only about 10 percent of manufacturing employment in the Federal Republic of Germany was in firms with less than 10 employees, and Germany is noted among the Western developed countries for the vitality of its small-scale manufacturing. The corresponding figure for Japan in 1975 was 19 percent. (Kaneda 1980, table 10). Probably about 7 and 14 percent respectively were in the very small workshops with 1–4 workers.

Indian figures derived from the 1971 census are held to have greatly understated the number of secondary workers, especially in household enterprises. Uncorrected figures show 36 percent of manufacturing employment in household enterprises. But a correction to allow for the underestimate and to make the figures more comparable with earlier censuses and other surveys raises this to 45 percent.[1] The nonhousehold, nonfactory sector accounted for another 32 percent, so that the total nonfactory sector comprised 77 percent of manufacturing employment. It can be guessed that workshops of 1–4 employees would have about three-quarters of the employment in the nonhousehold, nonfactory units,[2] so that 69 percent in cottage-shop establishments would be a reasonable guess to compare with 14–36 percent for Korea, 43 percent for Colombia, 63 percent for the Philippines, and 68 percent for Indonesia; and with rough estimates of 1 percent, 7 percent, and 14 percent for the United States, Germany, and Japan. Thus the Indian figures are not surprising, although it must again be emphasized that there is no good reason to suppose that the definitions used make these figures more than very roughly comparable.

At one time almost all manufacturing took place in households. In very poor countries in the 1970s about half of manufacturing may still have been in households, defined by a predominance of family labor and the absence of a workshop separate from the dwelling. However, this kind of manufacture is for technical reasons largely confined to a few industries. Thus in India,[3] three-quarters or more of total household employment in both rural and urban areas was in five two-digit industries: food, cotton textiles, textile products, wood and furniture, and nonmetallic minerals (Mazumdar 1983, part I, table 5).

Only in four industries (cotton textiles, textile products, wood and furniture, and leather and fur) was more than a third of the total employment in households (Mazumdar 1984b, table 4).[4] Finally, it should be noted that rural household employment was about five times that of urban areas (Mazumdar 1984b). Much the same industries figure elsewhere as the main preserve of household manufacture. Thus in the Philippines, textiles, clothing, footwear, and wood products (but no longer food) are important for household manufacture (Anderson and Khambata 1981, table 6-3).

In Colombia, only furniture, clothing and footwear, transport equipment, and nonelectric machinery (the last two presumably mainly repairs) had over 10 percent of employment in cottage shop units (Cortes, Berry, and Ishaq 1983, table 2-2). It is worth noticing that, in this rather more developed country, cottage shop employment in the production of textiles probably fell by over 80 percent between 1938 and 1964, (Cortes, Berry, and Ishaq 1983, table 2-25) a very different history from that of India. It would be tedious to give more country examples. With the exception of textiles, it is much the

1. The corrections are those made by Haan (1980, part II). See appendix to this chapter.

2. It will be recalled that a factory is defined as having 10 or more employees with power or 20 or more without power. According to the establishment tables of the 1971 population census, 60 percent of most nonhousehold employment was in units employing 1–4 persons. But the total number of workers reported by these tables was clearly an underestimate, and the degree of underestimation was probably higher for small units (see appendix).

3. Strictly, the figures are for Bihar, Haryana, Punjab, Uttar Pradesh, and West Bengal.

4. These figures are uncorrected for the underestimation of secondary workers.

same everywhere; and whether or not metal products or transport equipment are included depends more on definition than any real differences, that is, on whether vehicle repairs and services count as manufacturing, or whether garages count as households.

Small Establishments in the Nonhousehold Sector in India

Census data do not yield an estimate of the proportion of small establishments in the whole manufacturing sector. The ASI excludes establishments with less than 10 workers, and with less than 20 if not using power. The one-off SSI census of 1972 surveyed small units registered with the Directorate of Industries, "small" then meaning an undiscounted value of plant and equipment of Rs750,000 or less (Rs1 million in the case of ancillary units). But this excluded important manufacturing industries for which small units came within the purview of specialized boards outside the scope of the Small Industries Development Organization (SIDO). Many traditional industries were thus excluded, most importantly textiles of all kinds, but also grain milling, sugar, edible oils, and tobacco. (Further details are given in the appendix.)

Therefore, for an overall estimate, one has to use the *Census of Population* (India 1971). As already mentioned, this source severely underestimated the participation of secondary workers. After correcting for this, Haan (1980, tables 11 and 12) has estimated that total employment in nonhousehold manufacturing in 1971 was 12.1 million.[5] The Labour Bureau of the Government of India estimated total employment in the factory sector at this date as 5.1 million. Thus 7 million, or 58 percent of the total, were in workshops. Despite the limited coverage, some insights can be obtained by combining ASI data with the SSI census.

There are several difficulties in combining the data. The SSI data are for 1972, while the nearest ASI year with a plant size breakdown is 1974. We have made no adjustment for this. Since SSI employment almost certainly grew from 1972 to 1974, it follows that our estimate of the proportion of employment in small units is, on this account, too low.

Probably more important is the fact that there is an unknown degree of overlap. This is most likely to occur with the sample sector of the ASI (less than 50 workers), the mean size of an SSI unit being 13.1 workers. We assume therefore that there is negligible employment in SSI units with 50 workers or more.[6] A high estimate of the proportion of employment in small establish-

5. The details of the adjustment are summarized in Mazumdar (1983, pp. 4–8).

6. There are two industries where this assumption is probably unsound. In the RBI sample inquiry (referred to later) the mean employment in small assisted units was 12.8, almost the same

ments is then made by assuming no overlap, so that total employment in small units is obtained by adding SSI figures to employment in ASI factories with less than 50 workers, while total employment is the total of SSI and ASI. A low estimate is obtained by assuming total overlap, employment in ASI factories of less than 50 workers being subtracted from both numerator and denominator. The results for 115 three-digit industries are given in table 6-12. It can be seen that except in the case of some half-dozen industries the two estimates are close.

Table 6-1 summarizes the results for the 115 included industries at the two-digit level. It shows that 41–46 percent of total employment in the included industries was in small establishments (less than 50 workers). Table 6-2 likewise gives summary ASI data for 47 subindustries that are not included in the SSI and hence are excluded from table 6-1. It shows that only 7 percent of the total employment was in small establishments. If the two tables are taken together, it is seen that ASI employment in the 115 ASI industries that were also covered by the SSI (about 2.7 million) is only about 55 percent of the total ASI employment (about 4.8 million), though the former account for 69 percent of total ASI employment in small units (334,000 out of a total of 477,000).

Seven out of the sixteen two-digit industries in table 6-1[7] are small-establishment industries—apparel; wood products; leather and leather products; rubber and plastic products; nonmetallic minerals; metal products; and miscellaneous.[8] However, these account for no more than half the total SSI employment, since a number of them are small industries. For example, the two large-establishment industries, chemicals and nonelectric machinery and parts, give about the same SSI employment as the five small-establishment industries, apparel, wood products, leather and leather products, rubber and plastic products, and miscellaneous. The two large industries that are also small-establishment industries are nonmetallic minerals and metal products: together they account for almost a third of SSI employment (see table 6-1, seventh column).

However, at the more disaggregated level of three-digit industries, the small-establishment industries account for considerably more than half of

---

as in the SSI census. But in two two-digit industries the mean was far greater. In beverages and tobacco (22) it was no less than 99, which suggests that a very high proportion of small-scale employment was in units with 50 workers or more. However, these were probably making bidis (cheap cigarettes), and tobacco is excluded from our comparison. The other exceptional two-digit industry is nonmetallic mineral products (32), for which the RBI sample has a mean employment of 30 workers; this suggests that a significant number may be in units of 50 or more workers. In this case our estimates of the proportion of employment in small units may be too high.

7. Excluding 23, under which code the SSI included only one three-digit industry.

8. The criterion is 50 percent or more employment in small units on the higher of our estimates.

Table 6-1. *Summary of Estimates of Employment in the Small-Scale Sector, India, 1972–74*

Code and sector	Employment in the SSI[a] (1)	Employment in ASI units of less than 50 workers[a] (2)	Total ASI employment[a] (3)	$100\left[\frac{(1)+(2)}{(1)+(3)}\right]$	$100\left[\frac{(1)}{(1)+(3)-(2)}\right]$	$100\left[\frac{(2)}{(3)}\right]$	Percentage of total SSI employment	Percentage of total ASI employment
20–1 Food	131	20	217	43	40	9	8.0	8.2
22 Beverages	5	2	19	27	21	9	0.3	0.7
23 Cotton cleaning and baling only	1	25	89	29	2	28	0.1	3.3
26 Apparel	66	10	36	74	72	27	4.0	1.4
27 Wood products	92	19	37	86	83	50	5.6	1.4
28 Paper and printing	89	27	186	42	36	15	5.5	7.0
29 Leather and leather products	32	5	38	53	49	14	1.9	1.4
30 Rubber and plastic products	82	17	89	58	53	19	5.0	3.3
31 Chemicals	159	25	292	41	37	9	9.7	11.0
32 Nonmetallic minerals	203	30	251	51	48	12	12.4	9.4
33 Basic metals	110	22	193	44	39	12	6.7	7.3
34 Metal products	300	36	141	76	74	25	18.4	5.3
35 Machinery and parts	145	36	283	42	37	13	8.9	10.6
36 Electrical machinery	66	17	234	28	23	7	4.0	8.8
37 Transport equipment	84	15	380	21	19	4	5.1	14.3
38 Miscellaneous	39	10	53	53	47	18	2.4	1.8
97 Repair services	29	18	130	30	21	14	1.8	4.9
Total	1,631	334	2,668	46	41	13	100.0	100.0

a. Thousands of workers.

Source: Table 6-12.

Table 6-2. *Employment in ASI Manufacturing Industries not in SSI, India, 1974*

Code	*Big subindustries*	*Number of subindustries*	*Employment in units with less than 50 workers*[a] (1)	*Total employment*[a] (2)	*100 [(1)/(2)]*
20–1	Sugar, tea	7	68	366	18
22	Tobacco leaf, bidis	6	23	189	12
23	Cotton spinning and weaving	7	16	837	2
24	Synthetic cloth and wool spinning and weaving	9	16	146	11
25	Jute spinning and weaving	4	1	267	0.3
26–38	Steel, jewelry	14[b]	19	318	6
	Total	47	143	2,123	7

a. Thousands of workers.
b. Excludes aircraft, for which no figures were available.
Source: Table 6-13.

employment, especially in the SSI units (67 percent), but also in the ASI sample sector (53 percent). This is shown in table 6-3.

But there is no compelling reason why small-establishment industries should account for most of the total employment in small establishments. They did so in Taiwan (Ho 1980, table 3-6),[9] but not in Korea. In the case of the latter, shifting the definition of a small establishment from less than 50 to less than 100 workers, we find that small-establishment industries (still those where such small establishments account for half or more of the employment) accounted for only 39 percent of employment in all small establishments in 1975.[10] These figures, derived from a five-digit industrial breakdown, are not exactly comparable to those given above for India, partly because the Korean figures exclude workshops, and partly because the limit for a small establishment is 99 not 49 workers. A better comparison is however possible if the Indian industries are reclassified to comply more closely with the Korean definition of small,[11] and if the employment share comparison is then restricted to units with 10–49 workers, a size class which should correspond to ASI factories with less than 50 workers and one which is also identifiable in Korea. The results are given in table 6-4. It can be seen that in Korea small-establishment industries (the first two rows) account for

9. The definition of a small establishment in Ho (1980, table 3-6) is one with less than 100 workers.

10. That is, those with 5–99 workers. Establishments with 1–4 workers are excluded from the Korean figures (see Ho 1980, table 3-7).

11. There remains some difference. The classification of Indian industries now depends on the proportion of employment in all nonhousehold units with less than 100 workers. The Korean classification depends on the proportion of employment in enterprises with 5–99 workers.

Table 6-3. *Share of Employment of Industries of Different Configuration, India, 1972–74*

Percentage of workers in units of less than 50 workers[a]	Number of three-digit industries	Percentage of total employment		
		In SSI	In ASI (less than 50 workers)	All ASI plus SSI
75–100	24	25	18	13
50–74	37	42	35	30
25–49	29	29	37	38
0–24	25	4	10	19
Total	115	100	100	100

a. As given by seventh column of table 6-12.
Source: Table 6-12.

only 41 percent of employment in small factories, compared with 70 percent in the case of India.

In Korea, only 5 of the 175 small-establishment five-digit industries but 9 of the 198 large-establishment industries employ more than 5,000 persons in establishments of less than 100 workers. Of the 9, 4 are in clothing and 3 in textiles. One of the latter (synthetic cloth) employs twice as many persons in small establishments as any other industry, whether small- or large-establishment-oriented. It might therefore be tentatively suggested that some of the difference between India and Korea stems from the omission of textiles from the Indian data and from the far greater development of the clothing industry in Korea. But as against this, large-establishment-oriented industries do not provide most small-establishment employment in Taiwan, where the textile and clothing industries are as prominent as in Korea.

Staley and Morse (1965) categorized the factors making for small establishments in their classic study. Their eight categories comprise different varieties of locational, process, and market influence. Ho partitioned Taiwan and Korean industries into the Staley-Morse categories. The interest of this is limited by the fact that such industries are, as we have seen, not always of great importance in providing employment. The industries which were both small-establishment-oriented and provided much employment (over 5,000 persons) in Korea were grain milling, rice wine, sawmills, building materials, and printing. Except for rice wine, the same industries featured in Taiwan. But further comparison is difficult because the level of aggregation of industries was much greater. In India, the same industries as in Taiwan were both small-establishment-oriented and large in terms of employment: but several others could be added such as (see table 6-12 below for a list of all sectors with code numbers) cotton cleaning and baling, various metal-product subindustries (codes 340, 343, 348), and machinery producers (354, 359). Lastly, however, it is worth noting that, even at the five-digit level at which Korean manufacturing data were available, small establishments existed in almost all

Table 6-4. *Share of Employment by Industries of Different Configuration, India and Korea*

	India, 1972–74			Korea, 1971		
Percentage of workers in units of less than 100 workers	*Number of industries*	*Percentage of employment in ASI units of less than 50 workers*	*Percentage of employment in all units*	*Number of industries*	*Percentage of employment in units of 10–49 workers*	*Percentage of employment in all units*
75–100	35	30	22	114	17	5
50–74	34	40	30	61	24	10
25–49	28	23	31	86	39	27
0–24	18	7	17	112	20	58
Total	115	100	100	373	100	100

Sources: Table 6-12 for India; Ho (1980, table 3-7) for Korea.

industries,[12] and in only 47 of the 373 industries did small establishments (less than 100 workers) provide less than 10 percent of the employment.

Nonhousehold Manufacturing in Six Indian States

The establishment tables of the Indian population census give useful information for all sizes of establishment, excluding those classified as households. Potential differences between states made it desirable to look at each state separately, but a shortage of resources compelled a restriction to six states of North India.

It has been a common observation that medium-size firms are relatively thin on the ground in developing countries, particularly in India. It is easy to imagine why this might be so: large foreign or public firms with modern technology embodying economies of scale are grafted onto a very small-scale traditional base and may tend to inhibit a more organic growth that would include many medium-size firms that had grown from small beginnings.

In 1961, Dhar and Lydall wrote: "A peculiarity of the Indian distribution is that, while it has a high concentration of *establishments* in the lower size group, it has a high concentration of *employees* in the highest size group. Indian industry tends to be either on a very small scale, or on a very large scale, and it is somewhat thin in the middle. . . . But, whatever may be the origin of the peculiarity, it suggests that there may be scope for encouraging the development of more medium sized firms in India. This is rather an important conclusion, for the medium sized firm (say 50–499 employees) is very often close to the optimum size, especially in the lighter manufacturing trades" (pp. 30–31).

Since we later adduce developing-country evidence not available to Dhar and Lydall that tends to confirm their feeling for the relative efficiency of the medium-size firm (50–499 employees), it is worth examining the alleged peculiarity more closely. Table 6-5 is arranged to throw light on the matter by dividing establishments into tiny, small, medium-size, and large.[13] The top six rows of the table give percentages of employment in these categories for six North Indian states, and the bottom five rows for some other countries with which a rough comparison can be made.

The table shows that in the 1970s there is a trough in the size class 50–499 workers in Bihar, Uttar Pradesh, and West Bengal as compared with Korea, Taiwan, and the United States. But this is not so for the Punjab, for which

12. Only 6 of 373 industries had no small establishments (less than 100 workers). The list of 6 is somewhat bizarre: condensed milk, malt liquor, compound fertilizers, toothpaste, fuel oil, and gas or liquid supply meters (Ho 1980, table B-4).

13. It is worth noting, in the light of figures 6-1–6-4, that except for the tiny size, the class intervals are logarithmically equal (assuming that the largest extends roughly from 500 to 5,000 workers).

Table 6-5. *Nonhousehold Manufacturing Employment by Employment Size of Establishment*
(percent)

<table>
<tr><th>Location</th><th>Tiny
(1–4 workers)</th><th>Small
(5–49 workers)</th><th>Medium
(50–499 workers)</th><th>Large
(500 or more workers)</th></tr>
<tr><td>Indian states, 1971</td><td></td><td></td><td></td><td></td></tr>
<tr><td>Bihar</td><td>35.2</td><td>26.6</td><td>14.0</td><td>24.3</td></tr>
<tr><td>Haryana</td><td>26.1</td><td>23.1</td><td>25.6</td><td>24.2</td></tr>
<tr><td>Madhya Pradesh</td><td>25.1</td><td>28.2</td><td colspan="2">46.7</td></tr>
<tr><td>Punjab</td><td>35.7</td><td>34.1</td><td>21.4</td><td>8.8</td></tr>
<tr><td>Uttar Pradesh</td><td>30.1</td><td>29.7</td><td>17.8</td><td>22.5</td></tr>
<tr><td>West Bengal</td><td>16.6</td><td>25.1</td><td>21.1</td><td>37.4</td></tr>
<tr><td>Other economies</td><td></td><td></td><td></td><td></td></tr>
<tr><td>Korea, 1958[a]</td><td>n.a.</td><td>54.0</td><td>34.0</td><td>12.0</td></tr>
<tr><td>Korea, 1973</td><td>7.9</td><td>22.0</td><td>31.6</td><td>38.5</td></tr>
<tr><td>Taiwan, 1961[a]</td><td>n.a.</td><td>42.0</td><td>25.0</td><td>34.0</td></tr>
<tr><td>Taiwan, 1971</td><td colspan="2">29.1</td><td>36.1</td><td>34.7</td></tr>
<tr><td>U.S., 1977</td><td>1.2</td><td>14.1</td><td>43.7</td><td>41.0</td></tr>
</table>

n.a. Not available.

a. The tiny size (1–4 workers for Korea; 1–3 workers for Taiwan) is excluded because data were not in the source.

Sources: For Indian states, Mazumdar (1983, part I); for Korea and Taiwan, Ho (1980, tables 3-1, D2, D3); for U.S., *Census of Manufactures*, U.S. Department of Commerce, Bureau of the Census, table 4.

the figures fall from tiny to large, nor for Haryana, for which they are almost flat. Of the other five economy-years in the bottom five rows of the table, only Taiwan in 1961 shows a similar trough, which became a peak ten years later. In Korea, a steep fall from small to large is replaced by a steep rise fifteen years later. In the proportion of employment in small and medium-size units taken together, these Indian states are closer to the United States than to Korea or Taiwan.

It is a pity that the Japanese size divisions available to us do not fit table 6-5. But Japan in 1971 had only 32 percent of workers in establishments of 300 workers or more: West Bengal, Taiwan, and Korea have more than this in establishments of 500 workers or more. At the other end of the scale, Japan had 52 percent of employment in establishments smaller than 100 workers, more than West Bengal though less than the other Indian states.[14]

The conclusion concerning the lack of medium-size establishments is that there is some evidence for this in Bihar, Uttar Pradesh, and West Bengal but not in Haryana or Punjab (nor in Madhya Pradesh, for which figures available to us did not break down establishments with 50 or more workers.)

The data for the Indian states are available separately for urban and rural areas and are given in table 6-6, which also presents a finer size classification

14. The Japanese figures are from Kaneda (1980, table 10).

Table 6-6. *Nonhousehold Manufacturing Employment by Employment Size of Establishment and Location, Six Indian States, 1971*
(percent)

State and location	*Employment in respective location*	*Size of establishment (number of workers)* 1–4	5–9	10–49	50–99	100–499	500 or more
Bihar							
Urban	56.6	27.2	10.4	15.5	4.2	8.4	34.2
Rural	43.4	45.6	8.2	19.2	6.3	9.4	11.4
Haryana							
Urban	58.0	32.3	10.1	15.7	3.2	13.6	23.2
Rural	42.0	17.5	3.9	15.5	12.0	25.7	25.5
Madhya Pradesh							
Urban	75.6	26.7	8.2	14.5	←	50.6	→
Rural	24.4	20.2	7.2	38.0	←	34.7	→
Punjab							
Urban	73.2	31.5	15.8	24.6	8.3	11.2	8.6
Rural	26.8	47.2	4.7	12.3	12.2	14.5	9.2
Uttar Pradesh							
Urban	50.2	30.0	12.6	16.1	4.5	10.1	26.7
Rural	49.8	30.2	9.2	21.4	11.7	9.4	18.2
West Bengal							
Urban	76.5	13.6	9.9	14.0	4.9	14.1	43.7
Rural	23.5	26.5	8.8	20.0	9.4	18.7	16.7

Source: Mazumdar (1983, table 6).

than table 6-5. Features that stand out are: the relatively high employment in large units in urban areas of Bihar; the low employment in the tiny class combined with very high employment in the large class in urban West Bengal; the relatively pronounced slant in favor of small units, especially the tiny, in both urban and rural areas of the Punjab; the high employment in tiny units in rural Bihar; and the low employment in tiny and small units in rural Haryana, combined with high employment in units of 100 or more workers. These features show that the general industrialization policies emanating from New Delhi do not prevent wide divergences of industrial size distribution in different states. We do not, however, have enough information to attempt a good explanation. Only a very limited attempt is made in the next section to see how far these features result from the development of different industries. Again there is a variety of experience. For instance, the prevalence of small plants in the Punjab is not explained by the development there of industries which favor small plants.

In addition to the above, it is worth noting that the relative dearth of medium-size enterprises (50–499 workers) is an urban phenomenon only; it applies to urban areas in all of the states except the Punjab and to none of the rural areas; indeed, in the rural areas of Haryana and the Punjab the distribution is humped in favor of medium-size enterprises. Unlike Indian urban areas (except the Punjab), there are not enough large factories to result in more employment in the largest size class.

We saw earlier that the bulk of household manufacturing is in rural areas. This cannot be said of the shop-size enterprises. The percentage of employment in workshops is higher in urban than rural areas in Haryana and Madhya Pradesh and about the same in Uttar Pradesh. The fact that total nonhousehold manufacturing employment is usually much less in rural areas (the unweighted mean proportion in rural areas in the six states is 35 percent; see table 6-6) further implies that the absolute number of workshops in rural areas is less than in urban areas in most states.

We have not made any international comparisons of differences between rural and urban industry but have used the few figures available to us, because definitions of rural and urban vary greatly, and comparisons would be as likely to mislead as to enlighten.

Industrial Structure in the Six States and Interstate Differences

In this section, we look at interindustry differences in the importance of small establishments and tentatively explore the extent to which such differences explain interstate differences. Shortage of resources made us confine the analysis to urban areas.

A few industries account for the bulk of employment in very small units (less than 10 workers) in the six states, as is shown in table 6-7. These are also small-establishment industries, meaning that half or more of total employment in such industries is in units with less than 50 workers.[15] They also account for the bulk of household employment, which however is excluded from discussion in this section. The two-digit industries that are large-establishment industries in most of the six states are cotton textiles (code 23), wool and synthetic textiles (24), jute hemp and mesta textiles (25), nonmetallic minerals (32), basic metals (33), machinery and parts (35), electrical machinery (36), and transport equipment (37).[16] There are a few exceptions. Cotton textiles was a small-establishment industry in Bihar. But Punjab is the main exception, with only three of the above industries (24, 32, and 36) giving most employment in medium-size and large units.

As it is thus evident that the unit-size distribution within an industry may vary from state to state, it is of interest to see how far differences in overall employment percentages in large units (more than 100 workers) are to be

15. But for all of India, food products have less than half of total employment in small units, although most of the three-digit subindustries are small-establishment industries. See the section above on small establishments in the nonhousehold sector in India.

16. Nonmetallic mineral products (essentially building materials) is a small-establishment industry for all of India; see table 6-1.

Table 6-7. *Employment in Urban Nonhousehold Manufacturing Units of Less than Ten Workers, Six Indian States, 1971*
(percentage of total employment)

	Industry (industrial classification number)					
State	*Food products (20–21)*	*Repairs (97)*	*Apparel (26)*	*Metal products and parts (34)*	*Wood products (27)*	*Total*
Bihar	24.7	17.5	13.9	6.0	8.0	70.1
Haryana	22.6	16.9	11.9	9.4	9.3	70.1
Madhya Pradesh	24.3	17.5	19.8	5.7	8.7	76.0
Punjab	14.8	15.5	12.7	8.9	7.7	59.6
Uttar Pradesh	20.3	13.4	12.8	10.7	8.3	65.5
West Bengal	16.4	11.1	14.1	9.0	7.3	57.9
Mean	20.5	15.3	14.3	8.3	8.2	66.5

Source: Mazumdar (1983, table 8).

attributed to differences between states in industrial patterns, and how far to differing intraindustry unit-size distributions. We investigated this in a limited way comparing Haryana, Punjab, and West Bengal pairwise with Bihar, industries being classified at the two-digit level. The method (known as decomposition) is to pretend that each other state has the same percentage of employment in each industry as Bihar. Then using the former's own figures for the percentages of employment in large units for each industry, one calculates a hypothetical overall percentage in large units. Subtracting this figure from the actual percentage for, say, the Punjab yields a measure of the difference due to a different industrial structure. The remaining difference (that is, the actual difference between Bihar and the Punjab less the difference due to a different industrial structure) is due to interstate differences in the percentages of employment in large units, for each industry, which we may call the intraindustry unit-size distribution effect. The results are presented in table 6-8.

The percentage of employment in large units varies widely from about 20 percent in the Punjab to about 58 percent in West Bengal. In Haryana and

Table 6-8. *Percentages of Employment in Large Urban Establishments, Selected Indian States (100 or More Workers), 1971*

State	*Actual percentage*	*Difference of actual percentage of state from that of Bihar*	*Industrial structure effect*	*Intraindustry unit size effect*
Bihar	42.6			
Haryana	36.8	−5.8	7.1	−12.9
Punjab	19.8	−22.8	1.1	−23.9
West Bengal	57.7	+15.1	15.0	0.1

Source: Mazumdar (1983, table 9).

the Punjab (especially the latter), the relatively low percentage is despite the industrial structure favoring a higher percentage than Bihar; in other words, the intraindustry unit-size effect more than accounts for the low percentage of employment in large units. By contrast, the large-size inclination of West Bengal is exactly accounted for by its industrial structure.

It is remarkable that the intraindustry unit-size distribution effect can be so very important for the urban Punjab (and, to a lesser extent, urban Haryana) when the Punjab's industry is subject to the same macroeconomic and industrial policy influences as other Indian states. This development might result from the Punjab's (and Haryana's) rapid agricultural growth. The fact that the Punjab's industry is about as urban as that in Madhya Pradesh and West Bengal, and more urban than that in Bihar, Haryana, or Uttar Pradesh need not invalidate this idea.

Child and Kaneda (1975) studied the backward engineering linkage of tube-well irrigation in Punjab Province in Pakistan. One might expect rapid agricultural growth to be more associated with a different industrial structure than to favor small units within industries. This would surely be the case if industries were sufficiently narrowly defined, but the two-digit classification is too broad to catch what may be significant changes in structure which favor small units. Thus the diesel-engine makers, situated in small and large towns, were typically small (5–9 workers), and the rapid growth of their subindustry would have helped to slant the nonelectrical machinery industry toward small units.

In contrast to Haryana and the Punjab, the fact that the high proportion of employment in large units in urban West Bengal is wholly accounted for by its industrial structure is less of a surprise. Some historical comparative advantage leading to a concentration there of large-scale industries such as steel and jute is probably the reason. However, we have clearly left a puzzle, which should be more fully investigated for more states, for rural as well as urban areas, and at the three-digit level, before any strong statements are made concerning the determinants of the manufacturing size structure, especially those that appear to be largely independent of policies for all of India.

The Overall Size Distribution of Manufacturing in India: Value Added and Employment

We have so far reported only on the size distribution of employment, but the distribution in terms of income generated is also of interest. The following table sets out the relevant figures for India in 1971. The figures on value added are derived from the National Accounts Office, which estimates that almost 63 percent of value added in the early 1970s arose in the unorganized sector, and from table 6-9. Relative labor productivity is defined as the percentage of value added divided by the percentage of employment in any size group.

<table>
<tr><th></th><th>Household enterprises</th><th>Small non-household enterprises (1–49 workers)</th><th>Medium-size enterprises (50–199 workers)</th><th>Large enterprises (200–499 workers)</th><th>Giant enterprises (500 or more workers)</th></tr>
<tr><td>Workers (million)</td><td>9.9</td><td>7.8</td><td>0.9</td><td>0.6</td><td>2.8</td></tr>
<tr><td>Employment (percent)</td><td>45</td><td>35</td><td>4</td><td>3</td><td>13</td></tr>
<tr><td>Value added (percent)</td><td colspan="2">38</td><td>7</td><td>10</td><td>45</td></tr>
<tr><td>Relative labor productivity</td><td colspan="2">0.47</td><td>1.8</td><td>3.3</td><td>3.5</td></tr>
</table>

These figures imply that labor productivity in the large enterprises sector is about double that in smaller enterprises. We shall see in chapter 7 that these differences are reflected in wage levels.

The Structure of the Factory Sector: India and International Comparisons

The size structure of factories for all Indian industries (it may be recalled that factories employ only about a quarter of India's industrial workers) is given in table 6-9. The salient features are that 55 percent of the employment is in establishments of more than 500 workers, which use 78 percent of the capital in producing 67 percent of the value added. The medium-size units have the highest capital productivity, producing 25 percent of the value added with 15.8 percent of the productive capital.

The question remains as to whether the size structure of the Indian factory sector is exceptional. International comparisons may throw some light, especially perhaps on the question whether there is a dearth of medium-size establishments, for which we found some evidence in three of the six North

Table 6-9. *The Size Structure of ASI Factories (Including Electricity, Gas, and Water Supply), India, 1976-77*
(percent, except for number of workers)

Number of workers	*Number of factories*	*Employment*	*Productive capital*	*Value added*
Less than 50	79.2	15.1	6.2	8.6
50–99	9.7	8.3	3.2	5.4
100–199	5.3	9.6	4.2	6.6
200–499	3.2	12.1	8.4	12.9
500–999	1.2	10.2	13.2	13.5
1,000–1,999	0.8	14.0	15.4	19.2
2,000–4,999	0.5	15.9	12.0	18.2
More than 5,000	0.1	14.8	37.4	15.8

Sources: ASI; Mazumdar (1983, part II, table 2).

Indian states when considering the whole size structure from single-worker units upward.

Ishikawa (1967) has already published some results. We have used his data and also updated some of the series to draw a picture of the size distribution of manufacturing employment (including electricity, gas, and water) for a few countries at different dates. The comparisons, which were possible only for employment in establishments of 20 or more workers, are presented in figures 6-1–6-7.

Figure 6-1 portrays the Indian size structure for 1956 from CMI data and for 1977 from ASI data. We have earlier produced some evidence for believing that there has been a trend in favor of establishments with less than 100 workers in the 1960s and 1970s. This shows up in figure 6-1, especially when electricity, gas, and water are excluded from the ASI figures. The adjusted numbers employed per size class rise less steeply in 1977 than they did in 1956. The extraordinary dominance of the largest size class (over 1,000 workers) in 1956 has been considerably reduced, though far from eliminated.

Comparing figure 6-1 with figures 6-2, 6-3, and 6-4, we see that the contrast between India in 1956 (or 1977) and Korea in 1957 or Taiwan in 1961 is very marked, while Pakistan is similar to India. Employment per standard class interval (an interval where the upper employment bound is twice the lower) was greatest in the range above 1,000 workers in India (in both years) and in the range of less than 50 workers in Korea. However, there has been an enormous change both absolutely and relatively in both Taiwan and Korea. Employment per standard class interval peaked in the range 100–500 workers in both Taiwan in 1971 and Korea in 1979. However, in neither did the size class of 500 workers or more account for as much as half the total employment.

Comparing figure 6-1 with figures 6-5 and 6-6, we see that Indian manufacturing is much more large-size-oriented than that of the United States, where employment per class interval peaks, as in Korea or Taiwan, in the range 100–500 workers. Since 1954, there have been employment gains in the United States in the range 50–1,000 workers and actual falls outside this range. Pakistani manufacturing is also more slanted toward giant establishments than U.S. manufacturing. The contrast between South Asia and Japan is even more marked. Japan in 1958 shows a pattern similar to Taiwan and Korea at about the same time, that is, a level of employment per standard class interval that falls with size. Even in 1977 this remained true up to 1,000 workers, and the size class 20–50 workers still showed the highest level of employment per class interval, unlike any other economy under examination in the 1970s. Finally, Colombia's size distribution is close to that of Korea and Taiwan and shows none of the large-plant dominance of India or Pakistan (figure 6-7).

The relatively rapid growth of medium-size and medium-size-to-large enterprises may owe something to India's policy of encouraging small industry

Figure 6-1. *Employment in India by Size of Establishment*

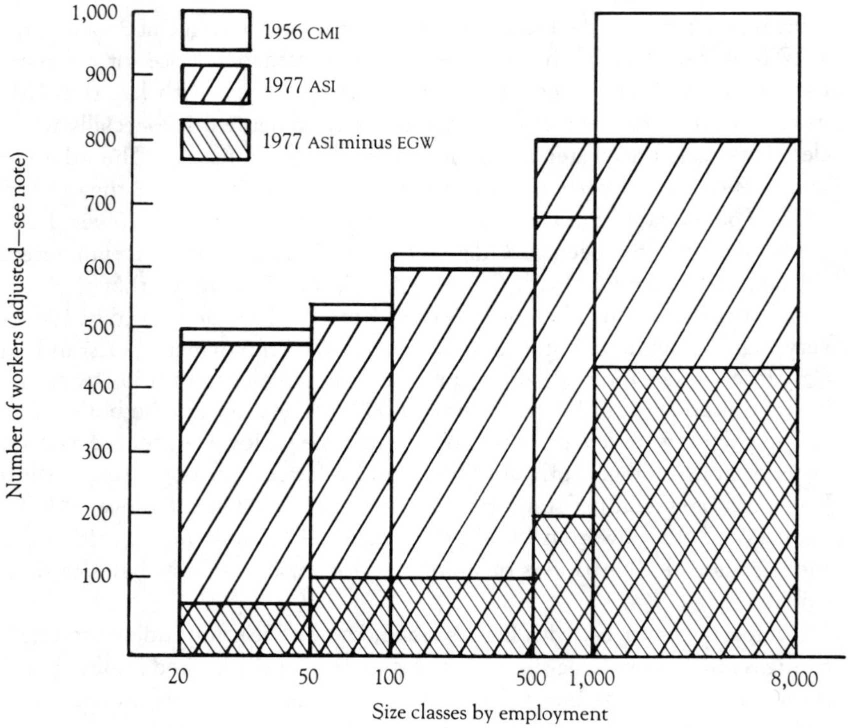

Note: The number of workers shown on the vertical axis is adjusted so that the relative *areas* of the columns correctly reflect the relative numbers employed in the size classes indicated on the horizontal axis. A standard size class is one in which the upper bound is double the lower bound (for example, 50–100 workers); for such a standard size class the total number of workers shown on the vertical axis is unadjusted.

The adjustment made is to multiply the number employed in a given size category by log 2 ÷ log (x_2/x_1), where x_2 and x_1 designate the upper and lower bounds of the size category. Thus any size is standard if the upper limit is twice the lower and the ordinate then gives the actual number employed. The largest size class is arbitrarily assumed to extend from 1,000 to 8,000 workers.

Figure 6-2. *Employment in Pakistan by Size of Establishment*

Figure 6-3. *Employment in Taiwan by Size of Establishment*

Figure 6-4. *Employment in Korea by Size of Establishment*

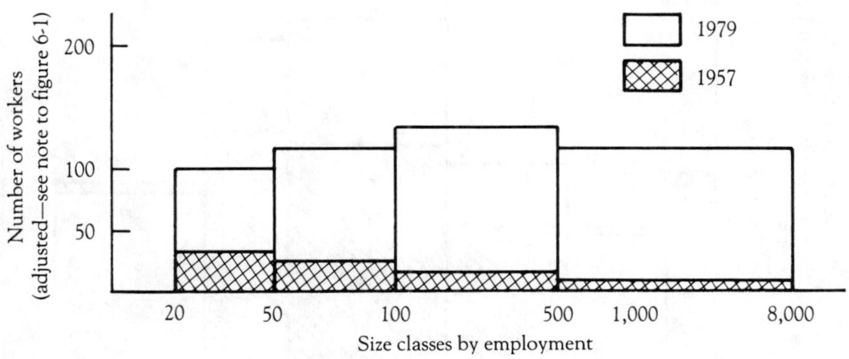

Figure 6-5. *Employment in Japan by Size of Establishment*

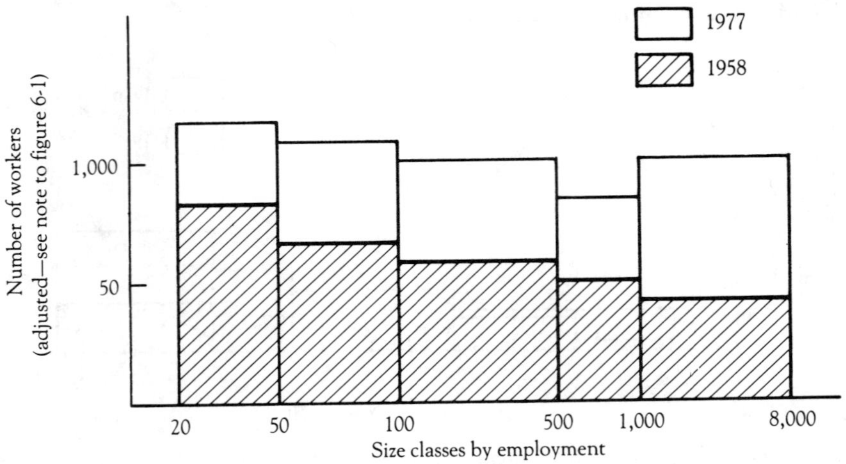

Figure 6-6. *Employment in the United States by Size of Establishment*

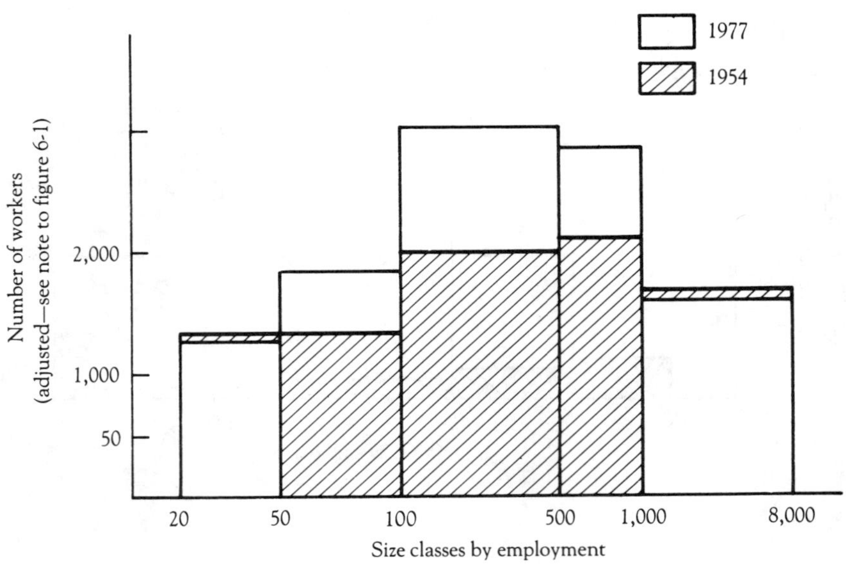

Figure 6-7. *Employment in Colombia by Size of Establishment*

but probably not very much, for several reasons. First, these policies operated for only about half the period. Second, many of the beneficiaries were the very small firms with less than 20 workers. Third, the size class of 500–1,000 workers is outside the range of the "small," and so also would be many enterprises in the range 100–500 workers. Fourth, the method of protecting small enterprises inhibits them from crossing the boundary lines, and thus their natural growth into the medium-size range.

Summary

The proportion of manufacturing accounted for by cottage shops, that is, households and workshops with 1–4 workers, has declined rapidly with development. It is about 70 percent for India and Indonesia, 60 percent for the Philippines, 40 percent for Korea and Colombia, 15 percent for Japan, and 7 percent for Germany. Such manufacture is largely in rural areas and is confined to a few industries, for example, food processing, textiles, wood products, and nonmetallic minerals.

In India in 1971, some 7 million people were employed in nonhousehold, nonfactory manufacturing establishments, and about 5 million in factories (10 or more workers with power, or 20 without). For a limited number of industries (excluding many where small units are mainly traditional, but also steel), a rough estimate can be made of the proportion of establishments with less than 50 workers. Overall the proportion is rather more than 40 percent, but from industry to industry it varies from about 20 percent in transport equipment to more than 80 percent in wood products.

The industries where small units account for most of the employment in that industry do not necessarily account for a high proportion of total small-unit employment (since some of them are small industries). They did so in India, but not in Korea, where such small-establishment industries accounted for only about 40 percent of employment in all small units. Some of the large-establishment industries in Korea, for example, synthetic cloth, are so large that they employ more people in small establishments than any small-establishment industry, which underlines the importance of industrial structure and exports in the generation of employment.

We examined the employment size distribution of manufacturing units separately for six North Indian states (Bihar, Haryana, Madhya Pradesh, Punjab, Uttar Pradesh, and West Bengal). There are considerable differences, despite the fact that all are subject to the same Indian macroeconomic and industrial policies. It has been observed in the past that India has relatively few medium-size establishments. Figures for nonhousehold manufacturing in the six states show that this is still true in that the proportion is well below those found in Korea, Taiwan, or the United States. But it is less true for Haryana and the Punjab than for the others, and in all the states it is an urban phenomenon only. We examined the question whether a different industry mix accounts for differences in the importance of small units. The Punjab stands out as the state with the highest proportion of small units, but this is not accounted for by its having an industry mix favoring small units. However, West Bengal has a high proportion of large units (more than 500 workers), which is accounted for by its industry mix. It is possible that the Punjab's rapid agricultural development has been a factor favoring the growth of small enterprises. Historical comparative advantage has made West Bengal the home of large-scale industries, for example, steel and jute.

Finally, we compared the size distribution of all Indian manufacturing establishments with more than 20 workers with that of similar establishments in Colombia, Japan, Korea, Pakistan, Taiwan, and the United States. India and Pakistan are similar. As compared with all the other countries they are relatively large-establishment-oriented. For instance, establishments with more than 500 workers account for a much lower proportion of U.S. employment than is the case with India; and while the medium range of 50–500 workers accounts for over half the employment in the United States and the East Asian economies, it accounts for less than a third in India.

Appendix: Sources on Size Distribution in Indian Manufacturing

The *Census of Population* provides two sets of tables which are relevant for the study of size distribution:

- The economic tables, which give a breakdown of total employment by two-digit industry groups between household and nonhousehold units.
- The establishment tables, which also provide figures on employment for each two-digit industry by employment-size groups, ranging all the way from 1-person units to units of 500 or more workers. The information is provided separately for household and nonhousehold establishments.

Table 6-10. *Indian Census Table Figures on Employment in Manufacturing, 1971*
(millions of workers)

Type of table	*Rural*	*Urban*	*Total*
Economic			
Household	3.8	1.3	5.0
Nonhousehold	3.2	7.2	10.4
Total	7.0	8.5	15.4
Establishment			
Household	2.9	0.9	3.8
Nonhousehold	3.0	5.9	8.9
Total	5.9	6.8	12.7

The figures of employment given in the two sets of tables do not agree. These data are compared in table 6-10.

The count in the establishment tables is lower, presumably because many workplaces were missed during the census enumeration. This problem is in addition to the well-known one that the 1971 census definition of secondary workers seriously underestimated the number who were gainfully employed. Compared with 1961, the participation rates of females dropped from 28.0 percent to 11.9 percent and of males from 57.1 to 52.5 percent. Other sources (for example, the National Sample Survey) give no indication of any change (let alone a drastic one) in participation rates during the 1960s.

Adjustments have been made to the 1971 census figures in order to correct for the underestimate of secondary workers. As an illustration we can give the revised figures prepared by Haan (1980). First, the female participation rate was increased from the observed 1971 census figure of 11.9 percent to 28 percent—the level of the 1961 census as well as the ten NSS surveys between 1959 and 1973. This increased the female labor force from 31 to 75 million. The male participation rate was also raised, but to a much smaller degree from 52.5 percent to 53.6 percent—adding 3.1 million males to the labor force. Second, the female work force was distributed between rural and urban areas according to the sex and residence distribution of the twenty-seventh round of the NSS (the 1971 census distribution could not be used because the underestimation of females in rural areas was relatively higher). The same procedure was adopted for males. Third, the work force had to be distributed over the industrial sectors. For this purpose the percentage distribution of workers by industry available from the 1971 census was used for males in both rural and urban areas, but for females only in urban areas. For females in rural areas, the percentage distribution by industrial category obtained from the 1961 census was used.[17] The revised figures, as calculated

17. Haan comments that, remarkably, the distribution of females by sectors remained more or less the same when the original 1971 percentage distribution of females by industry was applied to the total figure (1980, p.14).

by Haan for employment in the manufacturing sector in 1971, are presented in table 6-11.

The nonhousehold sector, of course, includes small workshops as well as registered factories. It is possible to get an estimate of employment in registered factories from the Directorate of Factories, Labor Bureau, Government of India. In 1971 this figure was 5.1 million, while total nonhousehold manufacturing employment was 12.1 million (see table 6-11). Thus factories accounted for about 42 percent of nonhousehold employment.

The establishment tables do give a breakdown of employment in the nonhousehold sector between registered factories and other workshops. We would expect these data to underestimate the proportion employed in unregistered workshops, which are easier to miss. This is indeed the case: factories account for 63 percent of total reported employment in the nonhousehold sector. In fact, employment in registered factories came quite close to the Labor Bureau figures—5.6 million against 5.1 million reported by the latter. Accepting the former would raise the estimated percentage of nonhousehold employment in factories from 42 to 46 percent.

The text has discussed the scope of the census of small-scale industry in India and the problems of comparing the data from this source with the data from the Annual Survey of Industries (ASI) to provide a picture of the size distribution of employment in different industries. This is the only way that information on size distribution at the three-digit level of industrial classification can be provided. But as mentioned in the text, many traditional industries were excluded from the SSI census.

In the two-digit industries, codes 20, 21, and 22, 13 subindustries were excluded. The two-digit textile industries, 23, 24, and 25, were wholly excluded except for one subindustry. Thus, industries in which small units were predominantly traditional were excluded, though powerloom units are exceptional in that they are both small and nontraditional in technique. In the remaining 14 two-digit industries (26 through 38, and 97), only 11 subindustries are excluded, of which steel is by far the most important. ASI employment in most of these excluded industries was overwhelmingly in units with more than 50 workers, but the principle of exclusion is obscure. Details of the 115 included industries are given in table 6-12, and of the 53 excluded industries in table 6-13.

Table 6-11. *Employment in Manufacturing, India, 1971*
(millions of workers)

	Rural			*Urban*			*Total*		
Sector	*Males*	*Females*	*Total*	*Males*	*Females*	*Total*	*Males*	*Females*	*Total*
Household	3.8	4.0	7.8	1.4	0.7	2.1	5.2	4.7	9.9
Nonhousehold	3.3	0.5	3.8	7.4	0.9	8.3	9.2	1.4	12.1

Note: Revised figures.
Source: Haan (1980, part II, table 11-12).

Table 6-12. *Estimate of Employment in the Small-Scale Sector, 1972–74*

Code and sector	*Employment in SSI (1)*	*Employment in ASI sector unit of < 50 workers (2)*	*Total employment in small scale (1) + (2) (3)*	*Total employment in ASI sector (4)*	*Total ASI + SSI (1) + (4) (5)*	*Revised total (1) + (4) − (2) (6)*	*Percent employment in small scale (3) − (5) (7)*	*Revised percent (1) − (6) (8)*	*Employment in ASI < 100 workers (9)*	*Total employment in small scale (1) + (9) (10)*	*Percent employment in small scale (10) − (5) (11)*
200 Slaughtering, preparation of meat	0.09	0.29	0.38	0.68	0.77	0.48	0.49	0.19	0.42	0.51	0.66
201 Dairy products	2.12	1.00	3.12	19.10	21.22	20.22	0.15	0.10	2.64	4.76	0.22
202 Canning and preservation of fruits and vegetables	4.38	0.88	5.26	5.61	9.99	20.22	0.53	0.48	1.88	6.26	0.63
203 Canning, preservation, fish processing and crustacea, and so on	4.15	1.19	5.34	4.72	8.87	7.68	0.60	0.54	2.85	7.00	0.79
205 Bakery products	20.32	3.25	23.57	14.79	35.11	31.86	0.67	0.64	4.38	24.70	0.70
208 Production of common salt	0.30	2.07	2.37	19.79	20.09	18.02	0.12	0.02	4.97	.5.27	0.26
209 Cocoa, chocolate, and sugar confections	4.95	0.75	5.40	3.43	8.38	7.63	0.68	0.65	1.18	6.13	0.73
214 Cashew nut processing, that is, drying, shelling, roasting, and so on	71.12	1.02	72.14	126.71	197.83	196.81	0.36	0.36	2.77	73.89	0.37
215 Ice	11.40	1.39	12.79	1.53	12.93	11.54	0.99	0.99	1.53	12.93	1.00
216 Prepared animal feeds	1.15	0.52	1.67	1.21	2.36	1.84	0.71	0.63	0.96	2.11	0.89
217 Starch	2.15	2.39	4.54	6.12	8.27	5.88	0.55	0.37	2.81	4.96	0.60
219 Food products not elsewhere classified	9.09	5.45	14.54	13.42	22.51	17.06	0.65	0.53	8.73	17.82	0.79
22 *Manufacture of beverages, tobacco, and tobacco products*											
220 Distilling, rectifying, and blending of spirits	0.44	0.57	1.01	7.89	8.33	7.76	0.12	0.06	1.96	2.40	0.29
221 Wine industries	0.23	0.05	0.28	1.26	1.49	1.44	0.19	0.16	0.14	0.37	0.25
222 Malt liquors and malt	0.31	0.24	0.55	3.6	3.92	3.68	0.14	0.08	0.52	0.83	0.21
224 Soft drinks and carbonated water industries	3.59	0.98	4.57	6.41	10.00	9.02	0.46	0.40	2.56	6.15	0.62

(Table continues on the following page.)

Table 6-12 *(continued)*

Code and sector	Employment in SSI (1)	Employment in ASI sector unit of < 50 workers (2)	Total employment in small scale (1) + (2) (3)	Total employment in ASI sector (4)	Total ASI + SSI (1) + (4) (5)	Revised total (1) + (4) − (2) (6)	Percent employment in small scale (3) − (5) (7)	Revised percent (1) − (6) (8)	Employment in ASI < 100 workers (9)	Total employment in small scale (1) + (9) (10)	Percent employment in small scale (10) − (5) (11)
23 *Manufacture of cotton textiles*											
230 Cotton cleaning and baling	0.87	25.16	26.03	89.21	90.08	64.92	0.29	0.01	53.43	54.30	0.60
26 *Manufacture of textile products (including wearing apparel)*											
260 Knitting mills	34.50	5.38	39.88	9.65	44.15	38.77	0.90	0.89	7.18	41.68	0.94
261 Thread, cordage	3.82	1.53	5.36	7.78	11.61	10.08	0.46	0.38	2.15	5.98	0.52
262 Embroidery and making of crepes, laces, fringes	4.39	0.97	5.36	5.18	9.57	8.60	0.56	0.51	2.46	6.85	0.72
264 Textile garments, weaving apparels	23.29	1.82	25.11	13.32	36.61	34.79	0.69	0.67	4.04	27.33	0.75
27 *Manufacture of wood and wood products, furniture and fixtures*											
271 Sawing, planing of wood	27.53	12.16	39.69	16.10	43.63	31.47	0.91	0.87	14.63	42.16	0.97
272 Wood and cane boxes, containers, baskets, and so on	9.82	2.04	11.86	3.51	13.33	11.29	0.89	0.87	3.14	12.96	0.97
273 Structural wood, wood beams, and so on	6.63	0.49	7.12	1.86	8.49	8.00	0.84	0.83	0.83	7.46	0.88
274 Wooden industry, goods and bobbins, and so on	2.99	1.16	4.15	4.19	7.18	6.02	0.58	0.50	2.18	5.17	0.72
275 Cork and cork products	0.31	0.11	0.42	0.22	0.53	0.42	0.79	0.74	0.11	0.42	0.79
276 Wooden furniture and fixtures	29.16	2.17	31.33	10.74	39.90	37.73	0.79	0.77	4.22	33.38	0.84
279 Wood, bamboo, and cane products not elsewhere classified	15.36	0.49	15.85	0.72	16.08	15.59	0.98	0.99	0.72	16.08	1.00
28 *Manufacture of paper and paper products, printing, publishing, and allied industries*											
280 Pulp, paper, and paper board	6.24	3.32	9.56	69.16	75.40	72.08	0.13	0.09	6.10	12.34	0.16
281 Containers and boxes of paper	10.98	2.71	13.69	8.92	19.90	17.19	0.69	0.64	5.72	16.70	0.84

283 Paper and paper board, not elsewhere classified	1.56	0.47	2.03	6.74	8.30	7.83	0.24	0.20	0.59	2.15	0.26
285 Printing and publishing of periodicals, books, journals, maps, and so on	48.95	9.89	58.84	67.08	116.03	106.14	0.51	0.46	15.84	64.79	0.56
287 Engraving, etching, block-making, and so on	2.25	1.06	3.31	1.64	3.89	2.83	0.85	0.80	1.17	3.42	0.88
288 Bookbinding	1.44	0.66	2.10	1.24	2.68	2.02	0.78	0.71	0.72	2.16	0.81
289 Printing, publishing, and so on, not elsewhere classified	17.73	9.36	27.12	31.04	48.77	39.41	0.56	0.45	13.67	31.40	0.64
29 *Manufacture of leather and leather and fur products*											
290 Tanning, curing, finishing, embossing, and japanning of leather	11.00	3.76	14.76	16.44							
291 Footwear excluding moulded rubber	15.61	0.75	16.36	20.50	36.11	35.36	0.45	0.44	1.17	16.78	0.46
292 Wearing apparel, for example, coats, gloves of leather	0.55	0.06	0.61	0.06	0.61	0.66	1.00	1.00	0.06	0.61	1.00
293 Leather consumer goods	2.28	0.35	2.63	0.72	3.00	2.65	0.88	0.86	0.51	2.79	0.93
299 Other leather and fur products	2.33	0.24	2.57	0.57	2.90	2.66	0.89	0.88	0.37	2.70	0.93
30 *Manufacture of rubber, plastics, petroleum, and coal products*											
300 Tyre and tube industry	9.88	2.14	12.02	26.68	36.56	34.42	0.33	0.29	3.64	13.52	0.37
301 Footwear of moulded rubber and plastic	4.98	1.49	6.47	5.75	10.73	9.24	0.60	0.54	2.27	7.25	0.68
302 Rubber products, not elsewhere classified	15.12	4.42	19.54	23.85	38.97	34.55	0.50	0.44	7.73	22.85	0.59
303 Plastic products, not elsewhere classified	40.88	8.03	48.91	22.74	63.62	55.59	0.77	0.74	12.63	53.51	0.84
305 Petroleum products, not elsewhere classified	10.27	0.36	10.63	1.59	11.86						
307 Other coal and coal-tar products, not elsewhere classified	0.56	0.43	0.99	8.34	8.90	8.47	0.11	0.07	0.72	1.28	0.14

(Table continues on the following page.)

Table 6-12 *(continued)*

Code and sector	Employment in SSI (1)	Employment in ASI sector unit of < 50 workers (2)	Total employment in small scale (1) + (2) (3)	Total employment in ASI sector (4)	Total ASI + SSI (1) + (4) (5)	Revised total (1) + (4) − (2) (6)	Percent employment in small scale (3) − (5) (7)	Revised percent (1) − (6) (8)	Employment in ASI < 100 workers (9)	Total employment in small scale (1) + (9) (10)	Percent employment in small scale (10) − (5) (11)
31 *Manufacture of chemicals and chemical products (excluding petroleum and coal)*											
310 Basic industry organic and inorganic chemicals	9.58	3.08	12.66	44.54	54.12	51.04	0.23	0.19	6.10	15.68	0.29
311 Fertilizers and pesticides	3.72	2.88	6.60	47.10	50.82	47.94	0.13	0.08	5.57	9.29	0.18
312 Paints, varnishes, and lacquers	12.64	2.91	15.55	26.31	38.95	36.04	0.40	0.35	5.23	17.87	0.46
313 Drugs and medicines	20.76	5.62	26.38	70.68	91.44	85.82	0.29	0.24	12.28	33.04	0.36
314 Toilet preparations, cosmetics, soap, and so on	29.02	2.20	31.22	18.39	47.39	45.19	0.66	0.64	4.31	33.33	0.70
315 Edible oils	1.86	1.84	3.70	9.45	11.31	9.47	0.33	0.20	4.86	6.72	0.59
316 Turpentine, plastic materials, synthetic fiber	1.86	1.35	3.21	28.94	30.80	29.45	0.10	0.06	1.88	3.74	0.12
317 Matches	31.25	1.91	33.16	18.07	49.32	47.41	0.67	0.66	3.96	35.21	0.71
318 Explosives, ammunition, fireworks	10.33	0.25	10.58	10.08	20.41	20.16	0.52	0.51	1.56	11.89	0.58
319 Chemical products, not elsewhere classified	37.98	3.00	40.98	18.06	56.04	53.04	0.73	0.72	5.47	43.45	0.78
32 *Manufacture of nonmetallic mineral products*											
320 Structural clay products	83.22	9.26	92.48	74.60	157.82	148.56	0.59	0.56	24.98	108.20	0.69
321 Glass and glass products	39.21	1.39	40.60	50.58	89.79	88.40	0.45	0.44	4.84	44.05	0.49
323 Chinaware and porcelain	10.47	1.64	12.11	25.06	35.53	33.89	0.34	0.31	3.80	14.27	0.40
324 Cement, lime, and plaster	8.63	1.20	9.83	46.36	54.99	53.79	0.18	0.16	2.15	10.78	0.20
325 Mica products	3.62	1.02	4.64	8.37	11.99	10.97	0.39	0.33	1.86	5.48	0.46
326 Structural stone goods, stoneware, and so on	19.60	6.60	26.20	11.15	30.75	24.15	0.85				
328 Asbestos cement and other cement products	9.07	0.46	9.53	9.27	18.34	17.88	0.52	0.51	0.69	9.67	0.53
329 Other nonmetallic mineral products	28.64	8.46	37.10	25.75	51.50	43.04	0.72	0.67	14.83	43.47	0.84

33 *Basic metal and alloys industries*											
331 Foundries for casting and forging iron and steel	97.15	18.74	115.89	156.07	253.22	234.48	0.46	0.41	30.62	127.77	0.50
332 Ferro-alloys	3.46	0.28	3.74	4.58	8.04	7.76	0.47	0.45	0.35	3.81	0.47
333 Copper	2.64	0.60	3.24	6.68	9.32	8.72	0.35	0.30	1.10	3.74	0.40
334 Brass	3.68	1.22	4.90	4.98	8.66	7.44	0.57	0.49	1.39	5.07	0.59
335 Aluminum	2.61	1.26	3.87	18.11	20.72	19.46	0.19	0.13	1.86	4.47	0.22
336 Zinc	0.08	0.28	0.36	2.07	2.15	1.87	0.17	0.04	0.28	0.36	0.17
34 *Manufacture of metal products and parts*											
340 Fabricated metal products	70.48	9.49	79.97	40.85	111.33	101.84	0.72	0.69	15.04	85.52	0.77
341 Structural metal products	25.27	2.35	27.62	10.38	35.65	33.30	0.77	0.67	3.84	29.11	0.82
342 Furniture and fixtures of metals	11.11	1.71	12.82	11.41	22.52	20.81	0.57	0.53	3.02	14.13	0.63
343 Hand tools and general hardware	79.86	11.15	91.01	47.35	127.21	116.06	0.72	0.69	17.21	97.07	0.76
344 Enameling, plating, and so on of metal products	8.25	1.88	10.13	6.72	14.97	13.09	0.68	0.63	2.86	11.11	0.74
345 Metal utensils, cutlery, and kitchenware	60.65	7.0	67.70	15.64	76.29	69.24	0.89	0.88	3.00	47.44	0.90
349 Other metal products	44.44	2.13	46.57	8.11	52.55	50.42	0.89	0.88	3.00	47.44	0.90
35 *Manufacture of machinery, machine tools and parts, excluding electric*											
350 Agricultural machinery and equipment and parts	5.91	4.56	10.47	20.09	26.00	21.44	0.40	0.28	6.52	12.43	0.48
351 Repair of heavy construction equipment	2.62	1.09	3.71	19.93	22.55	21.46	0.16	0.12	2.61	5.23	0.23
352 Prime movers, boilers, and steam generator plants	15.20	4.72	19.92	53.38	68.58	63.86	0.29	0.24	7.68	22.88	0.33
353 Industrial machinery for food and textile industries	23.01	7.86	30.8	54.81	77.82	69.96	0.40	0.33	13.63	36.64	0.47
354 Industrial machinery for other industries	28.34	3.40	31.74	26.01	54.36	50.95	0.58	0.56	5.87	34.21	0.63
355 Refrigerators, air conditioners, and fire fighting equipment and parts	3.56	0.93	4.49	15.97	19.53	18.60	0.23	0.19	1.80	5.36	0.27
356 Alterations and repair of nonelectrical machinery not elsewhere classified	9.40	3.91	13.31	37.34	46.74	42.83	0.28	0.22	6.42	15.82	0.34

(Table continues on the following page.)

Table 6-12 *(continued)*

Code and sector	Employment in SSI (1)	Employment in ASI sector unit of < 50 workers (2)	Total employment in small scale (1) + (2) (3)	Total employment in ASI sector (4)	Total ASI + SSI (1) + (4) (5)	Revised total (1) + (4) − (2) (6)	Percent employment in small scale (3) − (5) (7)	Revised percent (1) − (6) (8)	Employment in ASI < 100 workers (9)	Total employment in small scale (1) + (9) (10)	Percent employment in small scale (10) − (5) (11)
357 Machine tools, parts, and accessories	17.20	5.02	22.22	32.71	49.91	44.89	0.45	0.38	7.50	24.70	0.49
358 Office computing and accounting machinery and parts	0.58	0.58	1.16	6.03	6.61	6.03	0.18	0.10	1.31	1.89	0.29
359 Other nonelectrical machine equipment, accessories, and so on	39.51	4.15	43.66	16.92	56.43	52.28	0.77	0.76	6.60	46.11	0.82
36 *Manufacture of electrical machinery, apparatus, appliances, supplies*											
360 Miscellaneous industrial machinery apparatus and parts	13.48	9.71	19.17	102.71	118.17	110.48	0.17	0.12	11.07	24.53	0.21
361 Insulated wires and cables	2.57	1.98	4.55	25.48	28.05	26.07	0.16	0.10	2.92	5.49	0.20
362 Dry and wet batteries	1.82	0.42	2.24	16.19	18.01	17.59	0.12	0.10	0.84	2.66	0.15
363 Electrical apparatus, appliances, and parts	16.86	5.09	21.95	35.05	51.91	46.82	0.42	0.36	9.11	25.97	0.50
364 Radio and television transmitting and receiving sets	13.70	1.94	15.64	45.44	59.14	57.20	0.26	0.24	4.06	17.76	0.30
366 Electronic computers, control instruments, and so on	3.39	0.33	3.72	2.89	6.28	5.95	0.59	0.57	0.44	3.83	0.61
367 Other electronic components and accessories	3.74	0.36	4.10	2.54	6.19	5.83	0.66	0.60	0.87	4.61	0.74
369 Other electrical machinery apparatus, supplies	10.36	0.93	11.29	3.90	15.26	13.33	0.79	0.78	1.26	11.62	0.81
37 *Manufacture of transport equipment and parts*											
370 Ship and boat building and repairing	2.20	0.75	2.95	46.14	48.34	47.59	0.06	0.05	3.04	5.24	0.11
371 Locomotives and parts	1.32	0.25	1.57	31.65	32.97	32.72	0.05	0.04	0.25	1.57	0.05

372 Railway wagons, coaches, and parts	3.31	0.86	4.17	130.99	134.30	133.44	0.03	0.02	1.38	4.69	0.03
373 Other railroad equipment	0.15	0.24	0.39	8.92	9.07	8.83	0.04	0.02	0.48	0.63	0.07
374 Motor vehicles and parts	45.25	6.54	51.79	122.12	167.37	160.83	0.31	0.28	12.42	57.67	0.34
375 Motorcycles, scooters, and parts	7.61	1.10	8.71	14.03	21.64	20.54	0.40	0.37	1.47	9.08	0.42
376 Bicycles, cycle rickshaws, and parts	18.88	3.74	22.62	22.72	41.60	37.86	0.54	0.50	5.68	24.56	0.59
378 Bullock, push, and hand carts, and so on	3.90	0.08	3.98	0.08	3.98	3.90	1.00	1.00	0.08	3.98	1.00
379 Transport equipment and parts not elsewhere classified	0.87	1.11	1.98	3.77	4.64	3.53	0.43	0.25	1.75	2.62	0.56
38 *Other manufacturing industries*											
380 Medical, surgical, and scientific equipment	17.48	2.79	20.27	20.36	37.84	35.05	0.54	0.50	4.62	22.10	0.58
381 Photographical and optical goods (excluding chemicals, film, paper)	0.72	0.55	1.27	1.27	1.99	1.44	0.64	0.50	0.80	1.52	0.76
382 Watches and clocks	1.43	1.03	2.46	6.82	8.25	7.22	0.30	0.20	2.12	3.55	0.43
385 Sports and athletic goods	2.57	0.43	3.00	1.42	3.99	3.56	0.75	0.72	0.58	3.15	0.79
386 Musical instruments	0.61	0.08	0.69	1.93	2.54	2.46	0.27	0.25	0.08	0.69	0.27
387 Stationery articles, for example, pens, tags, and so on not elsewhere classified	7.39	1.76	9.15	6.82	14.21	12.45	0.64	0.59	2.90	10.29	0.72
389 Other miscellaneous projects, for example, costume jewelry	8.76	2.86	11.62	8.23	16.99	14.13	0.68	0.62	4.90	13.66	0.80
97 *Repair service*											
971 Repair of footwear and leather goods	0.26	0.12	0.38	0.12	0.38	0.26	1.00	1.00	0.12	0.38	1.00
972 Electrical repair shops	3.02	0.49	3.51	2.32	5.34	4.85	0.66	0.62	1.27	4.29	0.80
973 Repair of motor vehicles and motorcycles	23.50	15.98	39.48	97.88	121.38	105.40	0.32	0.22	36.58	60.08	0.49
979 Other repair enterprises	2.10	1.67	3.77	29.72	31.82	30.15	0.12	0.07	4.15	6.25	0.20

Sources: SSI Census, 1972; ASI, 1974.

Table 6-13. *Industries in ASI Census But Not in SSI Census, India, 1974*

Code and sector	ASI less than 50 workers (thousands)	ASI total (thousands)	Relative employment in small-scale establishments	ASI less than 100 workers (thousands)	Relative employment in small-scale establishments
204 Grain mill products	36.41	57.48	0.63	49.97	0.87
206 Manufacture and refining of sugar	0.08	132.05	0.001	0.35	0.003
207 Production of indigenous sugar and so forth from sugarcane and palm juice	9.06	43.68	0.21	19.40	0.44
210 Manufacture of hydrogenated oils, vanaspati, ghee, and so forth	0.11	14.86	0.01	0.92	0.06
211 Manufacture of other edible oils and fats	15.43	26.77	0.58	20.77	0.78
212 Tea processing	6.20	86.65	0.07	27.89	0.32
213 Coffee curing, roasting, grinding	0.16	4.82	0.03	0.72	0.15
223 Production of country liquor and toddy	0.33	1.37	0.24	0.59	0.43
225 Tobacco steaming and so forth for preparing raw leaf tobacco	1.30	95.42	0.01	4.43	0.05
226 Manufacture of bidi	17.10	65.44	0.26	38.35	0.59
227 Manufacture of cigars, cigarettes, and so forth	0.22	16.14	0.01	4.43	0.05
228 Manufacture of chewing tobacco, snuff	3.87	10.10	0.28	7.61	0.75
229 Manufacture of other tobacco and tobacco products	0.08	0.54	0.15	0.27	0.50
231 Cotton spinning, weaving and finishing, and so forth	3.45	793.71	0.0004	7.06	0.01
232 Printing, dyeing, and bleaching of cotton textiles	3.59	23.02	0.16	6.36	0.28
233 Cotton spinning other than in mills	0.23	0.34	0.68	0.34	1.00
234 Production of khadi	0.13	0.55	0.24	0.13	0.24
235 Weaving and finishing of cotton textiles in handlooms other than khadi	1.90	7.71	0.25	1.90	0.25
236 Weaving and finishing of cotton textiles in powerlooms	6.16	9.08	0.68	7.23	0.80
239 Other cotton textiles	0.51	2.65	0.19	1.06	0.40
240 Wool cleaning, baling, and pressing	0.79	4.08	0.19	1.06	0.40
241 Wool spinning, weaving, and finishing in mills	0.83	28.88	0.03	3.44	0.12
242 Wool spinning, weaving, and finishing in mills	1.42	5.57	0.25	1.73	0.31
243 Dyeing and bleaching of woolen textiles	0.56	0.70	0.80	0.70	1.00
244 Other manufacture of wool	0.25	4.30	0.06	0.70	0.16
245 Spinning, weaving, and finishing of silk textiles	2.74	13.76	0.20	3.53	0.26

246	Printing, dyeing, and bleaching of silk textiles	0.32	1.56	0.21	0.46	0.29
247	Spinning, weaving, and finishing of other textiles—synthetics, rayon, and so forth	7.90	72.16	0.11	10.48	0.15
248	Printing, dyeing, and bleaching of synthetic textiles	1.32	14.27	0.09	3.76	0.26
249	Silk and synthetic fiber textiles, not elsewhere classified	0.22	0.63	0.35	0.44	0.70
250	Jute and mesta pressing and baling	0.45	0.58	0.78	0.45	0.78
251	Jute and mesta spinning and weaving	0.02	262.67	—	0.12	0.0005
253	Preparation, spinning, weaving, and finishing of hemp and other coarse fibers	0.04	0.09	0.44	0.09	1.00
259	Manufacture of jute bag and other jute textiles not elsewhere classified	0.30	3.76	0.08	0.39	0.10
263	Weaving, carpets, rugs, and other similar textile products	0.54	4.46	0.12	0.76	0.17
284	Printing and publishing of newspapers	1.10	25.63	0.04	3.66	0.14
286	Printing of bank notes, currency notes, postage stamps, and so forth	0.06	9.25	0.01	0.20	0.02
304	Petroleum refineries	0.04	7.38	0.01	0.22	0.03
306	Production of coal tar in coke ovens	0.18	2.42	0.07	0.37	0.15
322	Manufacturing of earthenware and earthen pottery	0.05	0.36	0.14	0.05	0.14
330	Iron and steel industries	9.90	222.42	0.04	19.35	0.09
333	Copper manufacturing	0.60	6.98	0.09	1.10	0.16
334	Brass manufacturing	1.22	4.98	0.24	1.39	0.28
335	Aluminum manufacturing	1.26	18.11	0.07	1.86	0.10
339	Other nonferrous metal industries	1.16	6.04	0.19	2.37	0.39
365	Manufacturing and repair of radio x-ray apparatus and tubes and parts	0.14	0.20	0.70	0.20	1.00
377	Manufacture of aircrafts and parts					
383	Manufacture of jewelry and related articles	2.95	3.68	0.80	3.46	0.94
384	Minting of coins	0.06	6.14	0.01	0.06	0.01
400	Generation, transmission, and distribution of electric energy	2.0	512.11	0.004	4.84	0.01
410	Manufacture of gas in gas works and distribution through mains to household industrial, commercial, and other users	0.50	3.46	0.14	0.81	0.23
420	Water supply	1.38	7.91	0.17	3.34	0.12
741	Cold storage	2.10	2.53	0.83	2.42	0.96

—Not available.

Table 6-14. *Employment in Small Establishments Compared with the Total by Industry Group, India, 1971*

Industry code	*Employment in establishments with 1–50 workers* (1)	*Total employment* (2)	(1) ÷ (2) *as percentage*
20–21	968.20	1,547.71	62.6
22	479.01	729.98	65.6
23	478.11	1,309.57	36.5
26	474.09	552.80	85.8
27	308.47	356.15	86.6
28	164.69	318.96	51.6
29	98.64	136.36	72.3
30	62.42	137.23	45.5
31	116.41	334.71	34.8
32	204.58	618.13	33.1
33	26.68	148.88	17.9
34	339.23	502.40	67.5
35	159.92	430.62	37.1
36	43.5	209.97	20.7
37	50.34	298.29	16.9
38	254.50	408.46	62.3
39	428.44	571.84	74.9
Total	4,657.23	8,612.06	54.2

The census establishment tables give the size distribution of employment in all nonhousehold manufacturing units. The census includes all industries and is therefore more comprehensive in its coverage than the SSI. But the information is available only for the two-digit classification. Table 6-14 gives the proportion of employment in units employing 50 workers or less. These data can be compared with those given in table 6-1. With the inclusion of traditional industries, the proportion employed in small units increases significantly in all industry groups. This is in spite of the fact noted above that the establishment census underreported the total number of workers in manufacturing, and it is likely that the underreporting due to missed units would be more serious for smaller units.

7

Factor Intensities and Productivities

In the developing-economy context, the most compelling argument in favor of small-scale enterprises is their allegedly more efficient use of capital and labor. It is claimed that they usually produce a unit of output (value added) with less capital but more labor than large units. If only capital is scarce, the result is both more output and more employment. Even if this extreme hypothesis of surplus labor is rejected, it could still be true that, when properly valued, the cost of the extra use of labor by SSEs is more than outweighed by their economy in the use of capital.

The Theory

We shall first establish on certain assumptions an a priori pattern of relationships between size, wages, capital and labor productivity, and capital intensity. W will refer to the wage, K to the value of capital, and L to the number of employees, K and L being measured per unit of value added. We shall then confront these expected patterns with the facts.

First arrange all enterprises in order of labor productivity, that is, in decreasing order of the number of employees per unit of value added, thus:

$$L_1 > L_2 > \ldots \quad (7\text{-}1)$$

Next assume that the rate of profit is equalized. Thus:

$$\frac{1 - W_1 L_1}{K_1} = \frac{1 - W_2 L_2}{K_2} = \ldots \quad (7\text{-}2)$$

If wages rise faster than productivity, then $1 - W_2L_2 < 1 - W_1L_1$. In that case, the equality 7-2 implies $K_1 > K_2$. This says that greater labor productivity (value added per person) is achieved with less capital. If profitability is to be maintained, then for this to occur in a regular manner requires either that other price differences consistently offset the excess of wage differentials over productivity differentials or that increasing returns to scale are very generally associated with rising labor productivity. While bearing this in mind, let us assume (what is in fact usually the case) that wage differences are less than productivity differences. Thus

$$W_1 L_1 > W_2 L_2 \ldots \tag{7-3}$$

From 7-2 and 7-3 it follows that

$$\frac{1}{K_1} > \frac{1}{K_2} \ldots \tag{7-4}$$

that is, capital productivity falls. From 7-4 we derive

$$\frac{1}{L_1} \Big/ \frac{K_1}{L_1} > \frac{1}{L_2} \Big/ \frac{K_2}{L_2} \ldots$$

and therefore,

$$\frac{L_1}{L_2} < \frac{K_2/L_2}{K_1/L_1} \ldots \tag{7-5}$$

Given the assumptions, we thus have the pattern that, when enterprises are arranged in order of increasing labor productivity, wages will not rise as fast and capital intensity (K/L) will rise faster. Capital productivity falls.[1] There are two points to note about the above theory. First, it does not relate the patterns of variation of factor intensities and productivities to the size of the enterprise, however measured. It is, however, often assumed that size and capital intensity, and hence labor productivity, are positively related. There are no grounds for this assumption in the theory of production or of the firm. It is an empirical matter. There is a tendency to think in terms of containers (ships, tanks, furnaces, pipelines, and so forth) for which a larger size does not require much if any increase in the number of operatives who work outside the container. But there is a different model which may be more relevant in many industries. Thus, if a larger-size factory has more machines of the same type than a smaller factory and the same number of operatives per machine, then capital intensity will be lower since building costs rise less than in proportion to the number of workplaces. The essential difference is that the workers work inside the container, the size of which is determined by their number.

The second point is that output is value added. Enterprises may be producing the same or quite different things. Thus, since wages vary, they might be either at different points on the same production function or on different production functions. Nothing has been assumed about physical production functions, since "output" is value added. But we have assumed that more

1. The above algebra essentially derives from Tajima (1978, appendix 1), though the presentation is different.

value added cannot be obtained without either more labor or more capital. We have seen that insofar as enterprises can be said to be on the same production function, this could be falsified by increasing returns to scale; and, whether or not they are on the same production function, it could be falsified by monopolistic conditions which permit more value added for smaller inputs of capital and labor. These or other imperfect market conditions may thus invalidate the assumption of equal returns to capital. Widespread competition is thus suggested if we find that the foreshadowed patterns are observed.

Census of Production Analysis

This section analyzes the census of production figures for Japan, the developing economies excluding India, and India.

Japan

The aggregate of manufacturing firms in Japan conforms remarkably well to the above theory. Measuring size by employment, there is no exception to the empirical presumption that labor productivity rises with size.[2] Wages also rise, but less than in proportion, as hypothesized. Capital intensity also rises,[3] and faster than labor productivity. The only discordant element is that capital productivity *rises* from the smallest size class (1–9 workers) up to a peak in the range 20–50 workers, only thereafter falling monotonically as expected.

The above relations are largely preserved when manufacturing is split into four groups, although strict monotonicity begins to break down. In particular, capital intensity falls from size class 1 (1–9 workers) to size class 2 (10–14 workers) in three out of four cases. In the remaining case it falls in the size class of 20–29 workers. The main discordant element, that the smallest size has both the lowest labor and the lowest capital productivity, is true of all four groups of industries. This anomaly depends, however, on how size is defined. Kaneda (1980) reclassified size by capitalization instead of employment, and the anomaly disappeared, the K/L ratio rising and capital productivity falling (almost) monotonically, while other regularities were preserved.[4]

2. This refers to both 1957 and 1966 (the present tense is used for convenience). See Tajima (1978, table A-1).

3. The only exception was that K/L was lower in the size class 10–19 than 1–9 in 1957. K stands for the "value of assessment" of tangible fixed assets excluding land.

4. It is perhaps intuitively clear that K/L is more likely to rise regularly with size if size is defined by K rather than L. Putting Y for gross income, we have similarly Y/K more likely to fall (L and Y being closely correlated).

The Developing Economies Excluding India

We first consider figures for manufacturing as a whole collected for nine developing economies in Ohkawa and Tajima (1976) and in Tajima (1978). The figures exclude size classes of less than 20 workers. Labor productivity, capital intensity, and wages rise with size, with only a few anomalies.[5] But capital productivity often behaves in an erratic manner. We saw that in the case of Japan it rose, contrary to the hypothesis, from the smallest size class (1–9 workers) to somewhere in the range 20–50 workers and fell monotonically thereafter. A certain tendency for the curve to form an inverted U could also, perhaps, be observed for the Philippines (1960 and 1970) and Brazil (1960) (Ohkawa and Tajima 1976, tables 3 and 9).[6] Apart from the above, however, the figures are much more erratic than for Japan.

From the World Bank project we can add, for manufacturing as a whole, some census figures for Colombia (1976) and Korea (1968). In Colombia, if we exclude the smallest recorded size class of 5–9 workers, capital intensity and labor productivity rise monotonically. Wages also rise but probably less than labor productivity; but, contrary to expectations, the latter rises faster than capital intensity (implying an increase in profitability with size). The smallest size class shows relatively high capital intensity and labor productivity.[7] Capital productivity appears to behave erratically, but with some tendency to peak in the range 20–99 workers.[8]

Korean figures are given in table 7-1. Labor productivity rises from the smallest class, 5–9 workers, up to size class 200–499. Capital intensity rises almost monotonically up to the largest size; but contrary to the hypothesis, and as in Colombia, it rises less fast than labor productivity up to size class 5 (100–199 workers). It is notable that it is virtually constant in the range 10–200 workers. Capital productivity is virtually constant in the range 5–50 workers, but then *rises* dramatically in the range 50–200 workers, before falling back again (though not to the level of the range 5–50 workers), showing the inverted U already observed in Japan and elsewhere. Thus comparing size class 100–199 with 10–19, there is a rise of about 120 percent in both labor and capital productivity and virtually no change in capital per person. The average wage was 23 percent higher, while profitability was about two and a half times as great.

The above could result from strongly increasing returns. The figures also suggest that increasing returns vanish at about 200 workers and then go into

5. The worst irregularities are for India (1954), Sri Lanka (1952), and Pakistan (1969–70). We ignore these.

6. In Brazil, capital is measured by kilowatts; in the Philippines, by the book value of fixed assets including land.

7. This smallest class approximates the figures appropriate to a firm of about 50 workers.

8. This paragraph is based on tables 2-4, 2-6, 2-10, and 2-11 in Cortes, Berry, and Ishaq (1987).

Table 7-1. *All Manufacturing, Selected Economic Ratios, Korea, 1968*
(thousands of won)

Number of workers	Y/L	K/L	Y/K	W	$\frac{Y - WL}{K}$
5–9	196	296	0.66	63	0.45
10–19	248	375	0.66	83	0.44
20–49	259	388	0.67	84	0.45
50–99	315	387	0.81	92	0.57
100–199	553	380	1.45	102	1.18
200–499	607	520	1.17	125	0.93
500 or more	598	656	0.91	—	0.72

—Not available.

Note: Throughout the tables in this chapter, Y is value added; L is labor; K is the market value of tangible fixed assets including land; and W is wages, which are expressed in thousands of units of domestic currency. The fifth column is calculated.

Source: Ho (1980), tables D6 and D9, for the first four columns; p. 77 for the fifth column.

reverse. Thus the size class 200–500 shows sharply increased K/L, a 10 percent rise in labor productivity, and a 20 percent fall in capital productivity, while in the largest size class (more than 500 workers) a further rise in K/L and wages is accompanied by falls in both labor and capital productivity, and in profitability. If increasing monopoly rather than increasing returns were the explanation for rises in capital and labor productivity, these falls in the largest size class would be hard to explain. If increasing returns were the explanation, this should show up in many industries on disaggregation (see below in this section).

In table 7-2 figures are also given for small-establishment industries, that is, industries in which establishments with 5–50 workers account for half or more of industry employment. The results are similar, especially if one ignores the largest class (more than 500 workers), in which there could have

Table 7-2. *Manufacturing, Small-Establishment Industries, Selected Economic Ratios, Korea, 1968*
(thousands of won)

Number of workers	Y/L	K/L	Y/K	W	$\frac{Y - WL}{K}$
5–9	193	287	0.67	63	0.45
10–19	242	363	0.67	83	0.44
20–49	242	378	0.64	84	0.42
50–99	293	363	0.81	92	0.55
100–199	318	385	0.83	102	0.56
200–499	335	512	0.66	125 (200–499 and 500 or more)	0.41
500 or more	313	326	0.96		0.58

Note: Wages are in thousands of won and are assumed the same as for all manufacturing. The fifth column is calculated.

Source: Ho (1980) for the first four columns.

been only very few observations. Labor productivity rises, but much less strongly than for manufacturing as a whole. Capital intensity also rises almost monotonically, and about as fast as labor productivity over the whole range. But, as in the case of manufacturing as a whole, there is virtual constancy in the range 10–200 workers, capital intensity rising only 6 percent, while labor productivity and capital productivity rise by 31 and 24 percent respectively, the apparent increasing returns and rise in profitability being much less marked than for manufacturing as a whole, as might be expected.

Some aggregate figures are available for Thailand from a survey of small and medium-scale firms and are given in table 7-3. As in Colombia, capital intensity is relatively high in the smallest size class. It varies rather little until the size class 200 or more is reached, when it leaps. But both capital and labor productivity rise until somewhere in the range 50–200, so that smaller sizes are dominated. The inverted U of the capital productivity plot is plainly evident. This is similar to Korea, for which the size class 100–199 workers dominates the smaller.

So far we have discussed only manufacturing aggregates. Yet disaggregation is essential. For policy purposes, it is important to know whether small units survive and are labor intensive because they produce the great bulk of a limited range of things that are technologically unsuited to large-scale, capital-intensive production or whether a very wide range of things is produced in both small labor-intensive and large capital-intensive units. In the former case, significantly more small-scale production can be engineered only by a change in demand in favor of the sorts of things which can be efficiently made by small units; in the latter case, a large relative shift to small units may be both possible and desirable without such a change in demand. In narrowly defined industries, it is more likely that the products of large and small firms are competitive, though this is by no means always the case.

Table 7-3. *Manufacturing, Thailand*
(thousands of baht)

Number of workers	Y/L	K/L	Y/K
0–9	22.5	85.6	0.26
10–49	29.2	67.7	0.43
50–99	46.6	77.6	0.60
100–199	51.9	87.6	0.59
200 or more	80.1	216.1	0.37

Note: The third column is derived from the first and second and is not as given in the original table (the discrepancy is, however, slight).

The survey was intended to cover only firms in the range 10–199, so that the figures for those in smallest and largest classes may be less reliable than the others.

Source: Tambunlertchai and Loohawenchit (1981, table 5-17).

Ho's (1980) examination of the Korean census[9] under the World Bank research project is unique in its detail. It was found that the regularities prevailing in aggregated statistics very rarely carried over to industries defined at the four-digit level. For instance, monotonicity of rising capital intensity with size is found in only 2 of the 160 industries for which there are entries in at least four different size categories. More than this would occur by chance. It is almost as if there were something causing irregularity. The figures do not support the belief that small enterprises, as measured by numbers employed, are typically labor-intensive. The incidence of lowest capital intensity is almost even over the various size categories from 5 to 500 workers. Highest capital intensity occurs as often in the range 5–50 workers as in the range from 50 upward (Ho 1980, table 4-4). In contrast, capital productivity peaks in the range 50–500 in about two-thirds of 160 industries in which there were at least four size entries (calculated from Ho 1980, table D-9).

Both statistical weaknesses and reality doubtless contribute to the turbulence of the disaggregated figures. Capital measures are notoriously unreliable: in this census, capital comprises the book value of tangible fixed capital but excludes rented assets. So far as this goes, capital used is more likely to be understated for small enterprises than large ones. As against this, it is possible that the proportion of inventories to fixed capital is lower for small enterprises.[10] Value added is also likely to be very unreliable, most especially for the smaller size classes, many of which do not keep records.[11] Even where records are kept, value added may be falsified to deceive the tax collector. Turbulence is also present in reality. One enterprise, or even a size class of enterprises, may have a good or a bad year for reasons unconnected with production functions or normal market conditions. If we ignore the smallest size classes because of poor statistics and the largest because of few observations, comparability is likely to be improved. For this reason, we made comparisons between size classes 20–49 workers and 100–199 workers, which is also the range in which the aggregate figures suggested strong increasing returns. Out of 213 industries, 139 had entries in both these size classes. In 67 cases, capital intensity was lower and capital productivity higher in the larger size class. In 40 of the 67 cases, labor productivity was also higher. To this extent, the feature of the aggregate figures—nonincreasing capital intensity and higher productivities—carries through on disaggregation. As already

9. Ho also considers the Taiwan figures, but they did not permit the detailed size and industry breakdown of the Korean census (see below).

10. This was found to be the case in Colombia (see Cortes, Berry, and Ishaq (1987, table 2-17). But this finding is hard to reconcile with optimum inventory theory, which requires inventories to rise with the square root of inputs (see Baumol 1977, chap. 1).

11. See Ho (1980, table 5-5). Below 20 workers, less than half have enterprise accounts. From 20–50 workers, two-thirds have separate enterprise accounts.

mentioned, in view of the declining productivity in still greater sizes, economies of scale seem a more likely explanation than increasing monopoly. Before leaving the subject of capital intensity, it should be noted that interindustry variations are far greater than the unreliable average intraindustry variations by size group. The latter seldom vary by a factor of more than 3. The range of the former is over 100.

If one rejects the idea that only capital or only labor is scarce, then one cannot derive even a prima facie presumption of social efficiency from the basis of either capital or labor productivity alone. These measures must be weighted and combined to produce a measure of total factor productivity. The weights used for each enterprise may be either actual shares of capital and labor in that industry (which presumes that factor rewards are equal to their social opportunity cost) or shadow prices. Ho calculated total factor productivity for the different size classes in each Korean industry on both assumptions, using a 20 percent return on capital as the shadow price of capital while retaining the actual wage as a true measure of labor's social opportunity cost on the grounds that the labor market in Korea is as near perfect as makes no matter (an assumption he shares with several other authors and observers). We give results using the shadow price version.[12] Out of 138 industries,[13] 88 show greatest efficiency of factor use in the range 50–500 employees. Thirty-two are best below 50 workers, of which 7 are in the class of 5–9 workers. Eighteen are best in the range 500 or more. The rather few industries where small enterprises show up as most efficient are also rather small industries in terms of employment. For instance, the 16 industries in which enterprises with less than 100 workers account for half or more of the industry employment, and where the apparently most efficient size is 50 workers or less, account for only 7.3 percent of total factory employment (Ho 1980, p. 64).

Ho gave the same treatment to the 1971 Taiwan census but distinguished only between larger enterprises and those with less than 100 workers (because of data limitation). If the same method as in Korea were used, the two size classes could claim an equal number of wins when comparing 40 industries (more aggregated than in Korea) in which small enterprises accounted for half or more of the industry employment (Ho 1980, table D-12). Even for these industries the interindustry variation is much greater than the intraindustry variation.

Some size-differentiated figures for industries at the two-digit level are available for the Philippines for 1970.[14] There is little regularity. Out of 17

12. In most of the small minority of cases where a difference is made to the most efficient size class, the smaller is favored by shadow pricing.

13. The total number of industries was reduced from 213 by excluding 75 which had cases in less than 4 size categories, or for which there were indications of excessive heterogeneity of output.

14. From International Labour Organisation (ILO 1974); also quoted in Bautista (1981, table

industries with an entry in at least three out of four size classes (20–49, 50–99, 100–199, 200 or more workers), in only one (rubber products) does capital intensity increase monotonically and in one (printing) it decreases monotonically. In no single case does one find a regular association of higher labor productivity and lower capital productivity with higher capital intensity. Even the usually well behaved labor productivity is erratic. Small wonder that Bautista writes, "a mixed pattern is seen with respect to the size structure," and approvingly quotes the ILO report to the effect that what was needed were policies that encourage the development of the most efficient industries and the most efficient firms, regardless of size. Blunt policies that are strongly biased toward one size or another are not capable of doing this. (International Labour Organisation 1974, p. 57.)

India

The CMI covering twenty-nine industries, and aiming to include only power-using factories with more than twenty employees, gave data by employment size from 1953 to 1958. The figures for ten of these industries were analyzed for each of the six years by Sandesara (1966). He calculated rank correlations between size groups and various production attributes including capital intensity, capital and labor productivities, and wages. There was a positive and unusually strong but rarely perfect association for every year and industry between size and labor productivity. The association between size and wage levels was very similar. But most other expected relationships disappointed. Thus rank correlations between size and capital intensity were as often negative as positive and varied from year to year in most industries [15] and the correlations between capital intensity and labor productivity also jumped about from year to year in an apparently senseless manner. Capital productivity[16] tended to fall with capital intensity as expected, but unexpectedly tended to rise with size—so that there was some evidence of both higher capital and higher labor productivities in the larger sizes. Sandesara's conclusions were very negative for a policy of promoting small units.

But these conclusions were challenged by Mehta (1969), who pointed out that the K/L ratio for the smallest size group in the CMI (less than 20 workers, although the census aimed only at larger units) varied wildly from year to year and sometimes reached absurdly high figures. He suggested that the main reasons for this were excess capacity resulting from unfavorable policies in the 1950s, especially so far as raw material supply was concerned;

2-14). The ILO team made adjustments to convert book value of fixed assets to replacement values and expressed value added in international prices. These adjustments should make little difference to intraindustry size comparisons.

15. The CMI measure of capital was the book value of all real assets.

16. The product is (usually) *net* value added.

the presence in the smallest size group of large, sick firms with skeleton staffs; and the presence of new firms with capital installed but as yet little employment. These points are well taken and no doubt go some way toward explaining the relatively frequent poor showing of very small firms, while also rendering the figure unreliable and hard to interpret.

Mehta also presented calculations for 1960–63 from the Annual Survey of Industries, which succeeded the CMI. The ASI has a census section which enumerates all firms with more than 50 employees or more than 100 without power. Mehta was able to divide these into small, medium-size, and large firms on the basis of the book value of fixed capital, less than Rs500,000, Rs500,000–Rs2.5 million, and more than Rs2.5 million, these being the official definitions of small, medium-size, and large firms at the time. In the aggregate, the capital, labor, and value added ratios behaved as theory says they should have. However, when size is measured by the *book* value of fixed capital, the small will include units with fixed capital written down below its market value. There is thus a downward bias to the amount of capital need by the small, the opposite of the upward bias when size is measured by employment.

Disaggregated into 32 industries (between the two- and three-digit levels but nearer to the two-digit level), the small was invariably the most labor-intensive. This is not surprising, for this part of the small sector is selected in a very biased manner. They could be as well or better described as medium-size factories constrained to be labor intensive. The plants are constrained to have 50 or more, or 100 or more, employees and to have less than Rs500,000 of fixed capital. More interesting is the concurrent behavior of labor and capital productivity. In 105 out of 127 industry-years (one cell was empty), labor productivity rose, as capital intensity rose, from small to medium-size to large. In 98 out of the 127 cases, capital productivity fell, as in theory it should. Both productivities together behaved "correctly" in 74 cases. There was thus more regularity than we found, for instance, in the case of Korea. There are two main explanations: first, the size discriminator was fixed capital, not the number of employees; second, the level of aggregation was much higher in the Indian case.

The ASI also samples factories in the range of 10–49 (or 99) employees. For the same industries as before, Mehta compared this sample sector with the aggregate census sector, thus in effect comparing the aggregate of the small with the aggregate of the large, the dividing line now being employment of 50 (or 100 without power). The small were more labor intensive in 88 percent of cases (a case being an industry in a given year), had lower labor productivity in 83 percent of cases, and had a higher capital productivity in 73 percent of cases. For all industries taken together the ratios were as expected. Mehta concluded favorably for small enterprises, although he can hardly be said to have had a good basis for doing so. The level of aggregation was high, and there was no attempt to derive and compare total factor productivities.

In a reply, Sandesara (1969) compared for 1963 the sample sector with the small enterprises of the census sector of the ASI; this is a comparison of factories with 10–49 employees with those with 50 or more (or 100 or more) with less than Rs500,000 of fixed capital. In 19 out of 30 industries, the sample sector was more capital-intensive, which is not surprising in view of the bias already noted. Capital productivity was lower for the sample sector than the small census sector in 24 out of 30 cases, and lower than the medium-size census sector in 12 cases.

The total lack of data for establishments with less than 10 employees was partly remedied by the *Census of Small Scale Industries*, which collected data for 1972. The coverage of the SSI was discussed in chapter 5, and it will be recalled that the definition of small was Rs750,000 or less in original plant or machinery (Rs1million for ancillary units).

We compared the economic ratios for SSI units with the ASI in comparable industries (the latter averaged for 1974–77). This was done not only for all such industries taken together but also for four groups of industries classified by the degree of importance of employment in small-scale units. The results are presented in table 7-4. Taking all industries together, the capital, labor, and value added ratios behave "correctly" (as with Mehta's comparisons; see above).[17] The figures for wages and labor productivity are more notable. Labor productivity was 63 percent higher in the ASI enterprises, but wages were 117 percent higher, so that profitability was lower.[18] Lower profitability in large Indian factories is also confirmed over the ASI range (ten employees and upward) by the work of Shetty (1982; see below). Lower profitability in larger enterprises has been observed also for Japan and Malaysia (Tajima 1978), but higher profitability seems to be more the developing economy norm, for instance, in Brazil, Colombia, Korea, Philippines, and Taiwan (Tajima 1978, and above in this chapter).

Disaggregation even into only four groups of industries produces perverse results that are unfavorable to small-scale enterprises. In table 7-4, group IV, in which small factories (10–49 workers) account for less than half of factory employment, is well behaved, just as the aggregate was. But in groups II and III, in which small-scale factories account for 50–84 percent of the factory employment, the SSI enterprises, while remaining more labor-intensive, had both lower capital and lower labor productivity than the ASI factories. In group I, in which small factories dominate, accounting for 85 percent or more of total factory employment, the SSI enterprises have the same labor but lower capital productivity and are also more capital-intensive than the

17. Mehta did not have a breakdown by employment size and therefore used three size divisions by fixed capital.

18. It should be remembered that ASI sector employment is heavily weighted toward large units, with nearly 60 percent of the employees being in factories with more than 500 employees (see table 4-7).

Table 7-4. *Selected Economic Ratios for the SSI and ASI Sectors Broken Down by Categories of Industries*
(thousands of rupees)

Category of industry	Y/L		K/L		Y/K		W		WL/Y		(Y − WL)/K	
	SSI	*ASI*	*SSI*	*ASI*	*SSI*	*ASI*	*SSI*	*ASI*	*SSI*	*ASI*	*SSI*	*ASI*
Total	4.56	7.41	9.98	18.63	0.46	0.40	1.55	3.36	0.34	0.45	0.30	0.22
I	4.32	4.32	9.40	7.95	0.46	0.54	1.50	1.86	0.35	0.43	0.30	0.31
II	4.88	6.70	11.32	12.99	0.43	0.52	1.58	3.08	0.32	0.46	0.29	0.28
III	4.48	5.64	8.55	9.85	0.52	0.57	1.43	2.84	0.32	0.50	0.35	0.28
IV	4.54	7.82	10.49	20.71	0.43	0.38	1.65	3.50	0.36	0.45	0.27	0.21

Note: Wages (*W*) are in 1972 constant thousand rupees. The deflator for wages is derived from the general price index in the *Indian Labour Journal* 12, 5 (1980): 1723. The deflator for other money values is derived from the wholesale price index, National Accounts Statistics, Central Statistical Organization, Government of India.

For SSI, industries in which value added is negative are taken out of the calculations. These are industries 288, 300, 332, 344, and 971–79.

For ASI, some samples for which ratios take extreme values are taken out of the calculations. The criteria for extreme are: for Y/K, less than 0.20; for Y/L, more than 10; and for K/L, more than 50.

Groups I, II, III, and IV include industries in which, respectively, 85 percent or more, 70–84 percent, 50–69 percent, and less than 50 percent of workers are in small units.

ASI factories, which are still on balance larger. This is the only group in which SSI enterprises appear to be less profitable than ASI enterprises.

Table 7-5 presents the economic ratios broken down by employment size classes within the ASI sector, along with the data for the SSI sector. Several additional points of interest emerge. First, in the size classes up to 100 workers, capital intensity is virtually constant, but both labor and capital productivity increase. Capital intensity and labor productivity are higher than in the SSI sample, but capital productivity is somewhat lower. But second, in the medium-size class of 100–499 workers capital intensity and labor productivity rise sharply with only a small fall in capital productivity. Wages rise only a little more than labor productivity, which results in a small decrease in profitability. Third, the largest size class of 500 or more workers has the highest share of wages in value added and a significant fall in capital productivity associated with higher capital intensity. The profit rate is substantially lower. Fourth, wages rise monotonically through every size group, and every size group shows a higher share of wages in value added than in the SSI sample, albeit without any strong link to size within the ASI sector.

Table 7-5 gave some average ASI results for 1974–77. The ASI survey of 1978–79 has been analyzed by Shetty (1982). Table 7-6 has been compiled from this source, which gives figures only for fixed capital (but see note to table 7-6) but has a finer employment size breakdown for factories with 100 or more employees. The sharp rise in labor productivity and capital intensity appears to occur in the neighborhood of 200 employees, which is consistent with table 7-5. In table 7-5 capital productivity peaks in the range 50–99. In table 7-6 fixed-capital productivity also peaks in this range, though there is virtually a plateau from 50–500 workers. Since labor productivity is much greater from 200–499 workers than for smaller units, this size class may

Table 7-5. *Selected Economic Ratios in the ASI Sector by Size Groups, India, 1974–77*
(thousands of rupees)

Number of workers	Y/L	K/L	Y/K	W	WL/Y	$\frac{Y-WL}{K}$
10–19	3.84	8.81	0.44	1.53	0.40	0.27
20–49	4.06	8.47	0.48	1.71	0.42	0.28
50–99	4.42	8.68	0.51	1.97	0.45	0.28
100–499	6.57	14.46	0.45	2.75	0.42	0.26
500 or more	9.71	26.68	0.36	4.58	0.47	0.19
All ASI	7.41	18.63	0.40	3.36	0.45	0.22
SSI	4.56	9.98	0.46	1.5	0.34	0.30

Note: Wages (W) are in 1972 constant thousand rupees. The deflator for wages is derived from the general price index in the *Indian Labour Journal* 12, 5 (1980): 1723. The deflator for other money values is derived from the wholesale price index, National Accounts Statistics, Central Statistical Organization, Government of India.

The note in table 7-4 applies here.

Table 7-6. *Economic Ratios, ASI Sector, India, 1978-79*
(thousands of rupees)

Number of workers	Y/L	K_f/L	Y/K_f	W	$\frac{Y - WL}{K_f}$
0–49	7.07	9.40	0.75		0.28
50–99	8.08	9.29	0.87		0.24
100–199	8.76	10.60	0.87		0.24
200–499	15.86	18.45	0.86		0.35
500–999	15.76	28.46	0.55		0.19
1,000–1,999	17.40	32.79	0.54		0.19
2,000–4,999	14.98	21.36	0.70		0.19
5,000 or more	15.34	102.42	0.15		0.06
All ASI	13.18	[31.38]	[0.42]	6.37	[0.22][a]

Empty cells mean data were not available.

Note: K_f stands for fixed capital. Wages (W) are in thousands of rupees. Despite the text (p. 1614), it appears that Shetty must have used total capital for the ratios in brackets. Otherwise these figures are irreconcilable with table 7-5, with other published results, and even with the figures in the size breakdown.

Source: Shetty (1982). Derived from the text (Shetty).

almost be said to dominate. But this is not borne out by table 7-5. It is also worth noting that the small factories (0–50 employees, but mainly 10–50) have higher capital intensity and lower productivities than the size 50–99, this being yet another confirmation of Dhar and Lydall's results. The "rate of operating surplus to productive capital" (which is a better, if longer, description than the rate of profit) falls with size, with a watershed at about 500 employees, which also confirms table 7-5.

Table 7-6 sheds a little more light on the larger factories. The size class 500–999 (numbering 1,089 factories) shows a sharp rise in fixed capital per employee but an actual fall in both capital and labor productivity. In the class 1,000–1,999 (666 factories), labor productivity peaks, but the gain of 12 percent as compared with class 200–499 is bought at the expense of a fall of 37 percent in capital productivity. In the class 2,000–4,999 (377 factories), there is a fall in capital intensity matched by an expected fall in labor productivity and rise in capital productivity. This class is dominated by the 200–499 class, which clearly shows up with the best figures. The largest size class (80 factories) produces disastrous figures, but this is probably more because it largely consists of public enterprises than because of size as such. Indeed, the ASI figures are split into public sector, joint sector, and private sector. This shows that the public sector has 62 percent of the "productive" capital (fixed capital plus inventories), employs 27 percent of the labor force, and produces 30 percent of the net value added.[19]

Finally, the figures are split by age of the plant. This shows that the capital

19. To some extent, however, high capital intensity may be exaggerated by the inclusion in capital of the housing that very large enterprises may be required to provide for their workers.

intensity of Second Plan (1956–60) factories was more than double that of any other vintage. But fixed capital productivity has also been very low for the factories of 1971–78 vintages. The oldest (pre-1950) factories have higher labor productivity than both those of the second plan period and those built since the mid-1960s. They are also the least capital-intensive, though this may well be largely due to the survival of written-off buildings and equipment and to other assets not being written up with inflation.

The last compilation of evidence to be examined is that of the 1977 survey by the Reserve Bank of India (1979). Results are available for different industries and size groups (variously defined) but not for the two together. Since it was a general economic survey with no special emphasis on particular industries, it is more convenient to examine some of its results in the context of this chapter rather than in chapter 10.

Some of the financial results of this survey, and details of its coverage, will be presented in chapter 15. Here we note that the population surveyed was that of small industrial, mainly manufacturing, enterprises that had been assisted by a commercial bank, for which "small" took the current definition of an "original investment" not exceeding Rs1 million of plant and machinery (Rs1.5 million for ancillary units). There was no lower limit on size, and it was estimated that 46 percent of the population had a market value of fixed assets of less than Rs10,000 (Reserve Bank of India 1979, vol. I, table 3-1). In terms of employees, 29 percent of the estimated population of units had none; 22 percent had more than 10, and 12 percent had more than 20. The mean number of employees was 13, the same as the mean number of workers in the SSI (see chapter 5). But it is interesting that the *mean* number of employees in the size class of more than 20 was as high as 77 and that this class accounted for 71 percent of the employment in assisted small enterprises (Reserve Bank of India 1979, vol. I, table 3-4). In other words, much, perhaps most, of the employment in assisted small enterprises (by a capital criterion of small) was in medium-size enterprises (by an employment definition of medium as firms with more than 50 workers). We turn now to the evidence on factor intensity and productivity.

Table 7-7 gives the economic ratios for employment size groups, capital being measured as the market value of fixed assets. For the first time, for India, we have evidence, albeit aggregated for all industries, for different sizes within the group of the small. We see that capital intensity rises irregularly to a peak at 6–10 employees, then falls regularly. However, this could be the result of the nonrandom selection. As the employment size rises, so the probability rises of an enterprise being rejected for assistance as not being small by definition. Therefore, the capital intensity of the larger size groups is increasingly reduced. But this does not affect the manner in which capital and labor productivity should vary with capital intensity.

Labor productivity is irregularly flat up to 5 employees, after which it rises up to 21 employees before falling again. Capital productivity falls from 0–4

Table 7-7. *Factor Productivities and Intensity by Employment Size, India, 1977*
(thousands of rupees)

Number of workers	Y/L	K/L	Y/K
0	2.5	4.3	0.59
1	2.0	4.1	0.48
2	1.9	4.9	0.39
3	2.3	7.2	0.32
4	1.6	6.2	0.26
5	2.3	7.1	0.33
6–10	4.6	10.2	0.45
11–15	5.4	10.0	0.54
16–20	6.2	9.5	0.66
21 or more	4.9	5.8	0.85
All sizes	4.6	6.7	0.69

Note: Value added (Y) is not given in RBI table 3-3 (Reserve Bank of India 1979). It is here estimated in two steps. First, the value of materials consumed and services bought is subtracted from the value of manufactured products and services sold. Second, the resultant figure is adjusted down to conform with the figure given for gross value added in volume II, table 3.3. The same proportionate adjustment is made for each size group, so that the comparison between different sizes is unaffected. The difference between our value added and that of volume II lies in certain manufacturing costs which are not given in volume I, table 3.3.

Employees rather than workers are given in the RBI survey. We have added 1.5 persons per plant to allow for working owners or partners and unpaid family workers.

Source: Reserve Bank of India (1979, vol. I, table 3-3).

employees but then rises regularly both in the range in which capital intensity is rising and in which it is falling. The upshot of this rather discordant behavior is that size 16–20 dominates all smaller sizes, with 21 or more having still greater capital productivity and lower capital intensity, but at the cost of some reduction in labor productivity. The upshot is that the larger small are better than the smaller small.

The RBI survey enables us to group firms by a different size classification, that of the market value of fixed assets. The definitions of Y, K, and L remain the same. Results are given in table 7-8. Capital intensity now rises regularly until the market value of fixed assets reaches about Rs2 million, when it falls. Labor productivity after falling in the tiniest size classes (as in table 7-6) rises with only one anomaly up to the same largest class, when it falls. Capital productivity falls as expected, with only slight irregularity, up to the largest size class, where it rises. With this greater regularity of behavior, there are no dominant size classes.[20] A very clear lesson is that firms behave in the theoretically expected manner much better if they are arranged in order of capital size than if they are arranged in order of employment size. We have already had a hint of the validity of this proposition in the case of Japan and in Mehta's work on the ASI data in India (see above in this chapter).

20. In fact the class Rs100,000–Rs200,000 dominates its smaller and larger immediate neighbors. But this must be regarded as freakish, given the regularity elsewhere.

Table 7-8. *Factor Productivities and Intensity by Capital Size, India, 1977*
(thousands of rupees)

Size by market value of fixed assets	Y/L	K/L	Y/K
Less than 1	2.3	0.1	21.89
1–10	2.0	1.0	2.13
10–20	1.9	1.7	1.09
20–50	3.5	3.2	1.11
50–100	3.6	5.2	0.69
100–200	4.8	6.4	0.75
200–500	4.5	6.8	0.66
500–1,000	6.8	11.3	0.60
1,000–2,000	8.1	15.4	0.53
More than 2,000	7.2	12.3	0.59
All sizes	4.6	6.7	0.69

Source: Reserve Bank of India (1979, vol. I, table 3-3).

An alternative measure of capital is that officially used to designate small, medium-size, and large—the original value of plant and machinery. Table 7-9 presents the economic ratios, with some changes of definition, with size thus classified. It adds three more columns for the average wage, *W*; the share of wages in value added, *WL/Y*; and profitability, defined as *(Y − WL)/K*.

With only one anomaly, all three production ratios behave with theoretical precision. This seems to say a lot for the Indian mode of size classification and for using the market value of fixed assets plus inventories as the definition of capital. Wages rise monotonically and very fast, so that the level in

Table 7-9. *Various Economic Ratios by Capital Size, India, 1977*
(thousands of rupees)

Size by original value of plant and machinery	Y/L	K/L	Y/K	W	WL/Y	(Y − WL)/K
Less than 1	2.0	2.5	0.97	1.32	0.65	0.34
1–10	2.7	4.7	0.69	1.69	0.63	0.26
10–20	3.8	7.6	0.60	2.13	0.56	0.26
20–50	4.4	10.1	0.49	2.45	0.56	0.22
50–100	4.9	11.0	0.48	2.64	0.54	0.22
100–200	5.5	15.4	0.38	3.13	0.56	0.17
200–500	6.0	16.7	0.36	3.36	0.56	0.16
500–1,000	5.4	17.2	0.31	3.47	0.64	0.11
More than 1,000	7.2	21.6	0.26	5.08	0.70	0.08
All sizes	4.2	9.7	0.43	2.53	0.60	0.17

Note: Wages (*W*) are in thousands of rupees.

Value added (*Y*) is net, as defined in Reserve Bank of India (1979, vol. II).

Capital (*K*) is market value of fixed assets and inventories.

Labor (*L*) is as in tables 7-7 and 7-8. The wage bill (*WL*) has been inflated by ascribing the average wage to the 1.5 nonemployee workers (family workers or the proprietor) assumed.

Source: Reserve Bank of India (1979, vol. I, table 3-4; vol. II, tables 3-3, 4-3).

Table 7-10. *Economic Ratios in Different Surveys, India, Various Years*
(thousands of rupees)

Survey	Y/L	K/L	Y/K	W	WL/Y	(Y − WL)/K
ASI	7.4	18.6	0.40	3.36	0.45	0.22
SSI	4.6	10.0	0.46	1.55	0.34	0.30
RBI	4.2	9.7	0.43	2.34	0.56	0.19
RBI deflated	3.0	7.0	0.43	1.69	0.56	0.19

Note: Wages (*W*) are in thousands of rupees.
Source: See text.

the largest size class is 3.8 times that in the smallest. The share of wages dips in the middle of the range but is fairly stationary, with the result that profitability falls about as fast as capital productivity.

It is tempting to compare the overall RBI data with those for the ASI and SSI given in table 7-6. The last two rows of table 7-6 together with the RBI survey figures for all industries are given in table 7-10. But the reader should compare these figures only with great caution. The ASI and SSI figures are in 1972 prices and the RBI in those of 1976. This might reduce the RBI figures in the first, second, and fourth columns by about 28 percent to those shown in the last row. It is clear that the units in the RBI survey are less capital-intensive and have lower labor productivity than the SSI units. Although both supposedly cover the same field as far as size is concerned, it should be remembered that the SSI left out the industries in which the small units are predominantly of the traditional type, and operating without power.[21] This could well account for the lower capital intensity and labor productivity of the RBI sample. We cannot explain the apparently higher wages which together with the lower labor productivity account for the lower profitability of the RBI sample.

Apart from size, the RBI survey classified assisted units by whether or not they were registered and whether or not they used power. Seventeen percent of the estimated population was registered, although 22 percent employed 11 workers or more.[22] The average employment size of a nonregistered unit was 6, and that of a registered unit 48. The registered units dominate the unregistered having both higher labor and higher capital productivity, as table 7-11 shows (we have already seen in table 7-7 that the size group 16–20 dominated all smaller size groups).

Forty-five percent of units used no power, and they were a majority in the two-digit industries of beverages and tobacco, cotton textiles, other textiles, textile products, wood and wood products, leather and leather products, and

21. It should also be recalled that the RBI population consists only of units assisted by commercial banks, though it is unclear what difference this might make. A comparison between assisted and control units has been made by Sandesara (1982).

22. Those with 10 or more employees are supposed to register.

Table 7-11. *Economic Ratios in Different Categories of Enterprise, India, 1977*
(thousands of rupees)

Category	*Y/L*	*K/L*	*Y/K*	*W*	*WL/Y*	$\frac{Y - WL}{K}$
Registered	5.4	7.5	0.72	2.70	0.50	0.36
Not registered	3.4	5.6	0.61	2.22	0.65	0.21
Using power	7.4	9.2	0.81	2.97	0.40	0.49
Not using power	2.8	1.3	2.20	1.50	0.54	1.05

Note: Wages (*W*) are in thousands of rupees. Definitions are the same as for table 7-7.
Source: Reserve Bank of India (1979, vol. I, tables 3-1, 3-3, 3-4).

nonmetallic minerals. Not unexpectedly, table 7-11 shows them to be very labor-intensive and to have very high capital productivity and very low labor productivity and wages. These are the oft expected characteristics of very small units, but not using power is a much more reliable indicator than size. Indeed, the Indian government uses it to select enterprises to be subsidized, or otherwise favored, in cotton textiles, soap, and a few other industries.

Table 7-11, together with 7-7, may again be regarded as confirmation of Dhar and Lydall's main point, that among modern factories the larger very often have greater capital productivity, at least up to about 200 workers, while small traditional units may also have very high capital productivity, but at the cost of very low wages and labor productivity.

Thus far, labor has been treated as homogeneous, which is far from reality. Table 7-12 presents the variation in the composition of the labor force by size,[23] and also the level of wages (including allowances) paid to the four employee categories distinguished. There is a tendency for the proportion of unskilled workers to rise with size, and a stronger tendency for the proportion of skilled workers to fall.[24] This latter remains true even if supervisors, whose proportion rises, are ranked as skilled workers. But the fastest rising proportion is that of managerial workers (including office staffs). In these circumstances, can one say anything about the skill composition of the work force as a whole? Perhaps the proportion of the unskilled worker wage bill to the total wage bill is the best measure. This shows no clear trend with size, a slight tendency for the proportion of unskilled workers to rise with size being offset by a slight tendency for their wages to rise relatively less fast than those of the other categories. The size of the wage rise for both skilled and unskilled, especially for the largest size class, is remarkable. For instance, the plant with over Rs1 million of original value of plant and machinery pays 77 percent and 43 percent more for skilled and unskilled workers than one with

23. We chose the original value of plant and machinery as the size discriminator, since it produced the most orderly results when evaluating the economic ratios.

24. As we shall see, this important point is confirmed by the World Bank's surveys, as discussed in chapter 10.

Table 7-12. *The Structure of Employment and Wages by Size Classes, India, 1977*

Size by original value of plant and machinery (thousands of rupees)	*Percentages of total employment*				*Wages and salaries (thousands of rupees per year)*			
		Production workers				*Production workers*		
	Managerial	*Supervisory*	*Skilled*	*Unskilled*	*Managerial*	*Supervisory*	*Skilled*	*Unskilled*
Less than 1	3	2	61	34	3.50	2.66	1.23	1.22
1–10	5	2	48	45	3.24	3.61	1.89	1.21
10–20	8	3	52	37	2.99	3.98	2.34	1.46
20–50	9	4	46	41	3.43	5.04	2.64	1.74
50–100	10	4	43	43	3.95	5.24	2.89	1.82
100–200	13	5	36	46	5.09	7.03	2.94	2.31
200–500	13	5	36	46	5.68	8.00	3.43	2.08
500–1,000	13	6	31	50	6.02	8.38	3.83	2.06
1,000 or more	19	9	28	44	6.75	10.19	5.19	3.31
All	9	4	45	42	4.76	6.35	2.43	1.79

Source: Reserve Bank of India, (1979, vol. I, table 3-4).

Rs100,000–Rs200,000. But of course we do not know anything about the quality of the labor force within the categories of skilled and unskilled.

Summary

By way of summarizing this chapter's examination of census and survey data, we present the conclusions in terms of the behavior of the factor intensities and productivities across economies, rather than economy by economy. There is no doubt about the main conclusion that can be drawn.

Once overall manufacturing figures are disaggregated into separate industries, observed overall regularities between size, as measured by the number of employees, and capital intensity, and capital and labor productivity, break down. The greater the disaggregation the more they disappoint, and great disaggregation is needed to know that small and large produce much the same things and so could be competitive.

It was found that, even for manufacturing as a whole, the smallest size group (less than 10 workers) was not the most labor-intensive in two of the economies examined (Colombia and Thailand). In India also, ASI data show labor intensity to be higher in the range 26–99 workers than in the range 10–19. In Japan, when industry is split merely into four types, the size class of less than 10 workers is not the most labor-intensive in three cases. There may, however, be special reasons for this, in that the very small size classes catch a number of larger firms (measured by capital) which employ very few workers because they are newly born, temporarily ill, or dying.

Disaggregating ASI data for India to the level of 32 industries, and averaging the results over four years, it was found that factories in the range 10–49 (10–99 if no power was used) were more labor-intensive than larger ones in a high proportion of cases. But the much greater level of disaggregation possible in Korea showed that industry-specific labor intensity peaked evenly for all size classes in the range from 5–500 workers. Moreover, the *highest* capital intensity was as often found in the range 5–50 workers as in the range from 50 upward. The moral is obvious. Even if one were seeking a policy that favored only labor intensity, it would be essential to discriminate finely between industries. However, no such policy ignoring output could be sensible.

The crudest possible measure which might yet have a claim to measuring efficiency in a labor-abundant economy is capital productivity. The results are not favorable for the very small, for the very large, or even for manufacturing as a whole. In all the economies examined except Korea, capital productivity peaks in the range 20–200 workers. In Korea the 500-or-more size class has the highest capital productivity, with the class 100–199 next best. But disaggregating the Korean figures produces the result that capital productivity peaks in the range 50–500 workers in two-thirds of the industries. Further analysis of the figures shows that capital intensity was lower and capital productivity higher in the range 100–199 workers as compared with the range 20–49 workers in about half the 139 industries for which this comparison could be made. In India, an RBI survey of small industries (defined by capital) suggested that the large-small (more than 20 workers) have higher capital productivity than the small-small.

Total factor productivity is a still better measure of efficiency in the use of scarce factors of production. Results based only on the Korean census show that in about half the industries examined this measure peaked in the range 50–500 employees. Only about 5 percent of industries peaked in the smallest size class. Moreover, the few industries for which small enterprises show up as most efficient are also small industries in terms of employment.

Finally, it must be noticed that if size were defined in terms of capital, the results could be very different. The limited evidence we have (from Japan and from the Indian RBI survey) suggests that the expected labor and capital productivities would be much more reliable. Unfortunately a classification by capital size is rare. Even in India, where small is defined for policy purposes by the original cost of fixed capital, census results are presented in terms of an employment size classification. But if the usual more readily available size classification by employment is in question, then the upshot of the results of this chapter is that it is the medium-size, not the small, that is beautiful. The reader will find that this is, by and large, confirmed by the World Bank survey results presented in chapters 8 through 12.

8

An Introduction to the Enterprise Surveys

This and the following four chapters are based on the results of sample surveys of manufacturing firms conducted between 1978 and 1980 in five Indian industries. The aim was to produce a set of microeconomic data covering unregistered firms and those employing less than ten workers, categories often excluded from industrial censuses, and to provide a sufficiently large and diverse set of firm-level observations to permit econometric modeling of production relationships and statistical tests of hypotheses within narrowly defined product categories.

As the results of chapters 5 through 7 demonstrate, studies of small-scale enterprises based on industrial census data are of limited value in answering some of the more interesting questions regarding the role of small-scale enterprises and their relative level of economic efficiency. Both the lack of coverage of very small units and the level of aggregation of census results make comparisons of relative factor intensities and levels of productivity problematic. In addition, archival data provide little or no information regarding such important enterprise characteristics as the background of the entrepreneur, the history of the enterprise, or the nature of its products. For these reasons, purposive samples of firms in the laundry soap manufacturing, printing, shoemaking, machine tool manufacturing, and metal casting industries were undertaken. Surveys of handlooms and powerlooms were also undertaken, as reported in chapter 4, but the econometric analysis of production reported in chapters 10 and 11 could not be undertaken since there was very little variation in factor proportions.

The following chapters present the results of these sample surveys and analyze such issues as the relative factor intensity, productivity, and economic efficiency of small-scale enterprises. The basis for much of the analysis is the theory of cost and production. This chapter summarizes characteristics of the sample survey and briefly discusses each of the industries surveyed.

The Sample

The five industry surveys were designed to provide information on the full size range of firms within narrowly defined sectors. They do not represent a random drawing from the population of all industrial firms in India. Rather, an attempt was made to select sectors in which small-scale firms played an important role and in which valid comparisons could be made across all size categories. Two of the sectors—shoemaking and soap manufacturing—represent traditional activities in which small-scale enterprises have played a predominant role and in which handicraft methods coexist with more modern factory processes. Two others—printing and machine tool manufacturing—are sectors in which modern small-scale factories play a major role but in which handicraft methods of production are not present. The metal casting industry is intermediate between the second two groups.

Most of the small and medium-size units surveyed in each of the five sectors were in or near Bombay, Calcutta, and Delhi, while large firms were drawn from all regions of India. An attempt was made to enumerate all larger firms (more than 200 workers) currently active in the registered (licensed) sector of each industry.

The distribution of firms by employment size is given in table 8-1.[1] In addition the table presents the distribution of two other surveys of industry in India. The first is that of the Reserve Bank of India, which was a random sample of enterprises with fixed capital (undepreciated value at historic cost) of less than Rs1 million that had received loans from commercial banks. This sample includes a few firms whose employment size ranges up to several hundred workers, but the vast majority of firms fall into the size range of less than 50 employees. (The mean for the whole RBI sample is 12.6 workers; means for individual branches range between 1.6 and 99). In general our samples show greater coverage at the upper end of the size range than do the RBI data, which are heavily weighted toward very small-scale firms in the size range 1–5 workers.

The second distribution is taken from the tabulated results of the Annual Survey of Industries for 1977. This sample consists of registered enterprises of all sizes throughout India. In general our samples (which are at a much greater level of sectoral disaggregation) show a higher proportion of firms in the size categories 6–10 employees and a smaller proportion in the size range below 5. With the exceptions of shoes and printing, large-scale firms (more than 100 employees) are slightly overrepresented in our sample. This reflects our attempt to obtain responses from all large-scale registered sector firms.

1. This is the distribution of the sample employed in the econometric work in chapters 10 and 11. For distributions of the full set of sample responses and discussion of the selection criteria for the econometric data set, see appendix 1.

Table 8-1. *Percentage Distribution of Firms in Sample and in Reserve Bank of India and Annual Survey of Industries Samples by Employment Size*

Industry	Number of observations	Number of employees: 0–5	6–10	11–20	21–50	51–100	More than 100
Shoes	99	24.2	25.3	33.3	10.1	0.0	7.1
RBI[a]		94.5	2.3	1.5	1.3	—	—
ASI[b]		21.1	8.9	21.1	29.3	10.6	8.9
Printing	66	9.1	39.4	28.8	15.2	3.0	4.6
RBI		51.2	28.0	14.4	6.4	—	—
ASI		15.5	17.6	30.5	19.2	7.1	10.1
Soap	57	15.8	35.1	22.8	3.5	10.5	12.5
RBI		56.3	28.7	11.1	3.9	—	—
ASI		13.1	8.8	20.7	34.7	11.9	11.2
Machine tools	78	0.0	5.1	35.9	30.8	11.5	16.7
RBI		28.8	38.2	13.5	19.5	—	—
ASI		10.6	10.9	40.1	26.8	5.3	6.3
Metal casting	45	6.5	32.5	25.2	25.9	2.3	7.6
RBI		19.0	40.2	20.5	20.3	—	—
ASI		12.5	30.2	20.2	27.1	5.5	4.5

— RBI survey does not report entries in the two largest cells.

Note: Frequency distribution and number of observations from World Bank sample refer to final data set used in econometric estimation. See appendix 1 for details of survey and selection criteria for econometric analysis.

a. RBI: Reserve Bank of India survey, 1977.

b. ASI: Annual Survey of Industries, 1977.

Size can be measured by total employment or by using estimates of the capital stock, total output, or value added. The advantage of using labor force data as the primary indicator of firm size is that it permits more direct comparison with other studies of small-scale enterprise, most of which use employment as the size criterion, and that it provides a more graphic indicator of firm size than values of capital or output. All the possible indicators of firm size are highly correlated in our samples. Table 8-2 presents zero-order correlation coefficients between employment and other indicators of firm size. All the variables are significantly correlated at the 1 percent level. Thus, total employment provides a good indicator of size of firm in our sample and is used throughout the following chapters as the principal indicator of firm size.

Data Generation

The major variables used in the empirical work are measures of labor inputs, capital, material inputs, and output. The variables collected in the survey instruments are listed and described in appendix 1. Many of the data drawn

Table 8-2. *Correlation Matrix: Indicators of Firm Size*

Industry	*Total labor*	*Capital stock*	*Output*	*Value added*
Soap				
Total labor	1.0			
Capital stock	0.9162	1.0		
Output	0.9743	0.9687	1.0	
Value added	0.9589	0.8803	0.9403	1.0
Shoes				
Total labor	1.0			
Capital stock	0.8752	1.0		
Output	0.9527	0.9345	1.0	
Value added	0.9994	0.8796	0.9466	1.0
Printing				
Total labor	1.0			
Capital stock	0.7753	1.0		
Output	0.8435	0.7867	1.0	
Value added	0.8815	0.8806	0.9056	1.0
Machine tools				
Total labor	1.0			
Capital stock	0.7273	1.0		
Output	0.8769	0.8633	1.0	
Value added	0.9158	0.8722	0.9412	1.0
Metal casting				
Total labor	1.0			
Capital stock	0.9500	1.0		
Output	0.9672	0.9926	1.0	
Value added	0.9733	0.9703	0.9898	1.0

from the records of the sample surveys required substantial manipulation, and in this section we discuss the generation of the data sets for the five industries surveyed, beginning with labor, continuing with capital, and finishing with our measure of output.

Labor and Wages

Labor inputs are measured in person-months and are presented for two aggregate skill categories. Skilled labor was defined by taking the occupational categories identified in survey responses as skilled tasks. In general, skilled tasks are those that involve substantial discretion by the employee over the pace and quality of work, or that involve the supervision of other workers. White collar labor, paid family labor, and working proprietors have been included in the skilled category, since they perform skilled labor and management functions in the industries surveyed.

Because the surveys cover a broad range of firm sizes, the types of workers classified as skilled change across the size spectrum. In small firms engaged in traditional craft-based activities, the majority of workers identified as skilled are craftsworkers responsible for a wide variety of tasks within the enterprise. In larger firms, skilled workers tend to be concentrated in specialized tasks or in the supervision of less skilled workers. For example, in shoemaking the majority of skilled workers in small traditional firms are cobblers who perform multiple tasks in the construction of the shoe. In larger enterprises that use factory methods, pattern cutters are classified as skilled because of their control over the quality of the product and the degree of waste, while machine-paced operatives, such as assemblers, are not considered skilled; although they perform some of the same functions as cobblers, they exercise little discretion over the pace or nature of their tasks. Thus, the unskilled group covers a wide range of tasks from semiskilled through casual labor and includes all workers classified by respondents as semiskilled, unskilled, and unpaid family labor.

The division into skill groups was cross-checked with reference to data on earnings. The average earnings premium of the groups classified as skilled relative to the remaining employment categories classified as semiskilled or unskilled is 36 percent. This is consistent with earnings differentials between skilled and unskilled labor reported in the Reserve Bank of India survey.

Labor inputs are measured in person-months based on the assumption of equal operating periods in all firms. Where firms reported seasonal changes in regular or casual employment, the person-month data have been adjusted to reflect these variations. No attempt was made to adjust the labor input data for variations in days worked per year. Although, where available, data from the survey indicated that on average registered sector firms worked a greater number of days than their counterparts in the unregistered sector, there were insufficient responses by unregistered firms to permit firm-specific adjustments to the labor input data.[2]

Wage data and information on labor earnings are drawn from two sources in the questionnaires. The record provides information on minimum and maximum monthly earnings at the level of detail of individual job classifications. The earnings data reflect variations in hours worked, in addition to variations in basic wages and in incentive payments, since many of the firms in the sample employed complicated systems of piece rates in addition to or in place of daily wages. The earnings data were used to estimate average monthly earnings for the two skill categories of labor by multiplying the midpoint of the range of earnings by the number of workers in each occupational grouping and aggregating into skill groups. An imputed total wage bill was then constructed for each firm by summing payments to skilled and unskilled labor. The imputed total was compared with data provided on total labor compensation. In general, the imputed wage bill exceeded actual payments to labor, which reflected both the imprecision of the averaging

2. The registered sector firms reported an average of 305 operating days in the financial year. The average for unregistered enterprises was 258 days.

procedure and the presence of unpaid family workers and working partners in the labor force. Thus, in the empirical work below, we use both imputed earnings information and data on actual expenditures.

Because data on payments to labor were required for some of the econometric work, it was necessary to construct a consistent estimate of earnings of skilled and unskilled workers and hence to reconcile the differences between estimates of annual earnings outlined above. The differences between the imputed wage bill and the reported level of payments to labor were particularly marked in smaller firms, the great majority of which had relatively high proportions of working proprietors and unpaid family workers in the total labor force.

After experiments with several alternative estimates of earnings by skill category, the following approach was adopted:

1. The reported value of total payments to labor was adjusted upward for the imputed earnings of unpaid family labor and working proprietors by adding the product of the number of such workers and the estimated annual earnings of employed workers (total wage bill divided by number of paid employees).
2. The adjusted total wage bill was then divided between skilled and unskilled labor in proportion to the share of the two skill categories in imputed labor earnings.
3. The estimates of adjusted total payments by skill category were then used to compute shares of payments to labor in total output.

Capital Inputs

The survey provided information on the historic cost values of fixed capital inputs including land, buildings, machinery, furniture and fixtures, and other fixed assets. Data on value of inventories in current prices of finished products, raw materials, and work in progress were also collected.

The historic cost values for fixed assets were converted to 1980 prices using the capital asset price deflators of the Indian National Statistical Office. Thus, although asset values are in constant prices, they do not reflect replacement cost, since no allowance has been made for the effect of obsolescence on the value of individual capital inputs. Rather, they consist of book values (before depreciation) at constant 1980 prices. Vintage effects are, therefore, still present in the capital stock estimates.

The capital stock measures were converted to capital services using annuity factors based on an assumed rate of return of 10 percent and the average expected economic lifetime of each asset type reported in the survey.[3] We experimented with the use of firm-specific rates of replacement for assets

3. The annuity factor is $r(1 + r)^t/(1 + r)^t - 1$, where r is the rate of return and t is the lifetime of the asset.

rather than industry averages to reflect the presence of differing maintenance and depreciation schedules across firms. Unfortunately, there were not sufficiently complete responses in all industries to pursue this alternative beyond the experimental stage; but in the one industry (soap manufacturing) in which we were able to use and compare both measures of capital services, the differences between the capital service estimates were slight for most firms and did not affect the results of the empirical work.

Output

Our principal measure of output is gross value added, defined as the difference between total ex-factory value of output and material inputs of both goods and services. Value added is used primarily to facilitate comparison with other empirical studies, which normally employ it as their measure of output, but it has the drawback of limiting our ability to treat material inputs as a factor of production and thus to examine the possibilities of materials-capital and materials-labor substitution.

Formally, the use of value added rather than gross output as a measure of output implies that material inputs are separable from factor inputs, such as labor and capital. Separability can occur either because no substitution possibilities exist between materials and other inputs or because materials are substitutable to exactly the same degree with all factor inputs. These are rather strong assumptions, but they have not deterred the use of value added functions as production functions in a wide range of empirical work. We have experimented with alternative specifications of the empirical work using gross output rather than value added; the results of some of these experiments are reported elsewhere (see, for example, Page 1984). The summary interpretation of the results is that our choice of value added does not fundamentally affect any of the analytical conclusions.

Characteristics of the Industries Surveyed

The following section briefly presents information on the structure and technological characteristics of the five industries surveyed. This description is based on qualitative results of the sample survey. Features of the policy environment within which the industries operate are also discussed.

Printing

Printing is an activity that appears to be dominated by small-scale enterprises. Official statistics covering enterprises registered under the Indian Factories Act enumerate approximately 3,500 firms with an average size of 41 employees. This is among the smallest of the average employment sizes for industrial activities. In addition to the registered enterprises, there are at least 60,000 units in the unregistered sector, the majority of which consist of firms with less than 10 workers.

In contrast to such industries as soap and shoes, the government does not restrict the capacity of registered units. Small-scale firms are eligible for the general fiscal subsidies available to other small-scale enterprises but enjoy no specific protection according to process stage or use of technology.

Marketing outlets are highly varied. About 30 percent of firms in the sample directly exported some of their production, while an additional 15 percent sold a portion of their output to firms engaged in exporting. Small-scale units in the sample mostly concentrated on job work servicing local markets, rather than on book production or printing of periodicals.

The printing industry in India uses two basic processes: letterpress and offset. Both processes are used for a large number of common products, notably books, and a number of firms use both processes on the same premises. Letterpress, the more traditional method, consists of setting up cast metal letters in a printing frame and imprinting the made-up image on the paper. Offset, in contrast, works from a photoengraving process in which an image of the desired printed form is made up, the image is photographed, and an engraved plate is etched.

There is some specialization of technique by product. Offset tends to be used to print wrappers, posters, product leaflets, calendars, maps and charts, and books and their jackets. These are products for which design is important and that are not well adapted to letterpress. Letterpress tends to be used for books, magazines, journals, and such commercial job work as bills, letterheads, and balance sheets.

Regardless of the process used, a printed product passes through three basic stages: preparation, production, and finishing. The production stage encompasses the actual printing process itself. Preparation is the design, setting up, and preparation of the material to be published. Finishing consists of cutting, binding, and packaging the printed material. Some specialization by stage of processing has occurred in India. Although virtually all printers engage in production, many do not have sufficient capacity to finish all their output. Specialized finishing firms have therefore developed and act as subcontractors to the printing presses.[4]

Shoemaking

Shoemaking is a traditional craft-based industry in India. Although factory assembly of leather footwear began in the mid-1930s with the establishment

4. Finishing firms are not represented in our sample. The use of value added, however, permits us to adjust for variations in the extent of vertical integration of enterprises, and exclusions of this segment of firms should not greatly affect the results.

of the Bata factory in Calcutta, the industry remains dominated in terms of both total output and employment by small- and medium-scale units; at least half of total employment in leather and leather products occurs in firms of less than 100 employees.

The Government of India mandated licensing of the industrial capacity of all enterprises with a book value of plant and machinery in excess of Rs2 million (Rs3.5 million in 1985). There are at present eleven such firms with a capacity of 10.2 million pairs of shoes per year. Actual production, however, falls short of the licensed capacity, and three of the registered enterprises are currently closed.

The Ministry of Industry has enforced a complete ban on the expansion of large-scale production of leather shoes for the domestic market. With the domestic market reserved for the small-scale sector, recent licenses for the installation of capacity in large-scale plants have been limited to exclusive production for export. Currently, 8 percent of the registered capacity of the shoe industry is licensed for export only, including 95 percent of the allowed capacity of two recently established firms.

In addition to the protection provided by quantitative restrictions on the output of large-scale enterprises, the central government provides an additional measure of protection via exemption from excise duty. Small-scale firms are exempt from duty on the first Rs1.5 million ex-factory value of output and pay duty at a reduced rate on an additional Rs1.5 million. The excise duty rate for all other enterprises is 10 percent ad valorem.

The principal market outlet for small-scale producers in the sample was sales to small retail shops. Approximately 45 percent of the small enterprises responding to the survey indicated that the major market for their products was small retail firms. Direct sales to households were the major outlet for an additional 23 percent of respondents, while sales to large retailers were the primary market for another 23 percent. Sales to foreign buyers and to exporting firms were the primary marketing channel for less than 9 percent of small enterprises in the sample, while less than 2 percent of firms indicated that sales to large-scale manufacturing enterprises constituted their primary market.

Among the large power-using firms in the sample, the primary marketing outlet was the firm's own retail outlets or other retailers. Two of the firms in the sample had been licensed for exports only (95 percent of output was intended to be sold on the international market). These firms sold directly to foreign buyers under licensing and purchase agreements.

There did not appear to be any systematic association between firm size and the household income of final purchasers among the firms contained in our sample. About 29 percent of the respondents indicated that some of their output was purchased by households with incomes of less than Rs300 per month, but only 9.5 percent indicated that their output was sold entirely to these households. An equal percentage of firms indicated that a portion of their output was purchased by upper income households, defined as those with monthly incomes in excess of Rs1,000.

The basic operations of shoemaking consist of cutting outer and lining materials to a pattern, stitching the cut pieces together to form the upper portion of the shoe, fitting the upper to a last, joining the lasted upper to the sole, and finishing the assembled shoe. Ancillary activities include materials handling, factory housekeeping, and packing.

In traditional shoemaking the upper and bottom leather is cut by hand, either with a pattern and knife on a cutting board or with a forged steel knife which has been shaped to the desired pattern. Stitching is performed by hand or by machine. The completed upper is put on a wooden last and shaped by hand before being hand tacked to the insole. Assembly of upper and bottom is by chemical adhesives, and the assembled shoes are then buffed and polished by hand.

Workers are specialized by process. The most common division of labor is for the owner to undertake cutting, which is the most highly skilled operation. The cutter aims to maximize the number of pieces cut from irregularly shaped hides and to minimize the wastage resulting from irregular thickness, coloring, and imperfections. Thus, the task requires considerable judgment to match pieces with regard to these characteristics. Stitching and closing are performed by a second worker specialized in these operations. Assembly of upper and sole is undertaken by a third worker, and finishing by unskilled or casual labor.

The most common type of machinery employed in traditional enterprises is a foot- or electrically powered stitching machine used in upper construction. Several firms in the survey also reported using manual stretching machines for lasting, and a few enterprises employed ancillary equipment for finishing and labeling.

A stylized intermediate technology represents the most common factory organization of shoemaking in India. Each enterprise varies in its mix of mechanized and manual processes, but all organized sector firms, and a number of the small- and medium-scale units in the unregistered sector, exhibited substantially increased specialization of labor and a marked increase in the level of mechanization when compared with the traditional technique.

The most important mechanized process added by nontraditional enterprises is mechanical pulling of the upper onto the last. Equipment designed for this process varies from simple mechanical pullers on which all tacking or adhesion is performed by hand to automated units which perform pulling and adhesion as a single operation. Firms choose among several worker-machine combinations which involve progressively reducing the number of manual operations. Machine pulling both increases the volume of production per unit time and permits greater uniformity in fitting the upper to the last, an important dimension of quality.

Cutting of upper and bottom material is also partially mechanized in nontraditional enterprises. Hydraulic or mechanical cutting presses of varying pressures are used in conjunction with forged steel pattern-shaped knives for preparation of uppers and soles. The machines do not, however, reduce the need for cutter operators to exercise considerable judgment with regard to the quantity and quality of pieces obtained from the hide.

A final mechanical process encountered in all modern enterprises is skiving—adjusting leather pieces to a uniform thickness. This permits greater interchangeability of parts in assembly and reduces the need to match pieces by natural thickness.

The advanced processes encountered in India differ from the intermediate technology primarily in the sophistication of the machinery used for lasting and in the shift from batch to machine-paced activities in assembly and finishing. Only three enterprises in the industry employ lasting machines which are fully capable of pulling and attaching the upper to the insole without some manual operations. Two firms machine-pace assembly by means of a conveyor which transports partially assembled pieces between individual assemblers.

Because of the separability of the process stages and the existence of a number of technical alternatives at each stage, substantial scope exists within the industry for the substitution of labor for capital. The greater labor intensity of manual, batch-type operations may be reduced by an increase in working capital requirements due to increases in process cycle time and, perhaps, by decreases in capacity utilization, when components are not closely coordinated; but it would not be surprising to encounter substantial variations in factor intensities among plants employing different combinations of manual and mechanical processes.

The impact of capital deepening on the skill composition of the labor force at the firm level is difficult to judge on the basis of the survey data. There is little doubt that the shift from traditional craft-based manufacturing processes to machine-based assembly methods involves the substitution of semiskilled labor for more highly skilled artisans in the main assembly process but it may not occur at the plant level. Mechanization appears to reduce the skill requirements in such tasks as cutting, upper assembly, and lasting. It generates additional requirements, however, for machine operators and skilled nonproduction workers.

Soap Manufacturing

In soap manufacturing, two sectors are recognized by the government. The organized or large-scale sector consists of firms using electrical power in the manufacturing process. These firms are included in regular statistical coverage of industrial establishments, and their installed capacity is subject to central government licensing. The unorganized sector of the soap manufacturing industry consists of small-scale and cottage industries which do not use electrical power. These establishments are mostly unregistered, and neither comprehensive output nor employment data for them are available. Approximately two-thirds of soap production, by volume, is undertaken in the unorganized sector (Goldar 1982). Nearly all of this output is laundry soap. Toilet soap is produced almost exclusively in the organized sector and accounts for approximately half that sector's output by volume.

The production of small-scale units in the sample was primarily marketed by small retailers serving local markets. This was followed by wholesalers, who were also the primary outlet for larger-scale units. Relatively few firms of any size engaged in direct sales to households.

Soap manufacturing is a simple process of mixing fatty materials with caustic soda (or caustic potash). The ensuing chemical reaction yields neat soap and a residue which contains glycerine. The traditional method of manufacturing soap is a batch process in which the fatty matter and caustic material are mixed in a cauldron. The mass is heated by burning coal or firewood and allowed to boil for a period of four to six hours. When soap is formed, a brine solution is added and the mixture is allowed to boil until lye is separated from the soap. These processes have a cycle time of approximately twenty-four hours. The quality of the soap may be improved by a series of washings with salt solutions, and when the neat soap has reached the desired degree of purity it is transferred to iron molds in which it is allowed to cool, dry, and solidify under atmospheric conditions. Cutting and packing are performed by hand. The entire process requires from six to seven days and is performed completely without power.

The processes followed in the modern sector involve varying degrees of substitution of mechanical for human power and replacement of batch processing with continuous flow processes. Saponification is achieved by boiling either in a kettle using open steam (the batch process) or in a sealed autoclave (the continuous process). For better recovery of glycerine and improved product quality, the soap is washed several times in brine solution. The neat soap is then mixed with the required quantities of builders (additives designed to improve washing quality), fillers (additives designed to increase bulk), and preservatives, and cooled and dried in either tubular driers or vacuum spray devices. During the final process the soap is milled, plodded, and extruded into a continuous bar which is cut, wrapped, and packaged. The soap thus produced is of a distinctly higher quality than that produced without power by traditional methods.

Public policy has limited the scope for factor substitution by drawing a distinction between manufacturing processes that employ electrical power and those that do not. The existing structure of excise tax legislation exempts firms which do not use electrical power from payment of the excise tax. In addition, the government has, through its industrial licensing policies, prohibited expansion of capacity by the power sector in soap manufacturing. The quantitative restriction on incremental output is intended to provide an additional measure of protection for nonpower enterprises.

One consequence of this distinction, based upon the use of electrical power, is that although there is apparently substantial variation in techniques of production within the (power-using) sector, there is almost no variation in the techniques employed in the nonpower sector, since the combination of mechanical with manual operations is prohibited. Increases in firm size give rise to replication of the initial series of operations, and thus within the nonpower sector there is no engineering relationship between firm size and capital intensity. Another of the consequences of the distinction, quality specialization by firm, is discussed in chapter 13.

Machine Tool Manufacturing

Machine tool manufacturing in India is dominated by small factory operations. More than 65 percent of all machine tool manufacturing firms employ between 10 and 50 workers. There are approximately 10 large-scale machine tool manufacturing plants, employing more than 200 workers. The largest, Hindustan Machine Tools, is a public enterprise, while the remaining large-scale firms are private. The large-scale firms frequently produce other engineering products, such as pumps, engines, and precision instruments, in addition to machine tools, while smaller-scale firms tend to concentrate exclusively on the production of machine shop equipment. Most large-scale firms have licensing and technical cooperation agreements with foreign machine tool manufacturers to obtain designs for more sophisticated types of equipment, while small and medium-size firms generally copy their designs from existing tools.

Production of simple machine shop tools such as bench grinders, lathes, and boring machines has been reserved for the small-scale sector. This limits the capacity of large-scale firms in these product groups to that currently installed and has had the effect of forcing larger-scale producers into more specialized product lines such as vertical turret lathes, milling and gear hobbing machines, and special purpose tools. The reservation policy has also had the result that larger-scale producers have attempted to enter export markets as a means of avoiding the capacity restrictions faced in the domestic market.

The great majority of firms in the machine tools sample marketed their products exclusively through wholesalers. Less than 10 percent of respondents reported sales of their products directly to households or producing firms, and there was no evidence of subcontracting by smaller firms. The extent of export sales among smaller firms in the sample was quite limited, with only 6 of 98 firms directly engaged in any export activity. Among the larger, registered firms, all respondents reported some export sales.

The production processes employed in both small and large-scale machine tool manufacturing firms are fundamentally similar. Cast bodies of individual tools are assembled with precision machined parts to form the complete product. In all cases the machining of the drive assembly and other precision components is undertaken by the manufacturing enterprise. Electrical components such as motors and controls are normally purchased.

Firms differ with respect to their degree of vertical integration in the manufacture of machine bodies. The larger-scale enterprises frequently have their own foundry for the casting of bodies and foundations, while smaller firms subcontract casting to specialized foundries.

There does not appear to be any sharp break in the choice of techniques employed by machine tool manufacturing firms of varying sizes. Small firms tend to use factory processes employing much the same equipment as their large-scale counterparts. Increases in size make it possible to add specialized items of equipment, such as gear hobbing machines, which make it possible either to manufacture, rather than purchase, inputs or to reduce manufacturing time on general purpose machines. There is also some variation in materials handling procedures between larger and smaller firms; larger firms generally tend to use more mechanized material handling methods.

Metal Casting

Much of the metal casting industry in India is centered in Howrah, a city across the Hooghly River from Calcutta. It grew up there in the nineteenth century in response to demand generated by construction of the railways. The industry is dominated by medium-scale firms; there are few enterprises engaged exclusively in casting employing more than 200 workers. A number of large-scale engineering firms, including machine tool and transportation equipment manufacturers, have foundries as part of their vertically integrated manufacturing operations.

The Howrah firms produce a wide variety of cast metal products varying in technical sophistication from manhole covers to high pressure vessels. A number of cast iron products are reserved for the small-scale sector, but there is no licensing of capacity. Metal casting firms primarily market their output through small and large retailers. About 20 percent of the sample, however, was engaged in the production of products such as machine foundations, vessels, and other job work which were directly purchased by large-scale manufacturing firms. A substantial proportion (11 of 51) of the firms in the sample engaged in the production of simple cast products including iron pipes and manhole covers for export.

All the firms contained in our sample use the sand floor casting process for gray iron. This consists of making a model (normally of wood), forming casting sand around the model, ramming the sand to achieve a compact mold in the form of the model, removing the model, and filling the mold with molten iron. In general, firms employ the open cupola process to melt the pig iron, which is the most primitive of metal heating processes.

Technical substitution in the industry occurs primarily in the ramming process, which can be performed by hand or by machine. The quality of the sand mold depends critically on the compactness of the casting sand, and machine ramming is intended to replace the skilled labor required to achieve uniform compactness of the mold. A second area where some technical substitution was observed was in the metal heating process. A number of firms had shifted from open cupola to more technically sophisticated methods of melting iron in order to achieve greater uniformity in the iron mixture and to reduce impurities.

The current Howrah technologies depend critically on skills available in the Calcutta area. In particular, the availability of skilled mold makers who are capable of achieving great uniformity of compactness by the hand ramming method is a major reason for the limited introduction of machine ramming techniques. In other regions of India, where these skills are not locally available, there has been substantial substitution of machine ramming for skilled labor, particularly in the vertically integrated foundries of large-scale engineering firms.

9

Firm Size and Economic Performance: Theory and Measurement

Much of the empirical work in this and the following chapters shares a common methodological framework based on the economic theory of production. By using explicit production models, we are able to treat such issues as the choice of technique, technical efficiency, productivity, and economic efficiency in a simple unified framework. This chapter outlines the methodology and discusses its uses and limitations.

Economists have traditionally summarized the relationship between inputs and output in terms of the production function—the set of technical relationships which govern the maximum quantity of measurable output that can be obtained from a given set of measured inputs. The concept of technology, therefore, represents the ability to transform inputs into output. With the level of technology given, there is a maximum amount of output that can be produced with a given amount of inputs, and production above that maximum is technologically impossible. The production function at a specified level of technology summarizes the series of maximum output levels corresponding to different levels of inputs. Thus, we define a production function for each of our industrial sectors giving output as a function of factor inputs and the level of technology:

$$X(s) = F[T(s); Z(s)] \tag{9-1}$$

where $X(s)$ is the level of output, $T(s)$ is the level of technology, and $Z(s)$ is the vector of production inputs for each observation s.

Figure 9-1 represents one possible production function relationship between a single input and output. Changes in output due to changes in input levels at a given level of technology are represented by movements along a single production surface, such as OA, while changes in output due to changes in the level of technology are represented as shifts in a production schedule, such as AA'. This conceptualization allows us to deal with a variety of issues such as factor substitution and production efficiency.

Figure 9-1. *Total Factor Productivity Differences of Two Production Functions*

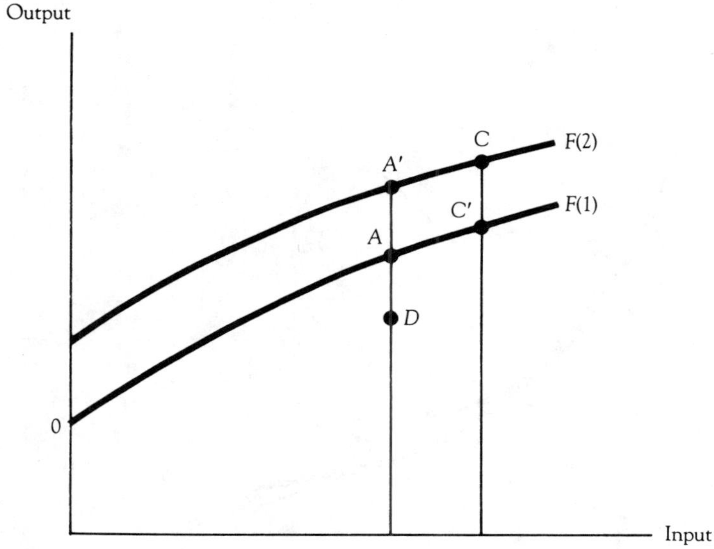

Capital-Labor Substitution: The Theory

At a given level of technology, producers make decisions regarding the use of inputs based on the technical possibilities embodied in the production function and on relative input prices. Factor intensities, therefore, are the consequence of economic decisions subject to the constraints imposed by technology. The simple stylization of this decisionmaking process is given in figure 9-2. The unit isoquant in the figure depicts all possible combinations of two inputs, capital and labor, which can efficiently produce a unit of output. No input combination to the left of the unit isoquant is technologically possible; combinations to the right of the isoquant are technologically feasible, but they represent inefficient choices in the sense that at least one input-output coefficient is larger than necessary.

The optimum ratio of inputs is given at point *A*, the tangency between the unit isoquant and the factor price line, and may be represented by the ray *OA*. Changes in relative factor proportions away from *A* increase costs for the given level of output. Thus, the point of tangency defines a position of producer equilibrium. In a riskless world in which all producers face equal relative prices, the conditions for producer equilibrium imply that all firms will exhibit equal relative factor intensities.

Reality, however, rarely conforms to this simple model of producer equilibrium. Cross-section studies of production encounter substantial variations in

Figure 9-2. *Choice of Techniques and Factor Substitution*

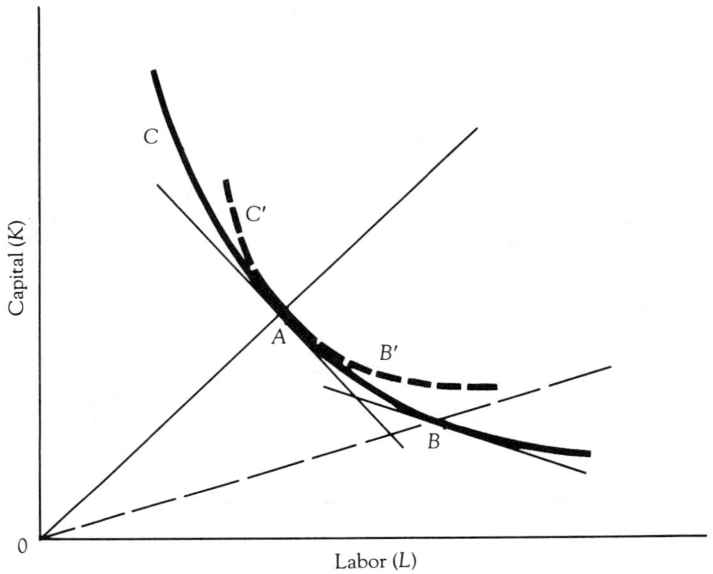

factor proportions among individual firms, and the literature on small-scale enterprises frequently emphasizes the observed differences in factor intensity between large and small-scale firms. In order to reconcile these differences in factor proportions with the notion of producer equilibrium, at least four kinds of differences may exist between firms in the variables affecting economic decisions.

First, firms may face different sets of market prices. Since the decision rule for profit maximization yields actual profits as a function of relative input prices, it is clear that two firms may choose different input combinations if they face different price regimes. In figure 9-2 both firm *A* and firm *B* fulfill the conditions for producer equilibrium at different input combinations since they confront two distinct sets of relative prices. Different input combinations in this case represent efficient substitution among inputs along a single production function.

The scope for efficient substitution along a single production function is defined by the degree of substitutability between factors of production. This is represented by the curvature of the unit isoquant in figure 9-2. As drawn, there is substantial scope for factor substitution in response to differences in relative factor prices. A production function which embodies less scope for factor substitution might be represented by the more sharply curved isoquant *C′AB′*. The reduced scope for factor substitution implies that even substantial differences in relative factor prices faced by firms will be reflected in only small changes in relative factor proportions.

A summary measure of the potential for factor substitution is given by the elasticity of substitution, the percentage change in relative factor proportions for a 1 percent change in relative factor prices. Because the elasticity of substitution is potentially an important determinant of the scope for differences in factor proportions among firms of different size, we devote a substantial amount of empirical work in chapter 10 to measuring the elasticity of substitution among pairs of inputs and to examining the manner in which substitution possibilities vary with firm size.

Second, levels of technology may differ between firms with the consequence that different optimum input combinations are chosen at equal relative factor prices. This is represented in figure 9-3. Firms *A* and *C* face equal relative factor prices but have different levels of technology, as represented by the unit isoquants *ABC* and *A′B′C′*. As drawn, the technology embodied in the isoquant *A′B′C′* is dominated by that represented by the isoquant *ABC* in the sense that both more capital and more labor are required to produce a unit of output along the former. The isoquant *A′B′C′* is also, however, uniformly more labor-intensive than *ABC*; the optimum technique is more labor-intensive at all possible factor price ratios. Thus, at the equal relative factor prices indicated on the diagram, the optimum technique *C′* is more labor-intensive than *A*.

Movement inward of the unit isoquant is associated with technological progress. In general, an isoquant such as *A′B′C′* would represent a productive technology of an earlier vintage than that of *ABC*. The manner in which the isoquants shift in response to technological change determines any possible biases in factor use resulting from technological progress. In figure 9-3 the bias may be defined as labor saving (capital using), since a shift from one level of technology to the other results in an increase in capital intensity at constant relative factor prices. Neutral technical progress occurs when the capital-labor ratio remains unchanged at constant relative factor prices following a shift in technology. This is defined as Hicks neutral technological change.

If smaller firms use older vintages of technology, their relative labor intensity may reflect the presence of capital using technological change. In chapter 10, we attempt to test for the presence of different vintages of capital and for possible biases in the pattern of technological change.

Third, factor substitution possibilities may be a function of the level of output. In this case, the optimum technique may depend on scale, as well as on relative factor prices. When relative factor proportions are a function of scale, even a single level of technology cannot be represented by a single unit isoquant. Rather, unit isoquants for various levels of output will cross, as represented in figure 9-4. Firms facing equal factor prices will choose different optimum techniques (such as those at *A* and *C* in the figure) at different

Figure 9-3. *Biased Technical Change in a Unit Isoquant*

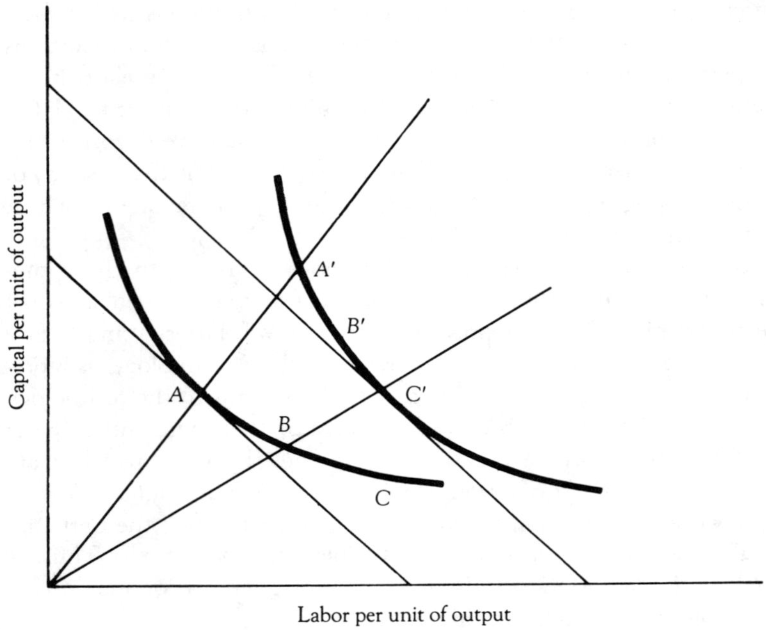

Figure 9-4. *Nonhomothetic Unit Isoquants*

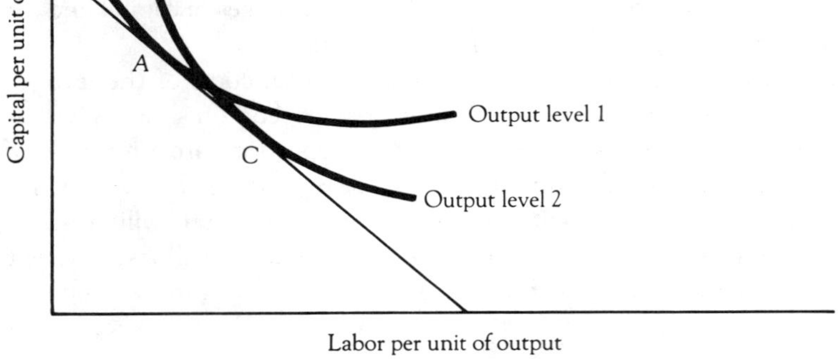

levels of output. The ability to represent a single technology by one unit isoquant, regardless of scale, is formally defined as "homotheticity." When the production function is homothetic, optimal factor proportions are independent of scale; when it is not, scale affects the optimal choice of technique. Clearly, the presence or absence of homotheticity can be a major explanation for observed differences in factor proportions between large and small firms, and it is therefore one of the characteristics for which we test the production function in chapter 10.

Fourth, firms may not maximize profits. Varying administrative and organizational structures among firms may lead to alternative decisionmaking rules. The techniques chosen under these rules need not correspond to those which would be predicted on the basis of profit maximization. Failures of profit maximization may occur as the result of incomplete knowledge of productive techniques or of factor prices, or as the consequence of nonprofit maximizing behavior by producers. In either set of circumstances, the result may be that firms having access to the same technology and facing unified factor prices will select different production techniques.

Measuring Capital-Labor Substitution: The Translog Production Function

In order to apply the above conceptual framework to economic data, we must have a form of the production function which is capable of statistical estimation and which provides a reasonably accurate picture of production technology. In the estimation of production models the conventional approach has been to impose rather strong restrictions on the properties of the technology to which the production function is an approximation. The Cobb-Douglas form, for example, despite its many attractive properties, restricts all elasticities of substitution between inputs to unity and is homothetic in input space. As the name implies, the Constant Elasticity of Substitution (CES) function restricts pairwise elasticities of substitution to be constant and equal for all input levels.

Flexible forms of the production function impose fewer a priori restrictions on the properties of the underlying technology. One such flexible form which has been used in recent empirical work is the transcendental logarithmic (translog) production function (Christensen, Jorgensen, and Lau 1971, 1973; Berndt and Christensen 1973a, 1984). In these production functions outputs are an exponential function of the logarithms of inputs. A two-factor version of the translog production function relating output, X, to capital, K, and labor, L, can be written as:

$$\ln X = a_o + a_k \ln K + a_l \ln L + B_{kk} \cdot \frac{1}{2} (\ln K)^2 + B_{ll} \cdot \frac{1}{2} (\ln L)^2 + B_{kl}(\ln K)(\ln L). \quad (9\text{-}2)$$

The logarithm of output is, therefore, a function of the logarithms of the individual factor inputs, the square of the logarithms of inputs, and an interaction term between the two factors.[1] The a and B terms are the production function parameters to be estimated. Readers familiar with the more traditional Cobb-Douglas form will recognize that the first line in equation 9-2 is the Cobb-Douglas, two-factor production function. The quadratic and interaction terms in the second line provide information on the curvature of the production function (expressed in logarithms) and hence on whether substitution possibilities significantly diverge from the unitary elasticity of substitution imposed by the structure of the Cobb-Douglas production function.

The elasticity of output with respect to any input is the percentage change in output which results from a 1 percent change in the level of inputs. Logarithmic differentiation of the production function in equation 9-2 shows that in the translog case the elasticities of output depend on input levels. For the case of labor input, for example,

$$\frac{d \ln X}{d \ln L} = a_l + B_{ll}(\ln L) + B_{lk}(\ln K). \quad (9\text{-}3)$$

In the Cobb-Douglas case, in contrast, the output elasticities are constant and are equal to the first term, a_l, in equation 9-3.

The necessary conditions for producer equilibrium under competitive conditions require that the elasticities of output with respect to each input be equal to the share of expenditure on the factor in total output. Thus, for the case of labor

$$\frac{wL}{pX} = \frac{d \ln X}{d \ln L} \quad (9\text{-}4)$$

where w is the wage and p is the price of output. In the case of the Cobb-Douglas production function, constancy of the elasticity of output implies constancy of the factor share. The translog function, conversely, is characterized by variability in the elasticity of output and in factor shares as input levels change.

The B coefficients, which are frequently called share elasticities, provide information on the structure of factor substitution possibilities implied by the estimated production function. When the share elasticity is positive, the value share of the factor increases as the quantity of the input increases (at constant factor and product prices) holding all other inputs constant. When the share elasticities are negative, the opposite case holds.

Changes in factor shares are related to changes in factor prices via the elasticity of substitution. In the case of two factors of production, the elasticity of substitution may be defined as the proportionate change in the capital-

1. A more general exposition of the translog production function is given in appendix 2.

labor ratio which occurs as a result of a 1 percent change in the wage-rental ratio. In the Cobb-Douglas case, for example, the unitary elasticity of substitution implies that factor shares are unchanged as a consequence of changes in the wage-rental ratio; changes in factor intensity precisely offset the change in factor prices. When more than two factors are present in the production function, the concept of the partial, or Allen, elasticity of substitution is used.[2] The partial elasticity of substitution indicates the proportionate change in the intensity of one pair of factors, holding all other prices constant, which occurs as a result of a 1 percent change in their relative prices.

Because output elasticities and factor shares vary with input levels in the translog framework, the elasticity of substitution is variable and is determined by input levels. This is an important advantage of the translog production function, since it permits evaluation of the way in which factor substitution possibilities vary with firm size.

Total Factor Productivity

Consider a firm such as that located at point *A* in figure 9-1 contrasted with that located at point *A′*. As drawn, the two utilize equal levels of measured inputs but achieve different levels of output. The two firms differ in their levels of total factor productivity (TFP), the efficiency with which all inputs are combined. As drawn, the two firms use different levels of technology implied by the production function relationships *F*(1) and *F*(2). The difference in TFP between the two observations can be defined as the proportional increase which would expand output by *A* to the observed level of *A′* (or, alternatively, the proportional decrease in the output of *A′* to achieve the level of output of *A*) controlling for the input level.

It is only very rarely possible to encounter two firms which utilize precisely the same levels of inputs. More generally the problem of determining relative levels of TFP consists in establishing the productivity differential between firms such as *A* and C which use different levels of inputs and produce different levels of output. The problem in this case is to separate the effect of different input levels on output differences from the effect of TFP differentials.

The key to sorting out the two effects empirically is the production function, which allows us to predict the maximum output which could be achieved by each firm if it employed the level of inputs actually used by the other. Given this information, we can then define TFP differentials as either the proportionate reduction needed to shrink output at C to its predicted level C′ on the basis of the production technology employed by firm *A* or as the proportionate increase which would expand output to the level implied

2. The formal definition is provided in appendix 1 and was developed by Allen (1938).

by the production technology of C, which is A'—given the level of inputs actually used by A. The two production functions summarize at every point different levels of maximum output obtainable by the two firms at each level of inputs, and differences between them represent differences in TFP.

An attractive feature of the translog functional form is that it can treat time symmetrically with other inputs. This permits us to examine, simultaneously, TFP change and the substitution possibilities among inputs with respect to such chronological variables as age of the enterprise, labor force experience, and vintage of technology. The estimated parameters of the production function which describe these characteristics are the share elasticities with respect to inputs and time.

The rate of TFP change is specified as the percentage change in output with respect to time, holding all factor inputs constant. If the production function in equation 9-2 is extended to include time, the rate of productivity change is given by

$$\frac{d \ln X}{dt} = a_t + B_{tt}t + B_{kt} \ln K + B_{lt} \ln L. \tag{9-5}$$

The coefficients in equation 9-5 have the following interpretations: a_t is the rate of TFP change at the point of expansion of the translog production function. Under normal economic conditions it should be non-negative; B_{tt} is the rate of TFP change and takes on a positive, negative, or zero value depending on whether there is acceleration, deceleration, or constancy in the rate of TFP change. The terms B_{kt} and B_{lt} are the share elasticities of each input with respect to time. These parameters define the bias in TFP change. If a share elasticity with respect to time is positive (negative), the corresponding value share increases (decreases) with time, and the bias is interpreted as input using (saving). If the share elasticity is zero, the share is independent of time, and if all share elasticities are zero, the pattern of productivity change is Hicks neutral, as defined in the section above on the theory of capital-labor substitution.

Technical Efficiency

Up to this point, we have discussed differences in TFP as though they were synonymous with differences in technology. The origin of this identification of TFP differentials with differences in technology is obvious. If both firms in figure 9-1 achieve their maximum potential output as defined by the production function, differences in TFP can only reflect differences in levels of technology. Firms need not, however, achieve their maximum potential output as defined by the production function. The efficiency with which existing techniques are used is also reflected in differentials in productivity between firms. Consider, for example, the productivity differential between firms *D* and *B* in figure 9-1. Firm *D* operates at a point below the production function. If by improvements in its use of existing techniques, firm *D* is able to reach its potential output, given its level of technology (in this case point *A*), the increase in output for the same level of inputs will be recorded as an improvement in TFP. The TFP differential between actual and potential (or "best practice") output is defined conventionally as technical inefficiency. The observed TFP differential between *B* and *D* may therefore be thought of as consisting of both differences in technical efficiency (*AD*) and differences in technological levels (*AA*′).

In order to define technical efficiency and to distinguish it from differences in technology, we must have a standard against which the productivity performance of any enterprise is measured. This standard is the best practice or frontier production function, which defines the outer boundary of all input-output combinations for a specified level of technology. Deviations from best practice are ascribed to technical inefficiency.

Farrell (1957) introduced the concept of technical efficiency along with that of the frontier or best practice production function, which defines for a set of observations the maximum output attainable from a given vector of measured inputs. Subsequent applications have focused on specification and estimation of the production frontier and on its use to measure and explain variations in efficiency in production.[3] The amount by which measured output is less than potential output, as given by the production frontier, is conventionally defined as the level of technical inefficiency.

Farrell's work was the first of a number of efforts to model efficient production behavior and to use information provided by the modeling exercise to draw inferences concerning the relative efficiency of economic units. Beginning with the pioneering work of Aigner and Chu (1968), substantial econometric effort has been focused on developing frontier production functions which can simultaneously serve as efficiency standards, provide information on the technical structure of the best practice production function, and generate measures of relative technical inefficiency.

The frontier production function represents the outer boundary in output space for a given range of input observations. Consider the production relation

$$X = g(Z;B) + j \tag{9-6}$$

where X is a vector of observations on output, Z is a matrix of observations on factors of production, B is a vector of production function parameters, and j is a vector of disturbances. Traditional estimates of the production function—such as those in chapter 10—specify j to be normally, independently, and identically distributed with zero mean and finite variance, and to

3. An excellent survey of the literature is contained in Forsund, Lovell, and Schmidt (1980).

thus provide an estimate of the average production function for the set of observations. Frontier estimators, conversely, specify j to have a negative expectation, which reflects the presence of technical inefficiency in production. In selecting a frontier production model, two principal issues arise. The first is the specification of the production function, that is, its ability to approximate the underlying functional relationship between inputs and output correctly. The second is the nature of the error added to equation 9-6, which determines the characteristics of the measures of technical efficiency generated by the estimates as well as the technique of estimation of the production function parameters.

The production function employed in the frontier estimates in chapter 11 is a three-factor translog production function. All deviations from the production function are constrained to be of negative sign, so that the parameters provide an estimate of maximum potential output for each observed vector of input levels. The estimation technique employed—which is an application of linear programming—minimizes the sum of the deviations of actual output from predicted output.[4]

A firm's actual output $X(s)$, given observed input levels, is equal to its predicted output $\hat{X}(s)$ only if the firm operates on the production frontier. Each firm is assigned an efficiency index (Farrell index) equal to the ratio of its actual to predicted output, $X(s)/\hat{X}(s) \leq 1$, which may be found as the antilog of the slack variable in the programming problem. The value of the index provides a measure of relative technical efficiency.

The concept of technical efficiency is closely related to that of total factor productivity. The amount by which actual output is less than potential output is formally equivalent to the difference between total factor productivity based on best practice and that based on actual practice (see Nishimizu and Page 1982). Since differences in technical efficiency between firms are equivalent to differences in total factor productivity, the production frontier provides a useful tool for analyzing the relative productive efficiency of individual economic units.

Note that the best practice frontier is defined relative to the actual performance of firms at a given technological level. In many industries, several vintages of technology may coexist, with the result that differentials in productivity between firms may reflect both differences in best practice and differences in technical efficiency. The empirical work in chapter 11 consists of an effort to define the best practice production function in each of the five industries in our sample and to explain apparent deviations from best practice. In doing so we consider both vintage effects, the possibility that firms operate at different levels of technology, and possible sources of divergence from best practice. In particular, we try to discover whether size acts independently of other variables in determining the level of technical efficiency.

4. A fuller description of the estimation technique is given in appendix 2.

Our discussion up to this point has treated output as though it is a single homogeneous unit. Products, even when they serve a basic purpose such as cleaning, are rarely completely homogeneous. They differ in physical characteristics, as well as in consumers' perceptions of their value. Our principal measure of output is value added, and variations in product quality are therefore summarized by variations in the value added price. In general, therefore, our approach has been to focus on factor substitution and relative efficiency in the generation of value added. But it should be noted that this approach assumes that a higher price reflects quality rather than market power.

Because firms exercise control over their level of product quality, quality choice and the optimal combination of factors of production required to minimize costs at a given level of quality are economic decisions which can vary from firm to firm. A drawback of our use of value added as the measure of output is that this interaction between quality levels and input use is obscured, and it is not possible to examine the relationship between the chosen technique of production and the production of characteristics which establish the quality level of output. Hence, we cannot determine, for example, if improved qualities of output use labor, skill, or capital.

The issue of firm size and product quality choice has been emphasized in a number of studies, generally with the result that large firms are found to produce higher-quality products, and that the higher product quality is associated with greater capital intensity. Three of our data sets provide some information on characteristics of products at the firm level. In chapter 13 we examine the relationship between size, product quality, and the choice of technique.

10

Factor Proportions and Substitution: Survey Results

One of the most widespread arguments for small-enterprise promotion activities has been the presumed greater labor intensity of production techniques in small-scale firms. In chapter 7 we examined the evidence available from Indian industrial survey data concerning firm size and capital intensity. The results were mixed. In many product groups the supposed monotonic relationship between size and capital intensity was not strongly supported by the data. Moreover, the survey data available provide us with little insight into the causes of observed differences in factor proportions and do not permit examination of patterns of substitution between labor of various skill levels and capital.

In this chapter we address the same set of issues using the data from our sample survey. Our primary objective is to examine the extent to which labor intensity in production changes with firm size within the industries in our sample. Because the skill composition of the labor force often changes as techniques of production vary, a major focus of the empirical work is to trace patterns of substitution between skilled and unskilled labor as well as between both categories of labor and capital.

The principal tool for analyzing these issues is a flexible form of the production function, which permits us to estimate substitution relationships between inputs at each data point. These substitution relationships are summarized by the partial elasticity of substitution between pairs of inputs, which are functions of the levels of inputs and therefore of size. A second objective of our empirical work is to use the production function to examine a number of characteristics of the structure of production, including the presence of economies of scale, homotheticity, and the effects of technical change and experience on patterns of factor use and substitution, which can be framed as testable hypotheses in econometric estimation.

The next section presents evidence from the surveys on firm size and factor intensity. Then we briefly discuss the nature and characteristics of the production function which is used in the econometric model of factor substitution. The main results on firm size and factor substitution are then discussed and some extensions of the analysis are presented in the final section of the chapter.

Evidence on Factor Proportions

Before we discuss the estimation and testing of the production function, this section presents results from the sample survey concerning factor proportions and skill intensity. In chapter 8 we discussed the definition and construction of the main variables used in the production analysis. We defined three aggregate factors of production: skilled labor, unskilled labor, and capital. Skilled labor was defined by reference to responses in the sample survey as those tasks which involve substantial discretionary control by the worker over the process of production or which involve the control and supervision of other workers. Hence it includes proprietors, managers, supervisors, white collar workers, and skilled production and ancillary workers. Unskilled labor is a residual category which includes both semiskilled and unskilled production and ancillary workers. Our labor force aggregates, therefore, differ from those usually encountered in industrial surveys in that the skill categories are defined by task rather than by function. In most industrial census returns, the classification of workers is made by functional breakdowns, for example, managers, white collar workers, production workers, and ancillary (nonproduction) workers. Since each of these functional groups includes both highly skilled and unskilled workers, clear evidence on skill composition is difficult to derive.

Capital is defined as the flow of capital services from both fixed and working capital at undepreciated historic cost in constant 1980 prices. Because we are able to include working capital in our estimates of capital services, the capital measure is more comprehensive than that usually encountered in industrial survey results. In addition the presentation of the capital stock values at constant prices differs from most industrial census tabulations. Output is defined as gross value added.

Capital-labor ratios by skill category and the ratio of skilled to unskilled labor by employment-size category of firm are presented in table 10-1, together with the standard deviation of each cell mean. The most striking feature of the data is the diversity among industries in the relationships between firm size and capital intensity. None of our industries shows a strictly monotonic relationship between firm size, as measured by employment, and capital intensity. Within-group variation in the capital-labor ratio is large in almost every employment-size category in each industry.

In shoemaking there is a sharp and statistically significant increase in the capital intensity of firms employing more than 100 workers over that of enterprises with less than 50 employees. The capital-labor ratio in large firms

Table 10-1. *Capital-Labor and Skill Ratios in Four Indian Industries*
(thousands of rupees per worker)

		Printing				Machine tools			
Size category[a]		K/L	K/skilled	K/unskilled	Skilled/unskilled	K/L	K/skilled	K/unskilled	Skilled/unskilled
N < 5	Mean	2.354	2.711	441.755	217.00	—	—	—	—
	S.d.	(1.108)	(1.878)	(306.564)	(116.164)	—	—	—	—
	Obs.	6			6	—	—	—	—
5 < N < 10	Mean	6.891	12.906	21.389	2.783	3.689	4.242	28.387	6.683
	S.d	(7.381)	(17.847)	(21.774)	(2.075)	(0.970)	(1.105)	(8.104)	(0.495)
	Obs.	26			26	4			4
10 < N < 25	Mean	9.596	16.888	578.461	91.252	5.418	5.971	70.984	9.518
	S.d.	(10.165)	(19.951)	(1,772.499)	(267.903)	(11.676)	(12.429)	(187.023)	(4.517)
	Obs.	19			19	28			28
25 < N < 50	Mean	6.338	11.845	2,847.938	334.614	3.140	4.100	16.545	4.60
	S.d.	(4.186)	(10.798)	(8,933.490)	(1,041.981)	(2.086)	(2.738)	(11.397)	(2.906)
	Obs.	10			10	24			24
50 < N < 100	Mean	10.470	19.728	24.315	1.206	3.445	5.094	15.807	3.483
	S.d.	(3.653)	(1.565)	(15.038)	(0.667)	(1.641)	(2.983)	(16.597)	(3.224)
	Obs.	2			2	9			9
100 < N	Mean	14.274	32.337	25.793	0.893	7.402	11.736	21.508	1.956
	S.d.	(12.513)	(31.321)	(20.516)	(0.149)	(7.307)	(11.476)	(20.227)	(0.888)
	Obs.	3			3	13			13

Size category[a]		Soap				Shoes				Metal casting			
		K/L	K/skilled	K/unskilled	Skilled/unskilled	K/L	K/skilled	K/unskilled	Skilled/unskilled	K/L	K/skilled	K/unskilled	Skilled/unskilled
N < 5	Mean	8.174	22.068	15.856	1.333	0.635	1.056	3.025	2.860	3.000	15.000	3.750	0.2500
	S.d.	(7.426)	(24.098)	(11.964)	(1.093)	(0.537)	(0.988)	(6.482)	(4.587)	(0.000)	(0.000)	(0.000)	(0.000)
	Obs.	9			9	24			24	1			
5 < N < 10	Mean	5.766	21.721	8.647	0.524*	0.490	0.834	1.86	2.114	30.285	67.259	106.917	2.875
	S.d.	(3.548)	(18.188)	(5.067)	(0.292)	(0.581)	(0.933)	(3.86)	(1.779)	(8.726)	(38.180)	(69.271)	(3.095)
	Obs.	20			20	25			25	4			
10 < N < 25	Mean	5.665	25.075	8.198	0.397**	0.448	0.846	1.633	1.660	35.044	57.434	293.699	5.213
	S.d.	(2.857)	(20.019)	(4.947)	(0.199)	(0.403)	(0.519)	(2.966)	(2.137)	(49.066)	(66.049)	(696.259)	(6.519)
	Obs.	13			13	33			33	19			
25 < N < 50	Mean	5.685	33.152	6.848	0.213	0.719	1.700	1.355	0.855	21.455	36.919	154.211	2.850
	S.d.	(0.918)	(9.852)	(0.910)	(0.036)	(0.618)	(1.500)	(1.175)	(0.462)	(21.203)	(21.613)	(332.924)	(4.273)
	Obs.	2			2	10			10	7			
50 < N < 100	Mean	6.726	24.860	17.723	1.284	—	—	—	—	18.028	37.087	37.339	1.624
	S.d.	(4.884)	(18.570)	(19.446)	(1.258)	—	—	—	—	(20.175)	(46.982)	(33.026)	(0.809)
	Obs.	6			6	—			—	3			
100 < N	Mean	6.440	39.549	12.803	1.087	5.529	16. 529	8.694	0.629	35.108	102.399	63.263	0.925
	S.d.	(3.587)	(55.129)	(8.068)	(1.277)	(4.128)	(13.279)	(6.171)	(0.312)	(31.278)	(85.703)	(63.232)	(0.788)
	Obs.	7			7	7			7	11			

— Not available.

*Significantly different from the smallest size category at the 0.05 level.

**Significantly different from the smallest size category at the 0.01 level.

Note: N is employment size of firm; K is capital; L is labor; s.d. is standard deviation; obs. is number of observations.

a. Employment size, including working proprietors.

exceeds that in small firms by a factor of approximately 10 to 1, but within the size range below 50 workers there are no significant variations in factor intensity. The dichotomy between capital intensity in large-scale activities and in small and medium-scale firms is consistent with the impression gained during the survey that shoemaking was the industry in which traditional handicraft methods of production, and factory methods, are most closely related to firm size defined by employment. Although there is substantial variation in the techniques of production employed by large-scale firms, as shown by the high coefficient of variation of the capital-labor ratio, virtually all factory processes involved a substantial increase in the level of capital intensity. The nature of the shift from handicraft to factory processes is further demonstrated by the change in the skill ratio which occurs between large-scale and small and medium-scale firms in shoemaking. The skill ratio declines continuously from about 2.8 for very small firms (less than 5 workers) to 0.6 for large-scale firms. Thus the data show a process of substitution of capital and unskilled labor for skilled labor as firm size increases.

In soap manufacturing, which is the other traditional activity in our sample, there is no systematic variation with firm size in either the capital-labor ratio or the skill ratio. This lack of variation reflects the policy regime in soap manufacturing in India, which imposes substantial financial penalties on firms that use power in the production process. The result of this policy regime has been to encourage the growth of large-scale traditional (non-power) firms, which are essentially multiple replications of the basic production process employed in small enterprises. For this reason, the shift between factory and handicraft methods is not as closely associated with firm size in soap manufacturing as in shoemaking.[1]

In machine tool manufacturing, the average capital-labor ratio in firms employing more than 100 workers is approximately double that of firms employing less than 10, but the variance of the large-scale sample is so large (the coefficient of variation is approximately 100 percent) that there is no statistically significant difference between the means. Moreover, the size category with the second highest level of capital intensity and the greatest standard deviation is in the class from 10 to 25 workers. There is also a sharp apparent decline in the skill ratio in large-scale firms, but again the differences between means are not statistically significant (at the 10 percent level). Thus, although there is some evidence of a shift in techniques between large-scale and small and medium-scale firms, our evidence is not conclusive. The lack of a systematic pattern of factor intensities is consistent with observations made during the survey that among the product groups represented, machine-shop lathes, milling machines, and drill presses, there is no sharp break between workshop and factory methods of production associated with

1. Goldar (1982), however, finds capital intensity sharply increasing with firm size. He uses unit level data from the *Census of Small Scale Industry.*

firm size. Rather, small-scale producers tend to be small factory operations. Indeed, the size range of 10 to 25 employees exhibits both high capital intensity and high skilled labor intensity.

Factor proportions in the metal casting industry are very similar to those in machine tool manufacturing. Except for the smallest size category, which is not sufficiently large to allow generalizations, capital-labor ratios are roughly similar for firms with less than 25 employees and for firms with more than 100 workers. In the middle size range of 25 to 100 workers the capital-labor ratio is about half as large, but the within-group variation in each size class is so large that there are no significant differences among size classes. The skill intensity of production rises to a peak of more than 5 skilled workers per unskilled worker in the size category of 10 to 25 workers and then declines to less than 1 skilled worker per unskilled worker in the largest firms. Within-group variation is large, however, and the only significant pairwise differences in the skill ratio occur between the 10 to 25 and more-than-100 size classes. The decline in skill intensity reflects the shift in technology from hand ramming of floor castings to machine ramming and other semiautomated techniques described in chapter 8. What is of interest is the apparent substitution of skills within the labor input rather than substitution of capital for labor. It is also of interest to note that, as with machine tool manufacturing, the size class of 10 to 25 exhibits both high capital and high skill intensity of production.

The evidence on factor proportions from the printing industry is the most difficult to interpret. Capital intensity increases sharply for firms with more than 5 employees, and the difference between the smallest size group and all others is statistically significant. There does not appear to be any systematic relationship between capital intensity and employment size in any of the size categories exceeding 5 workers. There is great variability in the skill ratio. In three employment size categories, the survey results indicated that there were virtually no unskilled workers employed by firms in the sample. While this is not surprising for very small printing firms, in which the limited number of workers in the enterprise perform both skilled and unskilled tasks but are recorded as skilled, it is less probable among larger-scale firms. Our data, however, show high skill intensity in the size range from 10 to 50 employees, where this consideration should not apply. The results suggest that in printing our enumeration of tasks may not have provided a good basis on which to classify skilled and unskilled labor and that there is substantial inaccuracy in the employment data.

When firms are ordered by investment size rather than labor force size, somewhat greater regularity emerges in the patterns of factor proportions. Table 10-2 presents the mean and standard deviation of the capital-labor ratio for investment size classes ranging from less than Rs1,000 to more than Rs1 million. Very small firms (those with less than Rs1,000 in fixed investment) are significantly less capital-intensive than the majority of larger size

Table 10-2. *Capital-Labor Ratios in Five Indian Industries by Size of Fixed Assets*

Fixed assets[a]	*Printing*	*Machine tools*	*Soap*	*Shoes*	*Metal casting*
$K < 1$	—	—	—	0.09	—
				(0.11)	
$1 \leq K < 5$	1.26	—	—	0.26	0.30
	(0.46)	—		(0.19)	(0.00)
$5 \leq K < 10$	1.56	—	2.35	0.63	—
	(0.51)	—	(0.74)	(0.80)	—
$10 \leq K < 20$	1.82	—	2.39	0.78	0.85
	(0.65)	—	(0.78)	(0.30)	(0.53)
$20 \leq K < 50$	3.27	1.97	3.15	1.26	1.19
	(1.29)	(1.03)	(1.48)	(0.85)	(0.98)
$50 \leq K < 100$	4.09	2.27	7.14	1.74	2.59
	(3.31)	(0.94)	(5.24)	(0.50)	(2.14)
$100 \leq K < 200$	9.03	3.10	7.50	—	1.88
	(5.56)	(3.81)	(2.23)	—	(3.99)
$200 \leq K < 500$	10.98	4.70	3.92	—	3.30
	(11.16)	(2.55)	(1.29)	—	(8.51)
$500 \leq K < 1,000$	10.66	4.79	5.40	4.31	2.72
	(5.00)	(1.44)	(0.00)	(1.28)	(0.89)
$1,000 \leq K$	11.66	8.84	—	2.62	9.54
	(10.02)	(17.60)	—	(5.65)	(0.61)

— Not applicable.

Note: Standard deviations in parentheses.

a. K = thousands of rupees per worker.

categories (at the 90 percent level) in all five industries. These micro or cottage shop enterprises frequently have capital-labor ratios which are less than 10 percent of those for firms with fixed investment of Rs500,000 or more.

Interestingly, though, a monotonic relationship between investment size and capital intensity only exists in two of the five industries, shoemaking and machine tool manufacturing. In soap and printing, capital intensity peaks in the interval between Rs50,000–Rs100,000 and Rs200,000–Rs500,000, respectively, and then declines. In all five industries, the within-group variation in each investment size category is large and pairwise differences between groups are frequently not statistically significant.

The general pattern which emerges when size is measured by investment is that firms with capital stock below Rs20,000 tend to be somewhat, but not uniformly, less capital-intensive than larger-scale firms. In the interval Rs20,000 to Rs1 million, there is substantial variation in factor proportions, and no uniform monotonic relationship exists between investment size and capital intensity. In three of the five industries, shoes, machine tools, and metal casting, very large firms (those with Rs1 million or more fixed investment) are the most capital-intensive on average, but the within-group variation in shoes and machine tools is so large that these apparent differences are not statistically significant.

Measuring Substitution Possibilities

The estimates of factor substitution possibilities presented in this chapter are based on the transcendental logarithmic (translog) production function discussed in chapter 9. This section briefly outlines in a nontechnical way the features of the production function and some of the implications of using it to estimate patterns of factor substitution.

The translog production function is a flexible form of the production function in the sense that many of the technological characteristics which it summarizes are themselves functions of the levels of inputs and thus vary from observation to observation. For example, in a production function such as the well-known Cobb-Douglas form, the logarithm of output is a linear function of the logarithm of individual inputs. This implies that the elasticity of output with respect to any input (the percentage change in output for a 1 percent change in the input) is constant for all observations. In a Cobb-Douglas function the elasticity of substitution is also constant and equal to one. In the translog form the logarithm of output is a quadratic function of the logarithm of inputs. The elasticity of output and the elasticity of substitution are therefore both functions of the levels of inputs and vary from observation to observation.

The relationship between the Cobb-Douglas and translog forms is illustrated in figure 10-1. Observed relationships in logarithms between output and a single input, say labor, are represented by the circles. The production function relates the logarithm of output to the logarithm of the single input. The Cobb-Douglas form is linear in the logarithms and is estimated as the line *AB*. The translog form is the curved relationship *A′B′*. As drawn, use of the Cobb-Douglas function tends to overstate the level of output which could be obtained from the single input at both low and high levels of inputs and to understate it in the middle range.

The elasticity of output is the slope of the production function drawn in figure 10-1. It is clear from the diagram that the elasticity of the translog function varies with the level of the input, first exceeding the elasticity of output of the Cobb-Douglas function and later falling below it. The elasticity of substitution is not a relevant concept in the case of one-factor production.

The advantage of using a flexible production function for dealing with issues related to firm size should be clear. Firm size, via the levels of the inputs, becomes an important input into the estimate of the structure of technology. If technology differs with firm size, the translog function will summarize those differences and will permit us to observe how such impor-

Figure 10-1. *Relationship between Cobb-Douglas and Translog One-Factor Production Functions*

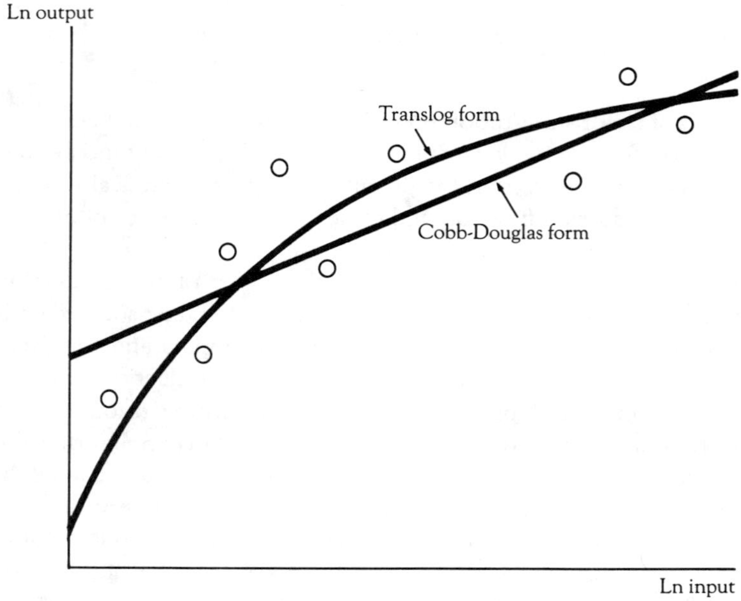

tant production function parameters as the elasticity of substitution vary with firm size.

There are costs associated with using flexible forms, however. The most significant of these costs is that the translog production function does not make the data define a function which has all the essential characteristics of a production function. Rather, it is necessary to test whether these characteristics are present in each set of data fitted to the production function, and hence, to test whether the translog form is appropriate to the data employed. The minimum properties which are essential to define a production function are summarized by the characterization of the function as "well-behaved."

A production function is considered to be well-behaved if it has positive marginal products for each input (monotonicity) and if the isoquants of the production function are convex (quasi-concavity). Since input and output levels are always positive, monotonicity is equivalent to requiring that the output elasticities be positive at each data point. Quasi-concavity is evaluated by examining the curvative of the estimated production function at each observation. A translog function need not fulfill these requirements globally, but if the conditions are met for a sufficient number of the observed levels of inputs and outputs the function is considered to be well-behaved.

If the production function is well-behaved, it can then be employed to explore a number of characteristics of the production technology. Three main characteristics of the production process are examined in this chapter:

homotheticity, constant returns to scale, and the elasticity of substitution between pairs of factors of production.

A production function is said to be homothetic if marginal rates of substitution between pairs of inputs are constant at constant relative levels of factor intensity. In the context of the translog function this may be tested by imposing a set of restrictions on the parameters of the production function. If the restrictions are accepted, the translog function takes the general form of a multifactor Kmenta (1967) approximation to the CES production function. While this approximation does not have constant elasticities of substitution, it is a homothetic form whose elasticities of substitution depend only on the ratios of the inputs.

Production under constant returns to scale implies an additional set of restrictions on the production function parameters which result in both output elasticities and value shares in output summing to unity. Again this may be tested directly. The parameter estimates of the translog function also provide sufficient information to estimate the partial, or Allen (1938), elasticities of substitution between all pairs of inputs at each data point.

The main empirical work in this chapter uses a three-input translog function with symmetry imposed. This may be written as:

$$\begin{aligned} \ln X &= a_0 + a_s \ln L_s + a_u \ln L_u + a_k \ln K \\ &+ \frac{1}{2} \cdot B_{ss} (\ln L_s)^2 + \frac{1}{2} \cdot B_{uu} (\ln L_u)^2 + \frac{1}{2} \cdot B_{kk} (\ln K)^2 \\ &+ B_{su} (\ln L_s)(\ln L_u) + B_{sk} (\ln L_s)(\ln K) + B_{uk} (\ln L_u)(\ln K) + j \end{aligned} \tag{10-1}$$

where X is value added, L_s is skilled labor, L_u is unskilled labor, and K is capital services. With the addition of a conventional stochastic error, j, equation 10-1 may be estimated directly by ordinary least squares (OLS).

A difficulty with the use of OLS is that the regressors are endogenous variables for an establishment within an industry. Thus parameter estimates derived from equation 10-1 will be biased and inconsistent. The problem can be overcome in the standard competitive setting by exploiting the properties of the elasticity of output. Since under competition the elasticity of output is equal to the share of expenditure on the factor input, the share equations, expressing each expenditure share as a function of input levels, can be employed to provide efficient estimates of the production function parameters (Berndt and Christensen 1973).[2] The production function and share equations are estimated jointly, following elimination of one of the share equations in order to obtain a nonsingular system.

In a single period cross section it is least likely that the equilibrium conditions underlying the capital share equation will hold. Indeed, capital may be

2. See chapter 9 and appendix 2 for definitions of the share equations.

truly a predetermined variable in a single production period. Thus an economically appropriate system to estimate consists of equation 10-1 and the labor share equations. In what follows, the disturbances are considered stochastically independent.[3] The disturbance vector of the system of equations is assumed to be joint and normally distributed with zero mean and a nonsingular covariance matrix.

Under the assumed stochastic specification with cross-equation constraints, the technique of multivariate regression will yield maximum likelihood estimates of the parameters on convergence. Hypothesis testing of the homotheticity and homogeneity (constant returns to scale) restrictions may be performed using the likelihood ratio test, $-2 ln \lambda$, where λ is the ratio of the likelihood of the restricted to the unrestricted system. The test statistic is distributed asymptotically as chi-squared with degrees of freedom equal to the number of independent restrictions imposed.

Because the translog production function is neither monotonic nor concave for any arbitrary range of inputs, it may be necessary to impose monotonicity and global concavity on the production structure in cases when the estimated function is not locally well-behaved. The most important of these restrictions is global concavity, which, because of the properties of the translog functional form carries with it the restriction of the production function to constant returns to scale. These restrictions rob the translog function of much of its flexibility, including the ability to approximate a nonhomothetic production structure, but they are necessary to ensure that the production function properly captures the relationship between time and levels of total factor productivity. The techniques for imposing these restrictions are discussed in appendix 2.

Patterns of Factor Substitution

Our analysis of the relationship between firm size and the pattern of factor substitution is based on estimates of partial elasticities of substitution derived from the production function. We estimate a three-factor translog production function using value added as the measure of output and skilled labor, unskilled labor, and capital as factor inputs. Time is excluded. Thus, we implicitly force all observations to conform to a single estimated technological level. In the next section, we shall employ several time-denominated variables to examine the extent to which the results of our cross-sectional analysis may require modification if firms are operating at different levels of technology.

3. The interpretation of this restriction is that technical and allocative inefficiency are uncorrelated. Schmidt and Knox Lovell (1980) consider the case of stochastic dependence between the errors.

Table 10-3 reports the results of fitting the three-factor translog production function to data from the sample survey. Parameter values and asymptotic standard errors are reported in the table for estimates of the single-equation production function by OLS and for the systems estimate.

The translog production function performs quite well. The R^2 values for the OLS single-equation estimates exceed 0.90, although the standard errors of the individual coefficients show evidence of the effects of multicollinearity among the right-hand-side variables. This is a common problem in attempts to estimate multifactor translog functions by single-equation methods (Greene 1980). Results of the system estimates show a marked improvement in the ratio of parameter estimates to standard errors; the maximum likelihood estimates were therefore taken as the basic parameters of the production function for the purposes of testing hypotheses regarding the form of the production function.

The testing strategy is outlined in figure 10-2. The unrestricted translog function is first evaluated for monotonicity and quasi-concavity. Tests of homotheticity and constant returns to scale are then conducted. For cases in which the restrictions cannot be rejected, the restricted translog production function is once again evaluated for monotonicity and curvature properties. Finally, the production function is restricted to the Cobb-Douglas form, which imposes equal unitary elasticities of substitution on all pairs of factors in order to test whether the curvature properties embodied in the translog functional form are appropriate for the data.

Results of the evaluation of the unrestricted translog form for monotonicity and quasi-concavity are reported in table 10-4. In all cases the production functions are monotonic in the neighborhood represented by the sample. The soap manufacturing sample, however, fails the check for convexity in 89.5 percent of the cases; thus, the unrestricted translog form is not well-behaved in the region represented by the soap manufacturing sample. The production function is satisfactorily locally convex in each of the other four industries, although the printing sample shows a relatively high percentage of observations failing the check for concavity.

These results suggest that the translog parameters in soap and printing may be unstable for the whole size range of the sample. There is little guidance concerning the effect of nonconvexities at some points on the production surface on the estimated elasticities of substitution at the point of approximation.[4] A recent simulation study indicated that translog estimates

4. A Chow test of the equality of parameters across the subsamples defined by splitting the sample at both 50 and 100 employees was conducted for each industry. It was not possible to reject the hypothesis of parameter equality based on the F test statistic for any industry. Thus, although it is not possible to detect a structural shift in the production function, the parameters which permit precise characterization of the curvature of the production function near the point of approximation may imply a lack of concavity at points removed from the geometric mean of the sample.

Table 10-3. *Ordinary Least Squares and Maximum Likelihood Estimated Parameters of the Unrestricted Translog Production Function for Five Indian Industries*

	Printing		*Machine tools*		*Soap*		*Shoes*		*Metal casting*	
Parameter	*Maximum likelihood*	OLS	*Maximum likelihood*	OLS	*Maximum likelihood*	OLS	*Maximum likelihood*	OLS	*Maximum likelihood*	OLS
a_o	5.049	5.041	5.451	5.479	4.956	4.914	4.022	3.992	5.372	5.425
	(0.067)	(0.095)	(0.069)	(0.085)	(0.065)	(0.077)	(0.070)	(0.087)	(0.075)	(0.092)
a_k	0.446	0.283	0.512	0.236	0.584	0.297	0.497	0.283	0.368	0.227
	(0.039)	(0.071)	(0.054)	(0.095)	(0.045)	(0.083)	(0.039)	(0.063)	(0.066)	(0.131)
a_s	0.266	0.356	0.424	0.859	0.155	0.312	0.310	0.780	0.459	0.623
	(0.013)	(0.109)	(0.027)	(0.155)	(0.011)	(0.077)	(0.014)	(0.129)	(0.096)	(0.277)
a_u	0.081	0.289	0.083	0.123	0.170	0.421	0.171	0.109	0.180	0.082
	(0.006)	(0.067)	(0.008)	(0.084)	(0.008)	(0.078)	(0.011)	(0.086)	(0.065)	(0.102)
B_{kk}	−0.032	−0.249	0.077	0.206	0.104	0.442	0.007	0.047	−0.097	−0.152
	(0.036)	(0.119)	(0.065)	(0.126)	(0.036)	(0.216)	(0.017)	(0.085)	(0.062)	(0.111)
B_{ss}	0.136	−0.235	0.095	0.652	0.138	0.221	0.154	−0.014	0.015	0.023
	(0.018)	(0.213)	(0.043)	(0.843)	(0.014)	(0.137)	(0.019)	(0.344)	(0.008)	(0.057)
B_{uu}	0.0196	0.083	0.032	−0.022	0.101	0.687	0.042	0.116	0.035	−0.076
	(0.003)	(0.034)	(0.006)	(0.192)	(0.012)	(0.207)	(0.012)	(0.109)	(0.012)	(0.132)
B_{ks}	−0.052	0.213	−0.057	−0.326	−0.039	0.090	−0.077	−0.003	−0.037	−0.103
	(0.014)	(0.129)	(0.038)	(0.343)	(0.016)	(0.155)	(0.011)	(0.163)	(0.096)	(0.086)
B_{ku}	0.003	0.035	−0.001	0.134	−0.019	−0.449	0.002	−0.043	0.003	0.055
	(0.006)	(0.059)	(0.011)	(0.126)	(0.014)	(0.167)	(0.009)	(0.080)	(0.010)	(0.059)
B_{su}	−0.012	−0.104	−0.040	−0.223	−0.049	−0.328	−0.027	−0.030	−0.041	−0.084
	(0.005)	(0.084)	(0.013)	(0.301)	(0.009)	(0.113)	(0.011)	(0.133)	(0.010)	(0.146)

Note: OLS = ordinary least squares. Standard errors in parentheses.

Figure 10-2. *Testing Strategy for Translog Production Function*

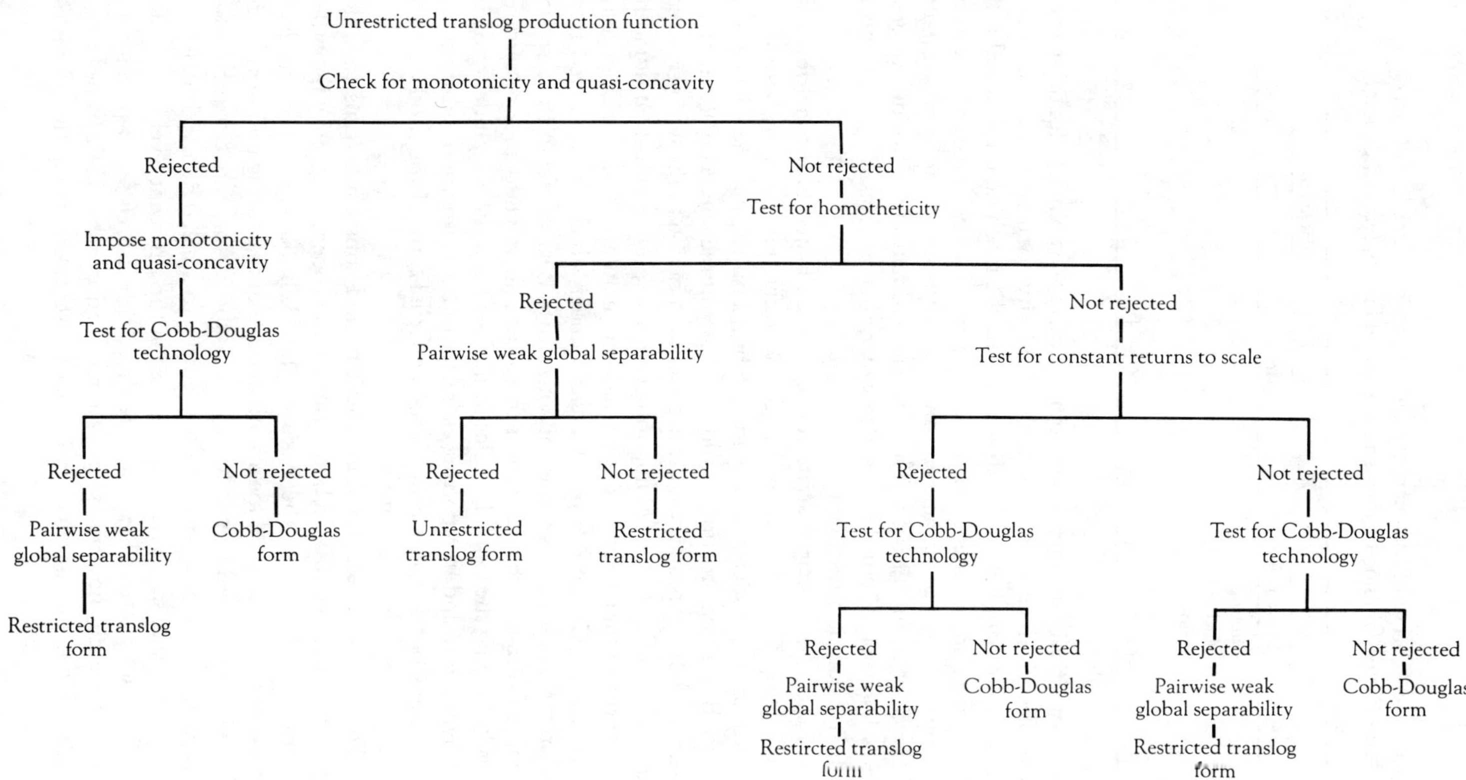

Table 10-4. *Tests of Monotonicity and Convexity for the Unrestricted Three-Factor Translog Production Function in Five Indian Industries*
(percent)

	Observations failing test	
Sample industry	*Monotonicity*	*Convexity*
Printing	12.1	22.7
Machine tools	0.0	1.3
Soap	8.8	89.5
Shoes	0.0	13.1
Metal casting	4.4	11.1

of elasticities of substitution are quite good at the point of expansion even for samples in which violations of the regularity conditions occur at a limited number of data points. By this criterion, the machine tools, metal casting, and shoe samples offer a sufficient number of locally convex observations to permit the use of parameters on the entire sample to estimate pairwise elasticities of substitution. The elasticity estimates for printing are presumably less reliable, but the extent to which they are biased by the nonconvex observations is not known.

In the case of soap manufacturing, the estimated functional form is not sufficiently well-behaved to describe a production function, and any estimates of the elasticity of substitution based on the parameters of the unrestricted translog function would be wholly unreliable. The alternative is to impose global concavity on the production function in the manner described in appendix 2. The cost of imposing concavity, however, is that homotheticity and constant returns to scale become maintained, rather than testable, hypotheses. Of the five possible cases under which the restricted translog form is globally concave, the only one which is successful in the case of soap manufacturing is the restriction of all own share elasticities to zero, effectively yielding the Cobb-Douglas form. Hence, in the remainder of our tests of hypotheses and in reporting estimates of the elasticity of substitution, we take the Cobb-Douglas form as the preferred form of the production function for soap manufacturing.

The results of tests for homotheticity and constant returns to scale are presented in table 10-5. In both printing and shoemaking, each hypothesis is decisively rejected on the basis of the likelihood ratio test, while both hypotheses are accepted for machine tool manufacturing and metal casting. Because the Cobb-Douglas form is homothetic, the only test performed on the soap sample was for constant returns to scale. The hypothesis of constant returns to scale cannot be rejected at the 1 percent level.

The results of our hypothesis testing imply a final set of parameter estimates for each industry, which are reported in table 10-6. In printing and shoemaking, the parameter estimates are taken from the unrestricted

Table 10-5. *Likelihood Ratio Test Statistics for Restrictions on the Translog Form, Five Indian Industries*

Parameters and industry	*Homotheticity test*	*Constant returns to scale test*	*Cobb-Douglas test*	
Number of restrictions	2	3	6	8
Chi-square critical value (0.01 level)	9.21	11.35	16.81	11.35
Industry				
Printing	25.52	33.52	95.54	
Machine tools	2.02	2.60	54.12	
Soap	n.a.	3.35	n.a.	
Shoes	31.08	31.10	80.44	
Metal casting	5.06	10.05	35.91	

n.a. Not available.

Table 10-6. *Parameter Estimates of the Three-Factor Translog Production Function in Four Indian Industries*

Parameter	*Printing*	*Machine tools*	*Soap*	*Shoes*	*Metal casting*
a_0	5.049	5.459	5.136	4.023	5.325
	(0.067)	(0.054)	(0.059)	(0.071)	(0.065)
a_k	0.446	0.500	0.670	0.497	0.359
	(0.039)	(0.033)	(0.021)	(0.039)	(0.040)
a_s	0.266	0.418	0.158	0.311	0.460
	(0.013)	(0.026)	(0.022)	(0.015)	(0.031)
a_u	0.081	0.082	0.172	0.171	0.181
	(0.006)	(0.008)	(0.014)	(0.011)	(0.020)
B_{kk}	−0.032	0.067	n.a.	0.008	−0.060
	(0.036)	(0.046)	n.a.	(0.017)	(0.080)
B_{ss}	0.137	0.089	n.a.	0.154	0.023
	(0.018)	(0.029)	n.a.	(0.019)	(0.019)
B_{uu}	0.020	0.031	n.a.	0.042	0.014
	(0.003)	(0.005)	n.a.	(0.012)	(0.002)
B_{ks}	−0.052	−0.062	n.a.	−0.077	−0.041
	(0.014)	(0.036)	n.a.	(0.011)	(0.030)
B_{ku}	0.003	−0.005	n.a.	−0.002	0.003
	(0.006)	(0.011)	n.a.	(0.009)	(0.006)
B_{su}	−0.012	−0.026	n.a.	−0.027	−0.035
	(0.005)	(0.009)	n.a.	(0.011)	(0.010)

n.a. Not available.
Note: Standard errors in parentheses.

translog system, while in machine tool manufacturing and metal casting the parameter estimates are based on the restricted linear homogeneous translog form. The final parameter estimates for soap manufacturing are those of the constant-returns-to-scale, Cobb-Douglas case.

The evidence from the survey with respect to homotheticity and constant returns to scale has some important implications for the interpretation of the technological relationship between small and large firms. In two industries, printing and shoemaking, the production function is neither homogeneous nor homothetic. Thus unit isoquants of firms of various sizes will intersect, which implies different optimal levels of factor intensity at equal relative factor prices. In machine tool manufacturing, metal casting, and soap manufacturing, conversely, it is not possible to reject the hypothesis that the production function is homothetic and therefore that relative factor proportions depend only on relative factor prices and are independent of scale.

The data do not support the hypothesis that there are significant economies of scale in any of the industries analyzed. In soap, metal casting, and machine tools, it is not possible to reject the hypothesis of constant returns to scale across the entire range of observations in the sample. Printing and shoemaking exhibit variable returns to scale; at the sample mean the sum of output elasticities is not significantly different from unity in shoemaking, while for the largest size group the fitted shares sum to less than unity.[5] In printing, the sum of factor shares at the point of expansion is significantly less than one.

Allen elasticities of substitution between pairs of factors were estimated on the basis of the parameter estimates in table 10-6. Because the elasticity of substitution in the translog form is a function of the levels of the inputs, it is possible to derive partial elasticities of substitution for each observation. Rather than do this, we present estimates for the sample mean and for the cell means of five employment size groups in table 10-7.

Capital and skilled and unskilled labor are all substitutes in each industry (except the largest size group of shoemaking) by the usual Allen definition, $\sigma_{mn} > 0$. The rankings of substitution possibilities are consistent across industries and are of some interest. Skilled and unskilled labor are the most highly substitutable factors, followed by skilled labor and capital. The partial elasticity of substitution between capital and unskilled labor is quite low. Thus, capital in these manufacturing industries appears to be more complementary with unskilled labor than with skilled labor.

This finding is at variance with the results of recent empirical work based on aggregate production functions for the whole manufacturing sector in

5. Goldar (1982) finds evidence of increasing returns to scale in his soap manufacturing sample. This is not inconsistent with the results reported here, since our sample covers a wider range of firm sizes (especially in the larger size groups) and exhibits first increasing and then decreasing returns to scale in size increases.

Table 10-7. *Allen Elasticities of Substitution by Size Group in Four Indian Industries*

	Printing						*Machine tools*						*Soap*		
Parameter	*Mean*	*N < 10*	*10 < N < 25*	*25 < N < 50*	*50 < N < 100*	*100 < N*	*Mean*	*N < 10*	*10 < N < 25*	*25 < N < 50*	*50 < N < 100*	*100 ≤ N*	*Mean*	*N < 10*	*10 < N < 25*
n_{su}	3.65	4.68	3.17	2.83	3.94	—	2.60	3.19	2.74	2.46	2.11	1.77	1.00	1.00	1.00
n_{sk}	2.37	2.50	2.29	2.38	1.94	—	1.45	1.44	1.44	1.53	1.33	2.20	1.00	1.00	1.00
n_{uk}	0.62	0.49	0.67	0.65	0.18	—	1.02	1.04	1.04	0.96	1.03	0.92	1.00	1.00	1.00

				Shoes						*Metal casting*					
Parameter	*25 < N < 50*	*50 < N < 100*	*100 < N*	*Mean*	*N < 10*	*10 < N < 25*	*25 < N < 50*	*50 < N < 100*	*100 < N*	*Mean*	*N < 10*	*10 < N < 25*	*25 < N < 50*	*50 < N < 100*	*100 < N*
n_{su}	1.00	1.00	1.00	3.20	4.17	2.80	3.54	—	4.05	3.09	2.87	3.35	3.20	2.95	3.38
n_{sk}	1.00	1.00	1.00	2.31	2.29	2.16	3.49	—	5.98	1.97	2.37	2.20	1.80	2.15	1.75
n_{uk}	1.00	1.00	1.00	0.37	0.23	0.42	0.27	—	−0.04	0.97	1.09	1.26	0.85	1.16	0.92

— Not available.

developed economies. These studies suggest that capital is more substitutable for unskilled labor than for skilled labor. The results are similar, however, to historical studies of productivity differentials between the United States and the United Kingdom that attribute the faster rate of capital deepening in the United States to a relative scarcity of skilled labor, which implies a higher elasticity of substitution between capital and skills than between capital and unskilled labor.[6]

Empirical studies of factor substitution in manufacturing industries in developed countries generally employ the functional labor force classifications of production and nonproduction workers. These functional groupings are associated with unskilled and skilled labor, respectively, because of the argument that direct production workers are normally machine-paced operatives while nonproduction workers are generally white collar and skilled ancillary employees. It is on the basis of this classification that skilled (nonproduction) workers are found to be more complementary with capital. Our skill groupings are directly based on tasks and permit both production and nonproduction workers to be allotted to each skill category. Since, particularly in small firms, there are substantial numbers of skilled production workers and unskilled ancillary workers, our classification is not strictly comparable with that employed in much of the literature.

In the previous section we noted that one pattern of factor use in several industries in our sample was a fall in the skill intensity of production as firm size rose. Our results, which suggest that capital is more substitutable for skilled than for unskilled labor, are consistent with the observed patterns of factor substitution across the size range of our sample. Most econometric studies based on developed-country industrial census data do not include observations at the small end of the size range, where (with the exception of machine tools) the substitution of skills for capital is most dramatic in our data set.

Substitution possibilities between skilled and unskilled labor and between skilled labor and capital do not appear to vary systematically with firm size. This is somewhat surprising in view of the greater division of labor and specialization of function encountered in factory enterprises as opposed to craft-based activities. It is important to note, however, that there is no dramatic shift in the structure of production across the range of firm sizes. Put another way, all the observations in our samples lie along a single production surface. Thus both factory-process and craft-based enterprises represent a spectrum of factor substitution possibilities along a single production function, which in general is characterized by greater substitution possibilities between skilled and unskilled labor and between skilled labor and capital than between unskilled labor and capital.

6. On the sectoral estimates of substitution elasticities, see, for example, Berndt and Christensen (1974); Habbakuk (1962) reviews the historical evidence.

Firm Size, Experience, and Factor Use

Our estimates of the production function in the preceding section were based on the assumption that the production relationship for all firms in the sample was adequately described by a single production surface. Implicitly this assumption constrained all observations to be at the same technological level, and although this is not an unusual assumption to make in work with cross-section data, it is in fact a testable hypothesis.

Among the data collected was a substantial body of information on the age of the enterprise and of its capital stock (the average age of which is the firm's capital "vintage"), as well as on the levels of employee experience. These data are not usually available in empirical studies using cross-section data, but they are appropriate for testing hypotheses concerning the rate and characteristics of total factor productivity change associated with capital vintage and experience.

The relationship between time and productivity change within the framework of the translog production function was summarized in chapter 9. The production function will permit us to establish whether there is a significant relationship between capital vintage (the level of technology) and/or employee experience and TFP. By including a measure of time (that is, the age of the enterprise) in the estimated production function, we no longer constrain all observations in the sample to exhibit the same set of production relationships in measured inputs. In fact the share elasticities with respect to time provide us with information on the manner in which vintage and experience affect factor use.

An important objective of the research reported in this section was to determine whether the experience–productivity relationship varied with establishment size. Some of the literature on small-scale enterprises suggests that levels of TFP in small firms may be below those encountered in larger ones (Page 1979; White 1978). Generally, these productivity differentials have been associated with the view that smaller firms employ techniques which are of older vintage and thus of lower productivity, but it is equally possible that firms of different sizes differ systematically either in their levels of employee experience or in the technological relationship between experience and productivity. The wide range of firm sizes present in the data set allows both these possibilities to be tested.

To estimate the production function, we use the three factors of production as previously defined: skilled labor, unskilled labor, and capital. Output, again, is defined as gross value added. The measure of experience employed in the production function estimates is average experience of employees within the firm. This variable was constructed by estimating the mean of the distribution of employees by length of experience in the enterprise. Since it reflects firm-specific experience only, the variable is an imperfect indicator of total employee experience, but it does successfully capture the generation of firm-specific skills as well as the effect of labor turnover.

Two variables reflecting the age of the enterprise were available from the survey. The first was the age of the firm from the date of its foundation to the year of the survey. The second was the representative capital stock vintage, which was estimated by taking a value-weighted average of the ages of individual capital components. In cases where there has been substantial renewal of or additions to the capital stock, the latter variable provides a much better estimate of the age of the plant than the former.

In cross-sectional data there are two potentially offsetting effects of age on TFP. The first is that, to the extent that the age of the enterprise acts as a proxy for accumulated experience in production, there may be a positive relationship between age and TFP. Alternatively, if the age-of-enterprise variable mainly summarizes the effect of vintage on TFP, and if newer plants embody higher productivity capital, the age of the enterprise may have a negative impact on TFP.

The restricted translog function was first estimated using each of the chronological measures (employee experience and vintage) independently in the estimating equations. It is of interest to note that the globally concave form of the production function was the same for each industry and reduced to a generalized Cobb-Douglas form with time terms.

The parameter estimates of the time variables provide the basis for testing hypotheses concerning the nature of the relationship between time and productivity. The data are not very informative with regard to the pairwise relationship between age of enterprise, employee experience, and productivity. The estimated time parameters are not significantly different from zero (at the 0.10 level) in nine of ten possible cases. The exception is machine tool manufacturing, for which the coefficient of the term a_t based on the average level of labor force experience is positive and significant. In all other cases it is impossible to detect any systematic variation of total factor productivity with either age of firm or employee experience.

Although the hypothesis tests for the age-of-enterprise variable fail to reveal a statistically significant relationship between chronological age and TFP, the sign of the age variable suggests that it may primarily reflect vintage effects of the capital stock on productivity. Firm-specific experience and the age of the enterprise were strongly correlated. These results, taken together, indicate that the production functions may suffer from specification bias caused by the omission of a variable reflecting the vintage of the capital stock. The capital stock estimates taken from the sample survey do not reflect productivity differentials embodied in new plant and equipment; hence, the capital services measures employed in the production function estimates treat all capital in use as equally productive. The results of this specification bias would be to obscure the experience–productivity and age–productivity relationships in each of the equations in which they are introduced singly.

Table 10-8. *Parameter Estimates of the Restricted Translog Production Function in Five Indian Industries*

Parameter	Machine tools	Printing	Shoes	Soap	Metal casting
a_0	5.478	5.149	3.979	5.066	5.027
	(0.64)	(0.090)	(0.077)	(0.074)	(0.065)
a_u	0.084	0.081	0.177	0.175	0.180
	(0.008)	(0.007)	(0.018)	(0.013)	(0.022)
a_s	0.433	0.277	0.323	0.157	0.465
	(0.032)	(0.037)	(0.016)	(0.016)	(0.039)
a_k	0.483	0.642	0.500	0.668	0.354
	(0.031)	(0.017)	(0.017)	(0.018)	(0.039)
a_t	0.049	0.015	0.044	0.053	0.039
	(0.014)	(0.023)	(0.023)	(0.018)	(0.012)
a_v	−0.030	−0.014	−0.017	−0.021	−0.019
	(0.010)	(0.008)	(0.008)	(0.009)	(0.009)
B_{tt}	−0.004	−0.014	0.001	0.001	0.003
	(0.002)	(0.006)	(0.004)	(0.004)	(0.001)
B_{vv}	0.004	0.000	0.001	0.004	0.003
	(0.003)	(0.001)	(0.001)	(0.002)	(0.003)
B_{ut}	0.000	−0.003	0.002	0.001	0.007
	(0.002)	(0.002)	(0.003)	(0.003)	(0.003)
B_{st}	−0.005	−0.003	−0.008	0.013	0.015
	(0.006)	(0.004)	(0.003)	(0.004)	(0.005)
B_{kt}	0.005	0.006	0.006	−0.004	−0.006
	(0.006)	(0.004)	(0.004)	(0.005)	(0.006)
B_{uv}	0.003	0.002	0.000	0.003	0.000
	(0.001)	(0.001)	(0.001)	(0.002)	(0.001)
B_{sv}	0.010	0.001	0.000	−0.002	0.001
	(0.005)	(0.002)	(0.001)	(0.002)	(0.003)
B_{kv}	−0.013	−0.003	0.000	−0.001	−0.006
	(0.010)	(0.002)	(0.001)	(0.002)	(0.002)
B_{vt}	0.002	0.002	−0.002	−0.006	0.001
	(0.002)	(0.002)	(0.002)	(0.003)	(0.005)

Note: Standard errors in parentheses.
The variable v is estimated as the average age of capital stock; t is estimated as the average experience of the labor force.

It is possible to respecify the translog production function with two time arguments, one reflecting the average age of the capital stock and the other reflecting employee experience. Parameter estimates of these production functions for each of the five industries are contained in table 10-8. The parameter estimates of the vintage variable conform to expected signs ($a_v <$ 0; $B_{vv} > 0$), and there is a marked improvement in the fit of the estimated production functions to the data.[7]

Hypothesis tests concerning the experience–productivity relationship are

7. The vintage variable should carry signs opposite to those expected for productivity growth with time. Thus the rate of change should be negative (productivity falls with age) and accelerating (the rate of fall increases over time).

Table 10-9. *Test Statistics for Experience of Labor Force and Vintage in Five Indian Industries*

Statistic	*Machine tools*	*Printing*	*Shoes*	*Soap*	*Metal casting*
Experience of labor force					
$a_t = 0$	3.530*	0.662	1.878	2.910*	3.357*
$B_{tt} = 0$	−2.027*	−2.155*	.364	0.324	−2.655*
$B_{ut} = B_{st} = B_{kt} = 0$	2.442	4.408	7.376	11.240	12.837
$a_t = B_{tt} = 0$	10.520	4.618	1.532		
$B_{ut} = B_{st} = B_{kt} = 0$					
Vintage					
$a_v = 0$	−2.919*	−1.697	−2.051*	−2.509*	−2.117*
$B_{vv} = 0$	1.338	0.617	1.400	2.309*	1.357
$B_{uv} = B_{sv} = B_{kv} = 0$	5.420	6.200	1.302	9.138	6.287
$a_v = B_{uu} = 0$	6.334	2.868	4.506	4.414	3.357
$a_v = B_{uu}$					
$B_{vv} = B_{sv} = B_{kv}$					

	Hypothesis			
	$t_0 = 0$ *(1)*	$h_{00} = 0$ *(2)*	$hs_0 = h_{u0} = h_{k0} = 0$ *(3)*	$t_0 = h_{00} = 0$ $(hs_0 = h_{u0} = h_{k0} = 0)$ *(3a)*
Critical values of test statistic[a]				
0.10	1.645	1.645	4.60	7.78
0.05	1.960	1.960	6.00	9.49
0.01	2.576	2.576	9.22	13.28

Empty cells indicate that no test was necessary.

*Significant at 0.05 level.

a. Test statistic for hypotheses 1 and 2 is the statistic with *df* equal to infinity. Test statistic for hypotheses 3 and 4 is the X^2 statistic with *df* equal to number of independent restrictions.

reported in table 10-9. The vintage productivity relationship is kept as a maintained hypothesis throughout. The overall level of significance is set at 0.10, and the individual tests are assigned a level of significance of 0.025. The results of the tests are summarized in table 10-10.

Four tests of the experience–productivity relationship were conducted. The first was for the presence of positive productivity change with increased experience. The second was for acceleration or deceleration of the rate of productivity change over time. The third was for the presence of biased productivity change with accumulated experience, and the fourth was for complete absence of productivity change associated with accumulated experience. These tests are performed sequentially by adding restrictions to the translog function. By setting the overall level of significance at 10 percent, we have accepted a 10 percent probability of falsely accepting the set of four maintained hypotheses.

Respecification of the translog production function with the vintage variable included results in a marked change in the estimated relationship

Table 10-10. *Summary of Test Results on Experience and Productivity in Five Indian Industries*

Sample industry	*Rate of productivity change with increased experience*	*Acceleration or deceleration of productivity change*	*Hicks neutrality of productivity change*	*Absence of all productivity change*
Machine tools	Positive	Negative	Accepted	Rejected
Printing	Not significant	Negative	Accepted	Accepted
Shoes	Not significant	Not significant	Accepted	Accepted
Soap	Positive	Not significant	Rejected	n.a.
Metal casting	Positive	Negative	Rejected	n.a.

n.a. Not available.

between TFP and accumulated experience. In three industries, machine tools, metal casting, and soap, the rate of productivity change is positive and significant at the point of approximation of the production function. In the other two industries, the rate of productivity change is positive but not significant. In all five industries, the rate of productivity change associated with an incremental year of average employee experience is equal to approximately 5 percent. Three industries, machine tools, metal casting, and printing, exhibit significant deceleration of the rate of productivity change, which indicates diminishing returns to experience. This result is of particular interest in printing (for which the rate of productivity change at the geometric mean of the sample is not significantly different from zero), since it appears to suggest that additional accumulated experience beyond the average level for this industry will have a deleterious effect on TFP.

Tests for the presence of biased productivity change indicate that only two industries, soap manufacturing and metal casting, show significant bias in TFP change associated with accumulated experience. The bias of productivity change is toward labor using and capital saving in both industries. This is consistent with the view that employee experience augments the productivity of labor relative to capital. The implication of a labor-using bias is that firms can increase their rate of productivity change by employing more skilled and unskilled labor relative to capital for a given level of firm-specific experience.

In four of the five industries, productivity deteriorations associated with an increasing average age of the capital stock are significant and fall in the range of 1.5 to 3.0 percent per year (see tables 10-8 and 10-9). These orders of magnitude are consistent with observed rates of technological change in manufacturing. The patterns of bias accompanying the vintage variable are also of some interest, because they further support the introduction of the average age of the capital stock into the regression as a control for vintage effects. The bias terms in all five industries indicate that as the average age of the plant increases the share of capital in total output decreases. This is equivalent to the presence of capital-using technical change and supports the view that the vintage variable captures the effect of embodied technical progress on productivity differentials.

When both vintage effects and the level of experience are included in the production function, the roles of time variables emerge clearly. Levels of technology differ among firms in the sample in four industries—machine tools, metal casting, shoes, and soap—and each additional year added to the age of the firm's capital stock decreases productivity by approximately 2 to 3 percent. Controlling for the age of the capital stock, additional experience in the labor force results in significantly increased productivity in three industries—machine tools, metal casting, and soap. Experience has little significant effect, however, on the patterns of factor utilization of firms in the sample. The share elasticities in most regressions are not significantly different from zero; with the exception of soap manufacturing and metal casting, it is not possible to conclude that factor proportions differ among firms as a result of different levels of employee experience.

The distribution (mean and standard deviation) of average employee experience and vintage of capital stock by size category of firm is given in table 10-11. In general, there is no significant relationship between firm size and either of the time-dimensioned variables. Exceptions occur in soap manufacturing, in which large-scale firms (employing more than 100 workers) have significantly higher average labor force experience and are significantly older than all other size categories, and in shoemaking, in which the vintage of firms with 10 to 25 workers is significantly less than that for both the largest and smallest size classes. Thus, although there is a significant relationship between vintage, experience, and productivity in the production function, it does not result in systematic differences in TFP between firms of various sizes, since in our data sets there is little systematic variation in either vintage or employee experience with firm size.

Summary

This chapter has presented the results of our effort to use data from the sample surveys to describe patterns of factor intensity and factor substitution in five industries. The general framework of analysis was the economic theory of production, and our principal tool was a three-factor translog production function, which expressed value added as a function of capital and two skill categories of labor.

We began our inquiry into patterns of factor use with a simple ordering of capital-labor ratios and skill ratios by firm size, as measured by employment. In general we found no systematic relationship between employment size and capital intensity. Variations in factor proportions within employment size groups were sufficiently large to make differences between the means statistically insignificant.

Table 10-11. *Distribution of Experience and Vintage by Size of Firm in Four Indian Industries*
(in years)

Firm size	*Soap*		*Shoes*		*Printing*		*Machine tools*	
(N = thousands of employees)	*Vintage*	*Average experience*	*Vintage*	*Average experience*	*Vintage*	*Average experience*	*Vintage*	*Average experience*
N ≤ 5	15.667	5.381	16.625	4.064	14.667	6.262	n.a.	n.a.
	(8.588)	(4.638)	(9.921)	(4.912)	(9.459)	(3.9)	n.a.	n.a.
5 < N ≤ 10	18.15	4.662	9.00	4.374	15.731	6.098	19.25	10.923
	(7.293)	(3.366)	(10.243)	(6.956)	(9.665)	(4.503)	(26.285)	(10.985)
10 < N ≤ 25	17.462	4.184	12.121	3.522	11.947	4.84	11.036	7.653
	(7.720)	(2.337)	(14.289)	(3.247)	(8.462)	(3.27)	(7.505)	(3.972)
25 < N ≤ 50	20.00	4.361	10.70	3.00	17.90	6.984	10.792	6.033
	(1.414)	(0.771)	(6.738)	(1.861)	(16.024)	(4.380)	(6.724)	(2.725)
50 < N ≤ 100	15.00	5.552	n.a.	n.a.	26.00	4.205	12.444	6.418
	(8.099)	(2.592)	n.a.	n.a.	(26.87)	(2.34)	(7.213)	(2.077)
100 < N	31.143	15.326	18.571	6.29	20.00	6.617	9.462	6.479
	(5.872)	(8.60)	(7.723)	(2.914)	(7.55)	(1.747)	(7.444)	(3.856)

n.a. Not available.
Note: Standard deviations in parentheses.

When size is measured by capital stock, rather than by the number of employees, somewhat greater regularity appears in the pattern of capital intensity, but not much. Very small firms, those with less than Rs20,000 in fixed assets, are for the most part less capital-intensive than larger firms, but within the wide range of fixed assets from Rs20,000 to Rs1 million there is substantial variation. In two of the five industries capital intensity first rises and then falls with firm size. The often sought monotonic relationship between firm size and factor intensity appears to emerge only in metal casting and shoes. In all cases, however, within-group variations are large and differences between means are frequently insignificant.

The translog production function allowed us to analyze patterns of factor substitution and other properties of the technology in three detailed ways. First, we were able to check the individual industry production functions for the characteristics of homotheticity and constant returns to scale. (Homotheticity is the production function property which means that at a given set of factor prices the optimal combination of inputs is the same at all levels of output.) Thus, in a homothetic production technology, differences in factor proportions among firms of different size should be due to differences in factor prices or to producers' errors, rather than to the nature of the technology. Two of our industries were not homothetic: shoemaking and printing. The remaining three, on the basis of our data, exhibit homothetic technologies.

The three industries with homothetic production functions—machine tools, soap, and metal casting—are also characterized by constant returns to scale. In these three industries, the whole production technology can be summarized by a single unit isoquant giving the quantities of inputs required to produce a single unit of value added. In the remaining two industries, shoes and printing, there are variable returns to scale. In these industries unit isoquants at various levels of output vary in terms of their position in input space both as a result of nonhomotheticity and as a result of changing returns to scale. There is no evidence of strongly increasing returns to scale, however, in any of the industries analyzed.

Second, we found that our estimates of factor substitution possibilities are similarly unrelated to size variations in the industries surveyed. Capital and skilled and unskilled labor were all substitutes, but capital was in general more substitutable for skilled than unskilled labor. This was consistent with our observation of capital-skills substitution at the enterprise level in many firms surveyed. These substitution possibilities do not, however, vary significantly with firm size.

Third, we attempted to analyze the extent to which productivity change in each industry might be responsible for variations in factor proportions. In order to do this we estimated an expanded translog production function that included two chronological measures, capital vintage and experience of the labor force. In four of the five industries we found that the vintage of the capital stock was associated with the level of total factor productivity; each year of additional age of the capital stock resulted in a loss of from 2 to 3 percent in TFP. Time had little significant effect on the pattern of factor utilization. Most of the estimated coefficients reflecting bias in productivity change were not significantly different from zero. Where it was significant, the bias was toward capital use. Because the age distribution of firms in the sample was not correlated with firm size, however, differences in factor proportions did not arise among size groups due to differing vintages.

11

Technical Efficiency and Productivity: Survey Results

The preceding chapter explored the relationship between firm size, factor intensity, and patterns of factor substitution in the five industries represented in our microeconomic data sets. This chapter examines the relationship between firm size and technical efficiency in production by using a three-factor frontier translog production function to measure technical inefficiency in our sample of five industries. The method of estimation permits the recovery of a firm-specific index of technical inefficiency. An investigation into the sources of technical efficiency differentials among firms in the sample reveals that firm size is positively associated with TFP in only one of the five industries, machine tool manufacturing. Several other variables, however, including the age of the enterprise, the vintage of the capital stock, the level of labor force experience, and the level of managerial education and training contribute to variations in productive efficiency in one or more industries.

Partial Factor Productivities

Before examining the technical efficiency of firms in our samples, we review in this section evidence from the surveys concerning variations in labor and capital productivity with firm size. Table 11-1 presents the mean and standard deviation of output-labor and output-capital ratios by employment size of firm for the five industries in our samples.

There is relatively little systematic variation in partial factor productivities with firm size. In soap manufacturing and metal casting, no statistically significant difference exists in the mean levels of labor or capital productivity between any pair of size categories. In soap manufacturing, there is no discernible trend in either labor or capital productivity with firm size; this lack largely reflects the limited variation in techniques. In metal casting, labor productivity rises with employment size, and the size group 50 to 100

Table 11-1. *Labor and Capital Productivity by Employment Size Class in Five Indian Industries*
(thousands of rupees)

Employment size category	*Shoes*		*Printing*		*Soap*		*Machine tools*		*Metal casting*	
	Y/L	Y/K	Y/L	Y/K	Y/L	Y/K	Y/L	Y/K	Y/L	Y/K
N < 1	5.23	14.28	8.92	4.86	10.80	2.47	—	—	3.20	10.67
	(3.27)	(13.57)	(5.28)	(1.11)	(1.84)	—	—	—	—	
5 < N < 10	5.61	29.87	12.15	3.51	9.33	2.20	4.16	1.11	13.91	4.97
	(4.28)	(33.72)	(6.47)	(2.47)	(2.83)	(1.22)	(1.61)	(0.17)	(10.90)	(3.98)
10 < N < 25	5.67	16.20	14.18	3.26	9.78	2.04	6.73	2.66	12.32	5.77
	(4.31)	(11.27)	(5.15)	(2.57)	(4.02)	(1.12)	(3.76)	(1.76)	(10.52)	(3.56)
20 < N < 50	6.30	13.13	14.56	3.51	11.19	1.98	7.84	3.28	14.44	7.63
	(3.29)	(10.20)	(7.53)	(3.13)	(1.19)	(0.11)	(2.43)	(1.78)	(16.66)	(7.60)
50 < N < 100	—	—	11.59	1.24	10.92	2.93	7.44	2.57	19.40	29.96
	—	—	(3.07)	(0.72)	(3.19)	(1.07)	(2.94)	(1.07)	(9.94)	(39.03)
100 < N	15.32	4.93	13.81	1.21	11.10	2.74	12.31	2.13	22.47	8.59
	(3.41)	(4.22)	(7.36)	(0.50)	(2.72)	(0.46)	(7.95)	(0.90)	(16.57)	(6.86)

— Not available.

Note: N = number of workers. Y is value added, K is capital, and L is labor. Standard deviations in parentheses.

workers in metal casting appears to have both higher average labor and capital productivity than all others. Within-group variation for each size category in both industries is large relative to differences between the mean output-labor and output-capital ratios of the size groupings.

In the printing industry, capital productivity falls sharply in the two highest size categories, and the mean values of the output-capital ratio are significantly below those for the smallest size group at the 1 percent level. There is, however, no accompanying rise in labor productivity, which suggests that firms in the largest size classes may be inefficient relative to their smaller-scale counterparts.

In shoemaking, there is a sharp break between the output-capital and output-labor ratios of firms employing up to 50 workers and those with more than 100 workers. Labor productivity rises and capital productivity falls significantly (at the 5 percent level) in the largest size class, compared with all others, which is the counterpart of the shift in factor intensities for the same size group reported in chapter 10.

Machine tool manufacturing is the only industry in our sample for which there is a statistically significant difference between the partial factor productivities of firms with 10 or fewer workers and those of large-scale enterprises. Both labor and capital productivities of firms with fewer than 10 workers are significantly (at the 5 percent level) below those for larger firm sizes. There is a tendency for both labor and capital productivity to rise with size up to the level of 25–50 workers and then to fall, but the differences between means in the higher size ranges are not in general statistically significant.

Table 11-2 contains output-labor and output-capital ratios ranked by investment size. As with the results for employment size groups, there is substantial variability in the partial factor productivities, despite the somewhat greater regularity of the pattern of capital intensity discussed in chapter 10.

Shoemaking is the only industry in which there is a regularly rising pattern of labor productivity with investment size and a regularly falling pattern of capital productivity. In all the other industries, there is some variation in both output-labor and output-capital ratios. In soap manufacturing, capital productivity first falls and then rises with firm size, while labor productivity remains virtually constant. In printing, capital productivity falls continuously with firm size, although within-group variation is large and differences between the means are not always significant. Labor productivity, however, rises up to Rs100,000, falls, and then rises again before finally declining above Rs1 million. A sharp increase in capital intensity above Rs1 million (see table 11-2) in machine tool manufacturing is reflected in a rise in labor productivity, but without a correspondingly marked decrease in capital productivity, which is not statistically different across the entire range of investment sizes. In the metal casting industry there is a significant rise in labor productivity above Rs20,000 in fixed capital but no corresponding reduction

Table 11-2. *Labor and Capital Productivity by Value of Fixed Assets in Five Indian Industries*
(thousands of rupees)

Size class	*Shoemaking*		*Printing*		*Soap manufacturing*		*Machine tool manufacturing*		*Metal casting*	
fixed assets	Y/L	Y/K	Y/L	Y/K	Y/L	Y/K	Y/L	Y/K	Y/L	Y/K
K < 1	3.77	42.30	—	—	—	—	—	—	—	—
	(2.15)	(36.94)	—	—	—	—	—	—	—	—
1 < K < 5	4.76	18.41	6.08	4.81	—	—	—	—	3.20	10.67
	(2.66)	(11.15)	(3.55)	(1.49)	—	—	—	—	(0.00)	(0.00)
5 < K < 10	6.13	9.72	8.38	5.40	9.86	4.20	—	—	—	—
	(4.11)	(1.79)	(3.09)	(1.16)	(2.46)	(1.24)	—	—	—	—
10 < K < 20	7.37	9.42	10.83	5.77	9.11	3.82	—	—	5.84	6.84
	(5.05)	(3.80)	(7.86)	(2.70)	(2.55)	(1.39)	—	—	(1.84)	(3.91)
20 < K < 50	10.84	8.64	13.02	3.98	8.39	2.35	6.25	3.17	12.74	10.72
	(7.26)	(2.77)	(5.31)	(1.91)	(2.43)	(1.19)	(3.09)	(2.00)	(8.03)	(19.46)
50 < K < 100	10.88	6.26	14.40	3.52	10.06	1.41	7.38	3.25	23.06	8.91
	(1.62)	(0.84)	(4.63)	(2.81)	(4.34)	(0.56)	(2.97)	(1.50)	(20.45)	(8.22)
100 < K < 200	—	—	13.00	1.44	12.08	1.61	6.45	2.08	12.12	6.45
	—	—	(4.91)	(0.54)	(4.18)	(0.42)	(1.56)	(1.09)	(8.35)	(5.91)
200 < K < 500	—	—	17.24	1.57	10.79	2.75	8.41	1.79	27.82	8.44
	—	—	(8.00)	(1.17)	(2.48)	(0.90)	(3.48)	(0.63)	(21.20)	(10.08)
500 < K < 1,000	14.45	3.35	15.13	1.42	13.03	2.41	10.63	2.22	21.26	7.83
	(5.71)	(1.72)	(1.94)	(0.45)	(0.00)	(0.00)	(4.69)	(0.99)	(8.31)	(3.33)
1,000 < K	15.98	6.11	12.71	1.09	—	—	15.61	1.80	25.96	2.72
	(0.48)	(5.41)	(6.40)	(0.47)	—	—	(9.08)	(1.11)	(7.13)	(0.58)

— Not available.

Note: K is capital, Y is value added, and L is labor. Standard deviations in parentheses.

in capital productivity. Within-group variations in each cell are large, and the apparently large pairwise differences in labor productivity above Rs200,000 are not statistically significant. Thus, the evidence from the firm-level surveys is quite similar to that presented from industrial census data in chapter 7. It is difficult to detect any systematic variation in labor or capital productivity with firm size.

The Measurement and Interpretation of Technical Inefficiency

We have selected as our measure of total factor productivity the index of technical efficiency discussed in chapter 9. Conceptually the measurement of technical efficiency is relatively simple. A frontier or best practice production function is used to predict the maximum output which could be obtained from a set of production inputs, based on the best practice actually observed in the sample. The difference between this predicted output and the actual output of the firm is attributed to technical inefficiency in production. The extent of technical efficiency is normally expressed as a Farrell index of the ratio of actual to predicted output. Firms performing best practice have Farrell indexes of one; those which are technically inefficient have indexes of less than one.

The Farrell indexes provide two measures of total factor productivity. The first is the extent to which actual TFP is below potential TFP for any enterprise. This is the economic loss due to inefficiency. The second measure is the extent to which TFP differs among individual firms. Because the Farrell indexes as we have estimated them are transitive, they can be used to compare the TFP levels of any pair of firms; the firm with the higher Farrell index has the greater relative TFP. Thus, comparison of Farrell indexes of technical efficiency can provide a picture of relative as well as absolute levels of total factor productivity.

The production function employed in the frontier estimates is the three-factor translog production function described in chapters 9 and 10. All deviations from the production function are constrained to be of negative sign, so that the parameters of the production function provide an estimate of maximum potential output for each observed vector of input levels. The estimation technique employed—which is an application of linear programming—minimizes the sum of the deviations of actual output from predicted output.[1] Each production function has the properties of monotonicity (positive marginal products) and concavity imposed.

A firm's actual output $X(s)$, given observed input levels, is equal to predicted output $\hat{X}(s)$ only if the firm operates on the production frontier. Each firm is assigned an efficiency index (Farrell index) equal to the ratio of its

1. A fuller description of the estimation technique is given in appendix 2.

actual to predicted output, $X(s)/\hat{X}(s) \leq 1$, which may be found as the antilog of the slack variable in the programming problem. The value of the index provides a measure of relative technical efficiency.

Interpretations of observed technical efficiency differentials among firms stress two major sources of deviations from best practice productivity: measurement error and the effect of nonmeasured inputs. Economic data, no matter how carefully collected and specified, inevitably omit some relevant inputs into production. In some cases this is due to the failure of measured data to reflect all relevant dimensions of an input, for example, the frequent lack of vintage information pertaining to capital stock measures or the absence of information on hours worked in labor force data. In other cases inputs are omitted because they are difficult or impossible to measure. Inputs of this type include the ability and motivation of managers, organizational and administrative characteristics of the enterprise, and the level of effort of the labor force.

If the principal source of deviations from best practice arises from incomplete specification of measured inputs, the role of the production frontier as an efficiency standard is questionable. The argument is most clearly demonstrated with respect to the role of the vintage of the capital stock. If plants of markedly different vintage coexist within a sample, the capital stock measures available may not reflect the productivity differentials embodied in successive generations of equipment. When technical efficiency is measured relative to a single industry frontier, enterprises employing older technologies will appear inefficient, even if they are achieving maximum output from their given vintage of capital. Similarly, the inability to construct input measures which encompass labor force skills or experience may give rise to variations in output for apparently equal input vectors. Technical efficiency differentials in this context are difficult to interpret. If inputs were more fully specified and variables more completely defined, much of the apparent variation in efficiency levels would presumably disappear.

Even in the most thoroughly specified empirical production function, however, some crucial inputs will remain unmeasured. Typically, these inputs are associated with the management practices of the firm. Entrepreneurs and managers differ in their ability, initiative, technical skill, and knowledge. These characteristics will affect the efficiency with which measured inputs are combined. The production frontier embodies the concept that these attributes make it meaningful to define the maximal output attainable from a given set of relevant inputs and to interpret deviations from the maximum as the consequences of suboptimal use of unmeasured inputs. Measures of technical inefficiency can be used to examine the relationship between such identifiable characteristics of the firm or industry as the level of protection or subsidization, the degree of competition, the background and training of entrepreneurs and managers, and the level of productivity.

Thus, empirical measures of technical efficiency should be interpreted quite broadly as embodying both the consequences of our inability to measure traditional inputs with ideal accuracy and the effect of nonmeasured inputs on the productivity of enterprises. Attempts to explain "sources" of inefficiency, therefore, generally consist of multiple correlation techniques in which enterprise attributes reflecting both characteristics of measured inputs—labor force skills, levels of capacity utilization—and variables thought to affect the quality or quantity of unmeasured inputs—levels of protection, entrepreneurial characteristics—are tested for their relationship to relative levels of technical efficiency.[2]

The effort to interpret sources of inefficiency in this chapter follows that tradition. The literature on small-scale enterprises suggests several reasons why small firms may appear technically inefficient relative to large and medium-scale enterprises in the same industry. First, the characteristics of measured inputs may differ across size classes of firms. Some authors have argued that smaller-scale enterprises tend to employ capital equipment which is older than that of medium- and large-scale firms. If capital asset prices fail to reflect fully the productivity differentials embodied in successive generations of capital equipment, smaller firms may appear inefficient relative to a single cross-sectional frontier defined by firms with newer capital equipment. A similar argument could be made with respect to hours worked by the labor force. It may be the case that family labor which occurs with greater frequency in small firms works longer and at greater intensity than hired labor. If this is so, a production frontier based on labor force data which fail to reflect hours worked or intensity of effort may indicate that smaller firms are relatively closer to best practice than large and medium-scale enterprises.

Second, if the production function is not characterized by constant returns to scale, imposing constant returns in order to ensure concavity of the estimated function will result in scale economies being reflected in the measured indexes of technical efficiency. Increasing returns to scale would imply that in the absence of other offsetting influences of size on efficiency, larger-scale firms will appear more efficient relative to the linear homogeneous frontier production function.

Third, small and large firms may differ systematically in the quantities and attributes of nonmeasured inputs. Varying administrative and organizational structures among firms may lead to alternative decisionmaking rules. The techniques of production chosen and the levels of productivity at which they are operated need not correspond to the principle of profit maximization. If small and large firms differ systematically in their decisionmaking criteria, these differences may be reflected in systematic differences in productivity levels.

2. This is the approach adopted in, for example, Shapiro and Muller (1977); Page (1980); Tyler (1979); Pitt and Lee (1981); and Martin and Page (1983).

Fourth, firms may also differ in their levels of knowledge of productive techniques. If large firms have more complete knowledge of the process of production, their scope for organizational and technical changes which enhance the productivity of existing plant may be greater. Small firms, conversely, may benefit from the greater opportunity for direct communication between employees and management and from corresponding improvement in worker motivation.

Empirical measures of technical efficiency capture the net effect of all the above factors simultaneously. They are not mutually exclusive, and distinguishing among them can be difficult. The strategy adopted in the section below on the sources of technical inefficiency is to control as far as possible for variations in characteristics of measured inputs among firms and then to test whether size has an impact on relative efficiency independent of variations in other measured enterprise attributes. Implicit in this strategy is the hypothesis that firms differ in a systematic way according to size in the quality or quantity of nonmeasured inputs.

Estimation of the Frontier

The estimated parameters of the frontier translog production function for each of the five industries in the sample appear in table 11-3. Three factor inputs are employed in each set of parameter estimates: capital services, skilled labor, and unskilled labor. Input indexes are expressed as ratios to the geometric mean of the sample in each industry. The estimated production frontiers have monotonicity, linear homogeneity, and concavity imposed.

The Allen partial elasticities of substitution between factor inputs along the production frontier at the sample mean are presented in table 11-4. Comparison of these estimates with those presented in table 10-7 provides some insight into the relationship between best practice and average practice

Table 11-3. *Parameters of Frontier Translog Production Function for Five Indian Industries*

Parameter	*Shoemaking*	*Printing*	*Soap manufacturing*	*Machine tool manufacturing*	*Metal casting*
a_o	5.291	6.197	5.307	6.468	6.268
a_k	0.391	0.121	0.394	0.300	0.330
a_s	0.387	0.632	0.207	0.650	0.520
a_u	0.221	0.247	0.399	0.050	0.150
B_{kk}	−0.115	−0.045	−0.252	0.000	−0.103
B_{ss}	−0.088	−0.064	0.000	0.000	−0.151
B_{uu}	−0.080	0.000	0.000	0.000	−0.078
B_{ks}	0.062	0.054	0.126	0.000	0.088
B_{ku}	0.053	−0.009	0.126	0.000	0.015
B_{us}	0.026	0.009	−0.126	0.000	0.063

Table 11-4. *Partial Elasticities of Substitution at the Production Frontier for Five Indian Industries*

Industry	*Skilled to unskilled labor*	*Skilled labor to capital*	*Unskilled labor to capital*
Shoes	0.76	0.71	0.61
Printing	0.96	0.52	1.21
Soap	2.75	0.08	0.73
Machine tools	1.00	1.00	1.00
Metal casting	1.86	0.97	0.60

technology in each of the five industries. In general the elasticities of substitution at the frontier are much lower than those for the average production function. This is consistent with the view that the best practice frontier represents a subset of the available technologies with less scope for factor substitution than is embodied by the average practice for the industry. Each pair of factor inputs is a substitute by the Allen definition, and the overall structure of factor substitution possibilities is not strikingly different from that represented by the average production function. The most notable differences in substitution possibilities occur in printing and soap manufacturing, for which the frontier estimates yield a higher Allen elasticity between unskilled labor and capital than between skilled labor and capital. Although the remaining three industries all have relatively higher Allen elasticities of skilled labor-capital substitution, the differences between the values of skilled labor-capital and unskilled labor-capital elasticities are smaller at the frontier than for the average technology. This implies that relative to average practice, frontier firms in the two industries use techniques of production which embody greater potential for the substitution of capital for unskilled labor.

Characteristics of Frontier Firms

The production frontier is defined by from three to five observations in each industry. Table 11-5 provides some descriptive statistics of these best practice firms and compares them with the average for all firms in each sample. The capital intensity of best practice firms is in general quite close to that of the average for the industry. In shoes, soap, and printing, the firms on the frontier are slightly less capital-intensive than the industry average; in metal casting, best practice firms are slightly more capital-intensive. The only striking difference in factor intensity occurs in machine tools, in which frontier firms are more than twice as capital-intensive as the average.

In four of the five industries (the exception is machine tools) best practice firms are more skill-intensive and smaller than the average firm in the sample. Differences are particularly striking in printing, in which firms at the

Table 11-5. *Characteristics of Best Practice Firms for Five Indian Industries*

	Mean value of best practice firms					
Industry	*Capital-labor ratio*	*Skill ratio*	*Number of employees*	*Labor force experience (years)*	*Entrepreneur's experience (years)*	*Age of enterprise (years)*
Shoes	0.8	2.5	14.7	3.9	13.7	5.6
	(0.9)	(1.9)	(69.3)	(4.0)	(20.8)	(12.4)
Printing	5.1	821.1	23.2	7.8	21.4	15.6
	(6.8)	(97.9)	(23.5)	(5.9)	(16.8)	(15.4)
Soap	4.2	1.1	7.3	7.4	10.3	13.7
	(6.3)	(0.8)	(56.4)	(6.1)	(17.4)	(18.9)
Machine tools	10.1	2.9	224.3	15.3	16.5	16.0
	(4.7)	(5.9)	(64.6)	(9.3)	(23.9)	(16.4)
Metal casting	2.9	6.7	49.0	11.2	—	23.5
	(2.7)	(3.1)	(112.2)	(8.8)	—	(22.8)

— Not available.

Note: Capital-labor ratio in thousands of rupees per worker. Skill ratio is ratio of skilled to unskilled workers. Entrepreneur's experience is for principal owner or manager. Sample means in parentheses.

production frontier are nearly ten times more skill-intensive than the average firm, and in soap, in which the best practice firm is only about one-eighth the size of the average firm. In machine tools best practice firms are on average approximately four times larger than the average firm.

In four of the five cases (the exception is shoes) the labor force in best practice firms has more average experience than that in average firms. This result is consistent with the results in the section on firm size, experience, and factor use of chapter 10, which indicated that firm-specific experience tends to increase levels of TFP. The evidence is more mixed with respect to the experience of the entrepreneur and the age of the enterprise. In printing, entrepreneurs of best practice firms have more experience than their counterparts in the average firm, while the opposite result holds for shoes, soap, and machine tools. In shoes and soap the firms on the frontier are substantially newer than the average firm in the sample, while in the other industries the average age of the enterprise is quite close to that of best practice firms.

Sources of Technical Inefficiency

The frontier production functions permit the recovery of a firm-specific index of technical efficiency at each data point. This Farrell index represents the percentage of predicted best practice output actually achieved by the firm. Under constant returns to scale, the Farrell index also represents the percentage by which unit factor input coefficients, and hence unit costs, could be reduced at the firm's existing level of output and (as noted in the above section on the measurement and interpretation of technical inefficiency) provides an index of differences in TFP between firms. The mean Farrell index for each industry, and for six size groups of firms within each industry, together with data on each industry's minimum index and variability are presented in table 11-6.

Actual output ranges from a low of 42 percent of predicted output for the average firm in shoes to a high of 69 percent in machine tools. These magnitudes are broadly consistent with other studies of technical efficiency in developing countries and indicate that there is substantial scope for improvements in the TFP of firms in the sample.[3] There is considerable variation in the firm-specific indexes of technical efficiency within each industry. The coefficient of variation of the Farrell index is greatest in shoes (54 percent) and least in machine tools (17 percent). The distribution of Farrell indexes becomes more compact, both in terms of the range and variance of observations, as the average level of efficiency improves.

Both the highest average levels of efficiency and the least variation occur in the two modern industries: printing and machine tool manufacturing. The results suggest that in these activities there is greater diffusion of technical knowledge and greater uniformity of efficiency among firms in the sample. There is also less variation in vintages of capital stock and in labor force characteristics than in the more traditional activities of shoemaking and soap manufacturing. It is also possible that establishments in the two more traditional industries show greater quality variations in output and cater to more localized commodity markets. Similarly, the traditional activities may exist within more isolated factor markets and take advantage of the limited mobility of local factors. Any or all of these factors would allow greater variation in measured levels of technical efficiency among firms in the more traditional activities and would allow less efficient firms to remain viable.

The evidence in table 11-6 regarding the relationship between firm size and relative levels of technical efficiency is mixed. In shoes very-small-scale enterprises—those employing less than 5 workers—have the lowest average level of technical efficiency, while those employing more than 100 workers have the highest average Farrell index. In soap small firms have lower levels of technical efficiency than firms with more than 100 workers but are similar to those in the size range of 6 to 100 workers.[4] In printing, very-small-scale firms have the highest mean Farrell index, while large-scale firms have the lowest. Machine tool manufacturing is the only industry in which there is a systematic increase for all size groupings in relative efficiency. The Farrell

3. Compare the results reported in Meller (1976); Page (1980); Tyler (1979); and Pitt and Lee (1981).

4. Using similar methods, Goldar (1982) finds a significant improvement in technical efficiency with firm size within the small size range. We found similar results for the size range < 5 workers but not the higher employment categories.

Table 11-6. *Farrell Indexes for Five Indian Industries by Employment Size of Firm*

Industry	*Measure*	*Overall*	$N \leq 5$	$5 < N \leq 10$	$10 < N \leq 25$	$25 < N \leq 50$	$50 < N \leq 100$	$N > 100$
Shoes	Mean	0.424	0.360	0.452	0.415	0.489	—	0.493
	Std. dev.	(0.228)	(0.215)	(0.267)	(0.200)	(0.274)	—	(0.167)
	Min.	0.069	0.069	0.104	0.137	0.175	—	0.346
	Obs.	99	24	25	33	10	—	7
Printing	Mean	0.645	0.797	0.569*	0.689	0.700	0.591	0.565
	Std. dev.	(0.179)	(0.101)	(0.151)	(0.178)	(0.214)	(0.229)	(0.157)
	Min.	0.373	0.629	0.373	0.414	0.405	0.429	0.444
	Obs.	66	6	26	19	10	2	3
Soap	Mean	0.579	0.578	0.548	0.602	0.647	0.560	0.688
	Std. dev.	(0.185)	(0.213)	(0.206)	(1.75)	(0.046)	(0.155)	(0.032)
	Min.	0.313	0.320	0.313	0.346	0.615	0.388	0.665
	Obs.	48	9	19	13	2	3	2
Machine tools	Mean	0.688	—	0.547	0.672*	0.638*	0.691*	0.773*
	Std. dev.	(0.119)	—	(0.036)	(0.129)	(0.077)	(0.125)	(0.137)
	Min.	0.500	—	0.516	0.500	0.553	0.563	0.610
	Obs.	78	—	4	28	24	9	13

— Not available.

Note: Std. dev. = standard deviation; min. = minimum; obs. = observations.

index rises continuously with firm size from 0.55 for firms with 10 or fewer employees to 0.77 for large-scale firms employing more than 100 workers.

Pairwise analysis of variance was applied between the smallest size class and all other size classes to test for the statistical significance of differences between the mean Farrell indexes of each size class within industries. The null hypothesis is that there is no significant difference in the mean levels of technical efficiency between pairs.

In three of the four industries, there is little statistically significant variation of efficiency with firm size. In shoes and soap it is not possible to reject the null hypothesis (at the 0.10 level) for any pair of size groups. In printing there is a statistically significant difference between the mean Farrell index for the smallest size category and that for the size class of 6 to 10 workers. Machine tools is the only industry in which there is a significant difference in the mean indexes between the smallest size category and all succeeding size classes. Thus, there is very little evidence to support the view that there is a systematic bivariate relationship between firm size and relative technical efficiency.

The absence of a bivariate relationship between firm size and relative efficiency is not wholly surprising. Size may act as a proxy for a number of attributes of the enterprise which may be offsetting in their effect on technical efficiency. For example, economies of scale based on the engineering characteristics of the production function may be offset by different organizational characteristics of large and small enterprises; the impact on productivity of differing vintages of plant may be offset by differing levels of labor force experience; or differences between small and large firms due to variations in managerial endowments or behavior may be masked by other sources of measured inefficiency.

The data set provided substantial information on a number of enterprise characteristics which might be related to the level of technical efficiency. These include measures of entrepreneurial education and experience, data on the age of the enterprise and vintage of the capital stock, information on labor turnover and employee experience, and a number of indicators of capacity utilization. A more complete test of the relationship between firm size and technical efficiency is therefore possible. The firm-specific Farrell indexes were regressed on a number of enterprise characteristics drawn from the record of the sample survey. The dependent variable in each regression is the logarithm of the Farrell index for the enterprise. The most successful of these experiments are reported in table 11-7.

Four regressions are presented for each industry. Regressions 1 and 2 provide information on the sources of inefficiency in each industry, exclusive of firm size. Having thus controlled as far as possible for the effects of other measurable contributions to the technical efficiency of firms, the final two regressions represent tests for the effect of firm size on relative technical efficiency. A dummy variable taking on the value of one for firms with 10 or

Table 11-7. *Sources of Inefficiency in Four Indian Industries*

	Leather shoes				*Printing*				*Soap*	
Source	*(1)*	*(2)*	*(3)*	*(4)*	*(1)*	*(2)*	*(3)*	*(4)*	*(1)*	*(2)*
Intercept	−1.001***	−0.968***	−0.924***	−0.099***	−0.635***	−0.538***	−0.512***	−0.463***	−0.906***	−0.782***
	(5.857)	(7.910)	(7.028)	(6.536)	(4.961)	(4.097)	(3.852)	(2.882)	(3.442)	(3.212)
Literacy	−0.310**	−0.305**	−0.277**	−0.265**						
(no = 1)	(2.320)	(2.411)	(2.129)	(2.019)						
Father's occupation										0.113
(same = 1)										(1.563)
Turnover problem	0.110	0.132	0.143	0.129					0.256*	
(no = 1)	(0.978)	(1.170)	(1.260)	(1.140)					(1.871)	
Lab < 10			−0.104				−0.071			
(yes = 1)			(0.923)				(1.104)			
Total employment				0.052				−0.028		
				(1.131)				(0.811)		
Experience of	0.040				0.085	0.078	0.081	0.074	0.177***	0.152**
entrepreneur	(0.661)				(1.440)	(1.505)	(1.559)	(1.418)	(2.771)	(2.449)
Age of firm		−0.082	−0.083	−0.080	−0.088				−0.215***	−0.183**
		(1.488)	(1.508)	(1.469)	(1.658)				(3.053)	(2.619)
Employee experience	0.152***	0.174***	0.171***	0.164***	0.117***	0.076	0.079	0.078	0.182***	0.206***
	(2.936)	(3.239)	(3.179)	(3.015)	(2.066)	(1.492)	(1.545)	(1.520)	(3.369)	(3.926)
Age of plant						−0.095***	−0.099***	−0.091**		
						(2.602)	(2.702)	(2.434)		
Capacity index					−0.144*	−0.203**	−0.173**	−0.215***	−0.165*	−0.185**
(deteriorated = 1)					(1.762)	(2.767)	(2.235)	(2.869)	(1.814)	(2.024)
$\bar{R}^2$	0.147	0.163	0.171	0.174	0.195	0.243	0.258	0.251	0.474	0.462
F	4.061	4.580	3.829	3.930	3.689	4.884	4.165	4.017	7.575	7.204
Observations	99	99	99	99	66	66	66	66	48	48

(Table continues on the following page.)

Table 11-7 *(continued)*

Source	Soap (3)	Soap (4)	Machine tools (1)	Machine tools (2)	Machine tools (3)	Machine tools (4)	Metal casting (1)	Metal casting (2)	Metal casting (3)	Metal casting (4)
Intercept	−0.907	−0.926	−0.344**	−0.287**	−0.269*	−0.446***	−1.153	−1.068	−1.617	−1.813
	(3.150)	(3.487)	(2.396)	(2.034)	(1.943)	(2.952)	(0.532)	(0.522)	(0.574)	(0.610)
Literacy (no = 1)										
Father's occupation (same = 1)										
Turnover problem	0.256*	0.243*								
(no = 1)	(1.836)	(1.759)								
Lab < 10	0.001				−0.160**				−0.272	
(yes = 1)	(0.083)				(2.126)				(0.377)	
Total employment		0.035				0.044**				0.084
		(0.798)				(2.474)				(0.091)
Experience of	0.177**	0.169**	−0.071	−0.074*	−0.073*	−0.053				
entrepreneur	(2.686)	(2.612)	(1.630)	(1.784)	(1.794)	(1.278)				
Age of firm	−0.215***	−0.229***	−0.085*				−0.230	−0.007	−0.001	0.015
	(3.053)	(3.198)	(1.832)				(0.229)	(0.251)	(0.258)	(0.253)
Employee experience	0.182***	0.185***	0.190***	0.160***	0.144***	0.127***	0.146	0.085	0.088	−0.003
	(3.369)	(3.448)	(3.825)	(4.508)	(4.079)	(2.893)	(0.241)	(0.236)	(0.237)	(0.255)
Age of plant				−0.091***	−0.083***	−0.089***				
				(2.867)	(2.659)	(2.893)				
Capacity index	−0.166*	−0.145						−1.618	−0.854	−0.836
(deteriorated = 1)	(1.657)	(1.519)						(0.571)	(0.494)	(0.492)
$\bar{R}^2$	0.474	0.482	0.404	0.251	0.294	0.309	0.027	0.117	0.131	0.139
F	6.063	6.365	6.307	8.254	7.615	8.149	0.506	1.549	1.276	1.370
Observations	48	48	78	78	78	78	39	39	39	39

* Significant at the 0.10 level.
** Significant at the 0.05 level.
*** Significant at the 0.01 level.
Note: Column numbers refer to regression analyses.

fewer workers is introduced into regression 3 with the expectation that its sign will be negative, if the hypothesis that small firms are less technically efficient is correct. In regression 4 firm size, as represented by total employment, is introduced as a continuous variable with the expectation that its sign will be positive.

The basic structure of the estimating regressions 1 and 2 is the same for all industries. In each case, three variables were tested for their contribution to the level of technical efficiency. These were the total experience of the entrepreneur in the industry (whether within the existing firm or in another), the average level of employee experience within the enterprise, and the vintage of the enterprise's capital stock.

Four other variables were found to affect the level of technical efficiency in at least one industry. Coefficient estimates for these variables—the literacy of the entrepreneur, the reported presence of a labor turnover problem, the entrepreneur's father's occupation, and the level of capacity utilization—are reported for those industries in which the variable is relevant.

The regressions are moderately successful in explaining variations in technical efficiency. *R*-squared values range from 0.147 to 0.482, which are comparable with other attempts at regression analysis of efficiency indexes, and the omnibus *F* statistics for all estimated equations are significant at the 1 percent level (see, for example, Tyler 1979; Pitt and Lee 1981; and Martin and Page 1983).

The explanatory variable which is uniformly most successful is the average experience of the labor force. In three industries—shoes, soap, and machine tools, the elasticity of the Farrell index with respect to employee experience is positive and significant at the 1 percent level. In printing, the estimated coefficient is positive but differs significantly from zero in only one of the four regressions.

The results with respect to experience, which support those in chapter 10, are particularly interesting in view of recent developments in the theory of industrial wage structures which stress minimization of the costs of labor turnover as an objective of wage-setting behavior by firms.[5] Despite substantial theoretical attention, relatively little evidence has been put forward to establish the impact of firm-specific, as opposed to general, experience of the labor force on the TFP of the enterprise. The presence of a significant positive elasticity of the level of technical efficiency with respect to experience provides direct evidence of a positive relationship between levels of experience within the enterprise and reductions in unit costs. The elasticity estimates are of roughly the same order of magnitude in all three of the industries in which they are significantly different from zero; a 10 percent increase in the

5. Stiglitz (1974) presents the basic theoretical model. Empirical examinations include Chapman and Tan (1980) and Knight and Sabot (1982).

average experience of the labor force results in approximately a 2 percent increase in the Farrell index (and hence in a 2 percent decrease in unit costs).

The data set did not contain a separate measure of the total experience of the labor force. This makes a test for differences between productivity-enhancing effects of general as opposed to specific human capital impossible, but unless total labor force experience and experience within the enterprise are perfectly correlated, which is unlikely in any of the industries, the results indicate that in three sectors with quite diverse production processes, firms have an incentive to retain employees rather than to hire workers with equivalent levels of experience from outside the firm.

It is not clear why the relationship between firm-specific experience and productivity should be less strong in the printing industry, in which the estimated coefficients are lower and often not significantly different from zero. Qualitative responses to the surveys suggest that skills acquired through employment in the printing industry may be more transferable between firms than in the other sectors surveyed. In addition there is some evidence to suggest that levels of educational attainment are higher among employees in printing than in the remaining industries. Both of these considerations may reduce the importance of firm-specific experience in determining productivity.

Managers in soap and shoes often cited high rates of labor turnover as a major management problem. For this reason a dummy variable, indicating the presence of a perceived labor turnover problem, was introduced into the regressions. Firms with high reported rates of labor turnover were expected to have lower levels of total factor productivity. Coefficient estimates in both industries were of the expected negative sign but were only significant (even at the 10 percent level) in soap manufacturing. The turnover variable is colinear with the level of firm-specific experience, since high rates of turnover reduce the average level of labor force experience in the firm, and this may explain the relatively high standard errors of the coefficient estimates.

The sample surveys provided information on the entrepreneur's (or senior manager's) total experience in his or her current position and in all previous positions within the same industry. Our measure of experience is therefore industry-specific, not firm-specific. Further, it does not distinguish between experience in management and experience in production. Results with respect to this experience variable are mixed. The logarithm of the variable is positive and significant at the 5 percent level in all four regressions in soap manufacturing, and the estimated value of the coefficient is quite similar to that for average employee experience. Soap manufacturing is the industry in the sample with the highest proportion of enterprises in which entrepreneurs engage directly in the process of production rather than primarily in the supervision of the enterprise. For this reason the results suggest that the entrepreneurial experience variable in soap manufacturing largely captures the impact of experience on direct labor inputs by the entrepreneur.

In none of the other surveys does entrepreneurial experience appear to have a significant effect on technical efficiency (at the 5 percent level) although in machine tools the coefficient estimate is of negative sign and is significant (at the 10 percent level) in two of the four regressions. It is not clear why increased entrepreneurial experience should reduce technical efficiency in machine tool production, nor why the relationship between experience and productivity in printing and shoes should be statistically insignificant. One conjecture is that years in related occupations are not an adequate measure of the acquisition of skills related to the effective management of an enterprise. As a check on the use of the total experience variable, the regressions were rerun using the age of the entrepreneur as a proxy for experience; the coefficient estimates of the age variable were similarly insignificant, and in the case of machine tools of negative sign. A possible explanation for the negative relationship between entrepreneurial experience and productivity in machine tools is that older entrepreneurs may be less familiar with new techniques of production and productivity improving innovations. In this case a measure of entrepreneurial experience is a poor indication of human capital; some skills are acquired over time but others are lost or outdated. Indeed, in each industry a similar argument could be made, and apparently only in the technologically most simple process, soap manufacturing, do management skills acquired over time fully offset the negative impact of those lost.

The samples are similarly uninformative regarding the effect of the entrepreneur's educational attainment on TFP. Five educational categories—illiteracy, primary school, secondary school, university, and postgraduate—were constructed for each industry. Entrepreneurs were assigned to the highest category attained; thus, for example, those who had some secondary school education, but did not necessarily possess a secondary school degree, were assigned to the secondary school level. In only one industry did a high proportion of entrepreneurs report themselves to be illiterate—shoes, in which the dummy variable representing illiteracy is negative and significant at the 5 percent level. Firms managed by illiterate entrepreneurs have lower levels of technical efficiency, with controlling for other enterprise characteristics, than those whose managers have achieved functional literacy; but there is no detectable improvement in total factor productivity with higher levels of educational attainment. In the remaining industries the index of technical efficiency rises with secondary school education and then falls at the post-secondary level, but the coefficients are not significantly different from zero. Thus there appears to be a substantial productivity gain associated with the achievement of literacy but no evidence of positive returns for further years of formal schooling.

This finding is of some interest. The literature on small enterprise development suggests that in industries characterized by craftsman-entrepreneurs, individuals with greater general skills in marketing and financial management will have greater success. All five industries in the sample have large numbers of craftsman-entrepreneurs. The level of formal educational attainment, however, does not appear to be a good indicator of general management skills. Once the hurdle of functional literacy has been achieved, additional years of formal schooling do not seem to equip small enterprise managers with the skills needed to increase the productivity of their firms.

A final entrepreneurial attribute which was tested was having a father in the same occupation. The expectation that this opportunity for acquiring skills would have a positive effect on technical efficiency was not supported by the data. The coefficient estimates, although of correct sign, were not significant for any industry; indeed, soap was the only industry in which the coefficient estimate exceeded its standard error.

Two variables reflecting the age of the enterprise are included in the regressions. The first is the age of the firm from the year of its foundation to the year of the survey. The second, which could be computed for only two of the five industries, is the representative age of the capital stock found by taking value weights and summing across the associated ages of individual capital components.

Two possible interpretations of the effect of the age of the enterprise on technical efficiency are possible, and each gives rise to a different predicted sign of the estimated coefficient. If age acts independently of employee experience as a measure of accumulated experience in production, the presence of learning by doing should be reflected in a positive and significant coefficient. Alternatively, if the age variable primarily reflects the impact of plant vintage on productivity, the coefficient should have a negative sign. The vintage variable, conversely, should primarily reflect the technological change–productivity relationship and therefore should carry a consistently negative coefficient. In all five of the industries, the coefficient of the age of the enterprise is negative when included with the measures of entrepreneurial and employee experience, and it is often insignificant. Hence the age variable appears to have little independent effect, and it probably reflects the impact of capital vintage on productivity.

The above relationship is confirmed in the two industries for which a separate estimate of capital vintage is available. In printing and machine tools, the coefficient of the vintage variable is negative and significant at the 1 percent level. Thus in three industries there is a significant negative relationship between age of plant and TFP. The vintage variables control for the effect of technological changes embodied in the capital stock and not reflected in the constant price estimates of capital inputs employed in the frontier production functions.

In printing, metal casting, and soap manufacturing, there was evidence of substantial reduction in capacity utilization during the recent history of many enterprises in the sample. Given the single-period, cross-sectional nature of the data, variations in capacity utilization will be recorded as changes in technical efficiency. To control for the impact of reported declines in capacity utilization on technical efficiency, a dummy variable—which took a value of one when capacity utilization in the survey year was below that reported for previous years—was introduced into the regressions. The coefficient estimates conformed to the expected sign (negative) and were generally significant at the 10 percent level or above.

The two tests for the effect of firm size (as measured by employment) on relative efficiency appear as regressions 3 and 4 for each industry in table 11-7. The size variable, when included with other enterprise attributes, is intended to summarize the effects of omitted variables associated with scale on the index of technical efficiency. The most obvious of these effects arises from the inability of the estimation procedure to allow for variable returns to scale. Increasing returns to scale will be reflected in greater measured technical efficiency, and thus firm size may be associated with improvements in measured technical efficiency, controlling for other attributes. Other variables possibly associated with firm size but for which independent measures do not exist include the level of technical knowledge of the entrepreneur, possible differences in managerial objectives and behavior, product differences, or differences in techniques not associated with the age of the establishment or the vintage of its fixed capital.

The results do not show a strong association between firm size and relative efficiency. In four of the five industries—shoes, metal casting, printing, and soap—neither the dummy variable nor the labor force size variable has a significant relationship with the Farrell index, nor do they significantly add to the explanatory power of the regression. In machine tools both the dummy variable and the labor force size variable show a significant positive relationship between firm size and technical efficiency.

The positive association observed between firm size and technical efficiency in machine tool manufacturing, controlling for other variables, could arise from plant-level economies of scale. The econometric analysis of average production functions in the preceding chapter does not, however, support the hypothesis of technical economies of scale within the range of plants observed in our sample. Thus, we are left to conclude that the superior technical efficiency of larger firms must be due to other organizational or economic attributes for which size acts as a proxy. Because machine tool manufacturing is the most technologically complex of the production processes studied, one possibility is that the superior total factor productivity of large firms is due to better organization of production and more complete mastery of the technical possibilities of the production processes. This explanation accords with our observations of plants in the sample survey and with the surveys of light engineering summarized in chapter 4.

Summary

In this chapter we have examined the relationship between firm size and partial and total factor productivity in the five industries contained in our sample. We were able to discover very little regularity in the patterns of labor or capital productivity when size is measured either by numbers employed or by value of fixed assets. There was, in fact, surprisingly less regularity in the pattern of labor and capital productivities than in capital intensity, which suggested that there were substantial variations in efficiency in production, as measured by total factor productivity.

The wide variation in total factor productivity was confirmed by our analysis of technical efficiency. Within each of the five industries technical inefficiency (measured by the difference between actual and predicted output) was substantial and variations in total factor productivity (indicated by comparison of the Farrell indexes across firms) were also marked. We were able to find relatively little evidence of a systematic bivariate relationship between employment size and technical efficiency, however. The only industry in which technical efficiency was correlated with firm size was machine tool manufacturing. In each of the other four industries there was no relationship between employment size and TFP level.

Estimates of technical inefficiency in cross-sectional data are often correlated to such enterprise characteristics as managerial endowments and experience, labor force stability and experience, variations in the level of technology, and capacity utilization. The surveys contained substantial information concerning a number of these enterprise attributes. Four variables—the average experience of the labor force, the age of the capital stock, the experience of the entrepreneur, and the level of capacity utilization—were identified as significant sources of variations in technical efficiency in one or more industries. With these characteristics controlled for, there remained little relationship between employment size, measured either as a continuous or discrete variable, and total factor productivity.

The only industry in the sample for which the regressions revealed a positive (or negative) association between size and technical efficiency was machine tool manufacturing. We have conjectured that this relationship reveals the effect of superior organization in production and mastery of technology by larger firms.

The absence of a significant relationship between firm size and technical efficiency means that neither a positive nor a negative case can be made for small firms on the grounds of superior or inferior economic efficiency. As we indicated in chapter 4, industrial policy in India insulates large firms from competition among themselves and with imports. It is therefore possible that our results reflect the fact that smaller firms exist in a more heavily competitive environment than large firms. In this case the impact of competition on cost reduction may offset the potential organizational and technical advantages of larger enterprises. Conversely, the frequently hypothesized advantages of smaller work groups may help to raise productivity in the small-enterprise sector, but not sufficiently to dominate larger enterprises.

12

Entrepreneurship, Growth, and Success: Survey Results

The literature on entrepreneurship has been mainly concerned with historical, sociological, and psychological theories of entrepreneurial supply. Moreover, most attention has been paid to large-scale industry. Tycoons have fascinated students. There has thus been rather little work in India or elsewhere on small business entrepreneurs and on the relation between their characteristics and success. Our research was not primarily concerned with this problem, but the questionnaires included questions on entrepreneurial attributes. Therefore in this chapter we describe these attributes and discuss their relationship to the efficiency (see also chapter 11), profitability, and growth of their firms. We compare the resulting crumbs with those produced by other inquiries.

Portrait of the Small Indian Entrepreneur

The Relationship between Entrepreneur and Enterprise

Nearly half of all the units in our combined samples had a single owner-entrepreneur. The differences between the industries were considerable. About 70 percent of the shoe units were of this type, but only 10 percent of the soap and 3 percent of the metal casting units, in which two or three working partners was the rule. The number of partners and working partners clearly tends to rise with size (except for shoes) and averages about 4 for units with over 50 employees. Details are given in tables 12-1 and 12-2. Figures for Colombia are similar. Half of all firms in the metal working sector surveyed had working partners, and 34 percent of those in food processing, and in both cases the number rises with size (Cortes, Berry, and Ishaq 1987, table 4-1).

The stereotype of the small business is that of the single owner-entrepreneur who, besides often working with his or her hands, combines

Table 12-1. *Percentage Distribution of Units by the Number of Working Partners by Industry, India*

Industry	*Nil*	*1*	*2*	*3*	*4*	*5 or more*
Machine tools	49.4	3.4	18.0	16.9	7.9	4.5
Printing	45.1	11.3	19.7	11.3	8.5	4.2
Powerlooms	27.0	2.7	48.6	5.4	8.1	8.1
Shoes	70.2	2.1	13.8	5.3	3.2	5.3
Soap	10.0	10.0	48.0	18.0	8.0	6.0
Metal casting	2.6	13.2	31.6	23.7	13.2	15.8

the more entrepreneurial functions of initiating the business, and making subsequent investment and pricing decisions, with the day-to-day managerial functions of overseeing production and organizing sales. The common presence of two or three partners when the size of the firm exceeds, say, 10 employees probably means that there is some delegation of function as the firm grows and that there is no sharp dividing line between different management structures. Conversely, the presence of working partners does not imply that the more entrepreneurial functions do not usually remain with a single boss. But we did not obtain details of the functions of partners and so cannot be very definite.

Table 12-2. *Mean Number of Partners and Working Partners in Different Industries by Size Groups, India*

Size group (number of workers)	*Machine tools*	*Powerlooms*	*Printing*	*Shoes*	*Soap*	*Metal casting*
1–5	0.0	0.0	0.0	0.4	2.3	—
	0.0	0.0	0.0	0.4	1.8	—
6–10	0.3	1.0	0.8	0.8	2.3	6.2
	0.3	0.3	0.7	0.4	1.9	4.2
11–15	1.6	2.7	1.8	1.4	3.4	3.5
	1.4	1.0	1.4	1.2	2.7	2.9
16–20	1.9	3.5	1.4	2.7	3.5	1.6
	1.7	2.1	1.3	2.5	2.0	1.3
21–25	1.1	3.7	2.7	—	3.5	3.0
	1.1	2.0	2.3	—	2.0	3.0
26–30	—	4.5	2.0	2.0	1.0	4.6
	—	2.6	2.0	1.5	0.0	1.3
31–40	2.8	—	1.0	3.0	4.0	3.3
	2.8	—	1.0	2.0	2.0	3.3
41–50	3.5	5.0	3.4	—	7.0	2.0
	2.5	4.7	3.0	—	3.0	1.0
51 or more	3.6	6.0	3.6	1.3	5.2	6.3
	3.2	1.0	3.4	1.3	3.8	1.9
Overall mean	1.6	3.3	1.7	1.3	3.0	4.7
	1.4	1.9	1.5	1.7	2.2	2.3

— No firm in the size class, or no response to the question.

Table 12-3. *Percentage Distribution of Native and Migrant Entrepreneurs by Age of Migrant at Migration, and Industry, India*

	Migrants (years of age)							
Industry[a]	*Under 15*	*16–20*	*21–25*	*26–30*	*31–40*	*41–50*	*51 or more*	*Natives*
Machine tools	3.4	5.6	5.6	2.2	1.1	36.0	2.2	43.8
Printing	8.4	4.2	4.2	15.5	23.9	7.0	1.4	35.2
Powerlooms	8.1	0.0	0.0	0.0	0.0	0.0	0.0	91.9
Shoes	20.2	9.6	6.4	5.3	2.1	2.1	0.0	54.3
Soap	32.0	24.0	24.0	10.0	0.0	2.0	0.0	8.0

a. Metal casting is excluded because out of 33 migrants only 4 recorded their migration age.

The Age and Provenance of the Entrepreneurs

The mean age of the entrepreneurs and their enterprises was 42 and 12 years, respectively.[1] The modal entrepreneurial age was over 35 in four out of five industries, shoes being the exception with a modal age of 26–30. The mean age is not very different from Korea, where a survey of small businesses in 1973 found it to be 46.

Age was considerably affected by migrant status, migrants (mainly from Pakistan) starting their businesses at a later age.[2] Migrants were in a majority in machine tools, printing, and soap. In powerlooms, almost all entrepreneurs were native while the rest consisted of those who had migrated as children. (Powerlooms were surveyed, but they could not be included in the production function in chapter 11.) Natives were also in a majority in shoes, but child migrants accounted for 20 percent. Young migrants (under 20 years at migration) accounted for a majority in soap. Older migrants, however, played an important part in the two modern industries—machine tools and printing (see table 12-3).

There was a significant positive but not very strong association between the age of the entrepreneur and that of his firm (see equation given in footnote 2). This was also found to be the case in Korea (Ho 1980, table 5-5).

The Education and Background of the Entrepreneur

Table 12-4 shows that the level of education is very different in the different industries. About 60 percent of the entrepreneurs in powerlooms and shoes had only a primary education or less. Most in machine tool manufacturing, metal casting, and soap had a middle school or high school education, while

1. The figures in this section exclude metal casting.

2. Significant variables for estimating the age of an entrepreneur (AE) when the business was started were age of firm (AF) and dummy variables nativity (N), powerlooms (P), and father in trade (FIT). The equation was $AE = 41.25 + 0.48AF - 11.57N + 6.47P - 2.36FIT$ ($R^2 = 0.31$). The estimated age of the entrepreneur at the inception of the business was, not surprisingly, 30 years.

Table 12-4. *Education of Entrepreneurs: Percentage Distribution by Educational Level, India*

Industry	*Below primary*	*Primary*	*Middle and high school*	*Inter-mediate*	*Bachelor's degree*	*Above bachelor's degree*
Machine tools	5.6	15.7	43.8	7.9	9.0	18.0
Printing	1.4	1.4	23.9	19.7	42.3	11.3
Powerlooms	64.9	0.0	16.2	5.4	10.8	2.7
Shoes	30.9	27.7	28.7	2.1	2.1	7.4
Soap	4.0	12.0	44.0	10.0	16.0	14.0
Metal casting	22.0	8.0	20.0	12.0	16.0	22.0

about 30 percent had college education or better. As might be expected, higher levels of education predominated in printing.

Previous job experience was poorly recorded in our surveys. Table 12-5 gives the results, such as they are. The most notable feature is that over half the entrepreneurs in machine tools were previously production workers. This echoes a very similar finding for the metalworking sector in Colombia (Cortes, Berry, and Ishaq 1987, table 4-4). A significant number of entrepreneurs in the printing industry (18 percent) had also once been production workers. It was noticed during the survey that as monotype gradually gave way to photo-offset processes, the monotype workers often set up their own enterprises which continued to use the older technology.

Some clue as to social background is provided by table 12-6, which shows both own and father's education for the total of our samples, and table 12-7, which gives the distribution of entrepreneurs by father's occupation for each industry. Primary and below; middle and high school; and intermediate or higher education each very roughly accounts for a third of present-generation entrepreneurs. But half the present generation sprang from fathers with less than primary education, and two-thirds from fathers with primary or less. A more general educational upgrading is also evident from

Table 12-5. *Percentage Distribution of Entrepreneurs by Previous Job in Different Industries, India*

Industry	*Partner or family worker*	*Production worker*	*Trader*	*Office worker*	*Farmer*	*Missing*[a]
Modern						
Machine tools	7.9	51.7	3.4	9.0	0.0	28.1
Printing	4.2	18.3	12.7	23.9	1.4	39.4
Soap	20.0	6.0	12.0	8.0	0.0	54.0
Traditional						
Shoes	11.7	14.9	4.3	9.6	0.0	59.6
Powerlooms	18.9	2.7	0.0	5.4	0.0	73.0
Metal casting	7.8	15.7	19.6	2.0	0.0	54.9

a. Includes those who did not have previous job experience.

Table 12-6. *Entrepreneur's Own Education Compared with Their Father's Education, India*
(percent)

	Own education						
Father's education	*Below primary*	*Primary*	*Middle and high school*	*Intermediate*	*University bachelor's degree*	*Above university bachelor's degree*	*Row percentages*
Below primary	34.1	23.8	26.8	4.9	9.8	0.6	49.7
Primary	0.0	12.5	58.3	14.6	6.3	8.3	15.5
Middle and high school	1.3	1.3	44.7	6.6	22.4	23.7	23.0
Intermediate	0.0	0.0	8.3	33.3	33.3	25.0	3.6
University bachelor's degree	0.0	0.0	11.8	11.8	35.3	41.2	5.2
Above university bachelor's degree	0.0	7.7	0.0	23.1	30.8	38.5	3.9
Column percentages	17.3	14.2	33.0	8.8	15.2	11.5	100.0

Note: The data in this table are based on five industries; metal casting is excluded because of poor response.

table 12-6. Table 12-7 shows that 40–60 percent of fathers in the more traditional industries, shoes, soap, and powerlooms, were proprietors or partners. It can be presumed that their sons inherited the business. In powerlooms, another 30 percent of fathers were production workers, presumably mainly weavers. Two of these industries, shoes and powerlooms, were strongly caste-based, and only rapid growth can have permitted the new entry, which was considerable, especially in shoes. Trade in some capacity figures quite strongly, and agriculture very weakly, as father's occupation.

Table 12-7. *Percentage Distribution of Entrepreneurs by Their Father's Occupation by Industry, India*

Father's occupation	*Shoes*	*Printing*	*Soap*	*Machine tools*	*Powerlooms*	*Metal casting*
Proprietor, family worker, and partner	58.6	4.2	44.0	24.7	39.3	2.0
Production worker	12.8	5.6	4.0	29.2	28.8	0.0
Trader or salesman in same industry	1.0	0.0	6.0	0.0	3.0	0.0
Trade-related occupation in other industry	13.8	36.6	40.0	27.0	15.7	56.9
Agriculture	3.2	7.0	4.0	12.4	0.0	2.0
Office worker	4.3	46.5	2.0	6.7	13.2	13.7
Missing	6.4	0.0	0.0	0.0	0.0	25.5

Table 12-8. *Percentage Distribution of Firms by Initial Employment Size, India*

Number of workers	*Machine tools*	*Printing*	*Shoes*	*Soap*	*Powerlooms*	*Metal casting*
1–5	37	48	71	60	27	16
6–10	32	34	13	18	38	16
11–15	14	11	6	12	27	4
16–20	8	1	2	2	8	6
21–25	1	4	2	2	0	6
26–30	1	0	0	0	0	2
31–40	2	0	0	0	0	0
41–50	3	0	0	0	0	4
51 or more	0	0	0	0	0	23
Missing	2	1	5	6	0	24

The new industry, printing, is exceptional in that office worker is the most common father's occupation, while inheritance plays almost no part.

In Colombia it was found that the entrepreneurs in small and medium-size enterprises were typically from a middle-income background. Our Indian surveys included much more traditional industry, and the somewhat scanty evidence presented suggests that a similar generalization would be unwarranted. It rather supports those who have claimed that the background of small-scale entrepreneurs in India is extraordinarily diverse despite caste affiliations (for example, Berna 1960).

How the Entrepreneur Started: Initial Size and Investment

A high proportion—64–84 percent—of enterprises started with less than 11 workers in all industries, except metal casting. Only in machine tools and metal casting did any firm start with over 25 workers. Table 12-8 gives the distribution. One can compare machine tools with metalworking firms in Colombia. In the latter, 18 percent started with over 20 workers, compared with about 8 percent in India.

The distribution of firms by initial investment is given in table 12-9. Metal

Table 12-9. *Percentage Distribution of Firms by Level of Initial Fund, India*
(thousands of 1970 rupees)

	Fund level									
Industry	*Under 20*	*20–50*	*50–100*	*100–200*	*200–300*	*300–500*	*500–1,000*	*1,000*	*Missing*	*Mean value of initial fund*
Machine tools	28.1	22.5	12.4	7.9	6.7	5.6	9.0	5.6	2.2	273.7
Printing	40.8	18.3	15.5	4.2	1.4	2.8	2.8	2.8	11.3	107.0
Shoes	41.5	26.6	4.3	4.3	3.2	1.1	1.1	0.0	18.1	44.7
Soap	38.0	30.0	18.0	8.0	4.0	0.0	0.0	0.0	2.0	49.6
Powerlooms	8.1	37.8	35.1	13.5	2.7	0.0	0.0	0.0	2.7	62.5
Metal casting	13.7	21.6	9.8	5.9	3.9	3.9	0.0	15.7	25.5	1,551.0

Table 12-10. *Percentage of Initial Fund Provided Out of Own Savings, India*

Industry	*0–20*	*21–40*	*41–60*	*61–80*	*81–100*
Machine tools	2.2	7.9	23.6	5.6	60.7
Powerlooms	62.2	8.1	10.8	13.5	5.4
Printing	12.7	14.1	11.3	4.2	57.7
Shoes	29.8	6.4	7.4	1.1	55.3
Soap	10.0	2.0	8.0	2.0	78.0
Metal casting	29.0	10.5	7.9	7.9	44.7

casting stands out as exceptional. As table 12-8 shows, about a quarter of the units had started with over 50 workers. This is echoed by the finding in table 12-9 that about 16 percent had an initial investment of over Rs1 million pulling up the mean value of the initial fund to about Rs1.5 million. In machine tools and printing there was also a handful of firms with an initial investment of over Rs1 million. But in all six industries, 45–70 percent of firms had started with less than Rs50,000 in 1970 prices (about $5,000). An employment size breakdown (not presented) shows the obvious tendency for the proportion of firms with low initial investment to fall as employment size increases.

After metal casting, machine tools show much the largest average initial investment of Rs274,000 (about $27,000) for an average initial size of 10 workers. The metalworking firms in the Colombia survey started with an average employment size of 14 and average initial capital of 2.6 million pesos (about $70,000) and were thus considerably more capital-intensive.

We regressed initial size (measured by both employment and capital) on various attributes of the entrepreneur.[3] The only pervasively significant association was with education. In machine tools, printing, soap, and shoes, higher education was strongly associated with initial size, and secondary education also significantly increased initial size in soap and shoes. Similar results were obtained in Colombia, especially in that those with university education had much greater initial investment (Cortes, Berry, and Ishaq 1983, p. 119). This relationship of education and initial size no doubt arises because the more educated usually have more money and are better connected with other sources of finance.

Most of the businesses surveyed were clearly eligible for loans from the various agencies supported by the government to help the small-scale sector. Nevertheless, as table 12-10 shows, own savings accounted for over 80 percent of the initial investment in 45–78 percent of the firms in all the industries except powerlooms. About one-third of the firms in the four industries apart from powerlooms and metal casting had obtained loans from friends or relatives or both, but the amounts were quite small compared with personal savings. In soap and powerlooms bank loans hardly figured. But they were of

3. See note 7 for the attributes used. The regression is not reported.

some importance for machine tools, printing, shoes, and metal casting, in which 14–19 percent of firms had used bank credit initially. At least in some sectors in India, bank loans now finance initial investment to a greater extent than in other countries (see chapter 15), though the amounts involved are still dwarfed by personal savings. Powerlooms was the odd industry; in it, neither personal savings nor loans from relatives or banks were important. It appears that traders were major suppliers of savings in this sector. Rather surprisingly, employment size breakdowns (not shown) add little or nothing to the above story. In particular, the high rate of personal savings to total initial investment is true of all size groups.

Growth of Firms

In considering growth, it cannot be too strongly emphasized that we could survey only survivors. Many inquiries bear witness to the high death rates of young firms and firms in the smallest size classes. The measures of growth must not be taken to indicate the true probabilities of growth of a cohort of newly established firms. This, of course, applies particularly to the smaller size groupings. This reservation must be borne in mind throughout.

The employment growth rates of firms in all the industries surveyed were high, as table 12-11 shows. There has been a large fall in the share of the two smallest employment size groups, 1–5 and 6–10 workers, in all industries except shoes and soap—and in the latter two, a substantial proportion of firms have grown from the 1–5 to the 6–10 size class. Larger size classes also show significant relative increases. In particular we should note the appearance of a significant proportion of firms in the over 50 size class when, except for metal casting, there was none at the time of establishment. The growth rate of employment itself has also been high, as these figures suggest. Details are given at the end of this section.

Tables 12-20 to 12-25 at the end of the chapter give the detailed picture of the employment size of the firms at the time of the survey, broken down by the initial size groups of these firms. Except for shoes and soap, over 80 percent of firms raised their size class.

Special interest attaches to the question how far firms starting very small have succeeded in growing. The information is presented in table 12-12, which gives the present size group of the firms whose initial size was 1–5 or 6–10 workers. Firms with initial sizes of 6–10 workers managed to do somewhat better than the 1–5 group. The proportion of the 1–5 group which reached 11 plus sizes is less than the proportion of the 6–10 group which reached 16 plus sizes, except in the case of printing. Shoes and soap stand out as the industries in which a high proportion of the smallest firms have stayed small.

This comparison of the present size with the initial size does not take into

Table 12-11. *Percentage Distribution of Units by Initial Size, Present Size, and Average Age, India*

Industry (total number of units)	*Unit size (number of workers)* 1–5	6–10	11–15	16–20	21–25	26–30	31–40	41–50	51 or more	Missing
Machine tools (78)										
Initial size	38.5	33.3	14.1	5.1	1.3	0.0	2.6	2.6	0.0	2.6
Present size	0.0	5.1	11.5	14.1	10.3	12.8	7.7	10.3	28.2	0.0
Average age (years)	—	6.0	12.8	7.8	11.3	12.3	13.3	20.3	17.8	—
Printing (71)										
Initial size	47.9	33.8	11.3	1.4	0.0	0.0	4.2	0.0	0.0	1.4
Present size	7.0	26.8	28.2	9.9	4.2	1.4	4.2	7.0	11.3	0.0
Average age (years)	15.8	15.4	12.3	13.2	5.3	2.0	12.0	8.0	16.8	—
Powerlooms (37)										
Initial size	27.0	37.8	27.0	8.1	0.0	0.0	0.0	0.0	0.0	0.0
Present size	5.4	16.2	8.1	21.6	16.2	21.6	0.0	8.1	2.7	0.0
Average age (years)	6.0	8.2	10.0	8.5	9.5	7.3	—	12.0	14.0	—
Shoes (92)										
Initial size	71.3	12.8	6.4	2.1	0.0	0.0	0.0	0.0	0.0	7.4
Present size	38.3	24.5	16.0	8.5	1.1	3.2	2.1	0.0	4.3	2.1
Average age (years)	13.3	9.6	13.5	12.9	38.0	4.0	11.5	—	11.0	18.0
Soap (50)										
Initial size	60.0	18.0	12.0	0.0	2.0	0.0	0.0	2.0	0.0	6.0
Present size	26.0	36.0	14.0	4.0	4.0	0.0	3.0	2.0	10.0	2.0
Average age (years)	17.8	19.2	19.6	6.0	16.5	—	21.0	6.0	24.2	6.0
Metal casting (51)										
Initial size	15.7	15.7	3.9	5.9	5.9	2.0	0.0	3.9	23.5	23.5
Present size	2.0	9.8	25.5	11.8	2.0	5.9	5.9	2.0	31.4	3.9
Average age (years)	18.0	33.6	12.4	21.5	20.0	20.0	24.7	14.0	26.9	28.0

— Not applicable.

Note: The present size distributions differ from those in table 8-1 for three reasons:

(1) In the samples used for chapters 8–11, and for the regressions in this chapter, a small number of large firms was added, using a separate survey and questionnaire, in order to get a better coverage of the whole size array.

(2) Rather more firms were subtracted because of lack of information for some of the variables.

(3) Size groups in table 12-11 are based on the number of permanent workers, this being more comparable for initial and present size than if family and transient workers had been included. But in chapters 10–12 the concern was with labor input, and the size groups there are based on the total number of workers including family and transient workers.

Reasons 1 and 3 affect mainly shoes and soap. Reason 2 affects mainly metal casting.

Table 12-12. *Present Size Distribution of Firms with Initial Size of 1–5 and 6–10 Workers, India*
(percent)

Industry	*Firm size (number of workers)* 1–5	6–10	11–15	16–50	51 or more	*Total number in sample*
Initial size of 1–5 workers						
Machine tools	0.0	13.3	23.3	56.7	6.7	30
Printing	8.8	41.2	29.4	20.6	0.0	34
Shoes	47.8	23.9	14.9	7.5	4.5	67
Soap	30.0	40.0	16.7	13.3	0.0	30
Powerlooms	20.0	40.0	20.0	20.0	0.0	10
Metal casting	12.5	12.5	25.0	50.0	0.0	8
Initial size of 6–10 workers						
Machine tools	0.0	0.0	3.8	77.0	19.2	26
Printing	4.2	16.7	33.3	29.1	16.7	24
Shoes	16.7	25.0	25.0	53.3	0.0	12
Soap	0.0	44.4	22.2	11.1	22.2	9
Powerlooms	0.0	14.3	7.1	77.6	0.0	14
Metal casting	0.0	0.0	85.7	14.3	0.0	7

account the age of the firms. It is possible that the tiny firms which have not moved up are very young ones. But table 12-13 refutes this hypothesis. In all industries, the great majority of small firms which stayed small have been in existence six or more years—and quite a lot have existed 11 or more years. The overall picture suggested is that in spite of the high mean growth rates of firms in terms of persons employed, our sample contains a sizable group of firms that have remained small over a long period of time.

It is frequently said that all firms start small. In table 12-14 we therefore present the initial size of the firms that now have over 50 employees. Apart from metal casting, 26 out of 37 such firms started with 20 employees or less. Metal casting is rather different in that half or more of the large firms started large, that is, with more than 50 employees.

Next we should note that age is strongly related to growth rates. Table 12-15 shows that the mean growth rate of the youngest surviving firms is much faster than that of the older firms. Also, the coefficients of variation of

Table 12-13. *Age Distribution of Small Firms that Stayed Small, India*

Industry	*Number of firms*	*Number of firms in size groups 1–10 which remained small*	*Number of firms in age group* 1–5 years old	6–10 years old	11 or more years old
Machine tools	78	3	1	2	0
Powerlooms	37	8	2	4	2
Printing	71	21	4	6	11
Shoes	94	44	12	13	19
Soap	50	25	2	3	20
Metal casting	51	2	0	0	2

Table 12-14. *Initial Employment of Medium-Size or Large Firms, India*

Number of workers	*Shoes*	*Soap*	*Printing*	*Power-looms*	*Machine tools*	*Total of five industries*	*Metal casting*	*Total with metal casting*
1–5	4	2	1	—	1	5	—	5
6–10	—	3	2	—	3	6	—	6
11–15	—	—	—	—	7	12	—	12
16–20	—	—	—	1	2	3	—	3
21–25	—	—	—	—	1	1	2	3
26–30	—	—	—	—	1	1	—	1
31–41	—	—	5	—	1	6	—	6
41–50	—	—	—	—	2	2	1	3
51 or more	—	—	—	—	—	—	8	8
Missing	—	—	—	—	1	1	5	6
Total	4	5	8	1	19	37	16	53

— No firms.
Note: Medium-size or large firms have over 50 workers.

Table 12-15. *Employment Growth Rate of Firms by Age of Firms, India*

Age group (years old)	*Machine tools*	*Power-looms*	*Printing*	*Shoes*	*Soap*	*Metal casting*
1–5						
Mean	41.2	46.9	71.2	47.6	18.6	4.1
SD/M	1.0	1.1	2.5	0.9	0.3	0.6
N	7	6	18	20	3	3
6–10						
Mean	23.7	9.1	16.7	12.5	11.4	6.3
SD/M	0.5	0.6	0.7	1.52	1.1	2.3
N	21	22	16	24	7	6
11–15						
Mean	16.2	6.7	9.6	4.3	7.4	3.5
SD/M	0.4	0.8	0.5	1.7	1.0	0.8
N	14	9	7	7	6	4
16–20						
Mean	8.8	—	9.0	6.2	7.6	6.5
SD/M	0.5	—	0.6	0.8	0.5	0.8
N	19	0	11	15	9	9
21–25						
Mean	10.2	—	0.8	3.2	4.1	1.6
SD/M	0.4	—	3.0	0.0	1.4	5.3
N	10	0	2	1	14	5
26 or more						
Mean	7.7	—	3.8	1.8	3.3	−11.8
SD/M	1.1	—	1.1	4.0	1.6	2.8
N	4	0	10	8	8	9
All						
Mean	17.5	14.7	27.4	18.6	7.1	0.7
SD/M	1.0	1.7	3.5	1.6	1.0	27.7
N	75	37	64	75	47	36

— No firms.
Note: SD/M = standard deviation divided by mean; N = number of firms.

the growth rates are higher for the younger. The table shows only employment growth, but sales growth exhibits the same features. It must be emphasized that the death rates for the younger firms would have been higher also. Thus declining *average* growth rates with age might be much less marked if those that declined to nothing could have been included. For survivors only, there is a clear parabolic relationship—the older the firm, the slower its growth. All the above applies if firms are grouped by initial size, though there are a few more exceptions.

The unweighted mean values of growth rates of employment and of sales (in real terms) in our firms are given in table 12-16. Except for soap and metal casting they are high. But there is a change in the relationship of the two sets of growth rates in the last three years. Over the lifetime of the firms the growth rate of employment was, except for metal casting, close to or higher than the growth rate of sales. The experience of the final three years is quite different, with the growth rates of sales exceeding the growth rates of employment by a large margin, except in the case of soap. We can only guess at the reasons for the high growth rate of labor productivity in recent years. But in view of the short period it must be mostly due to higher capacity utilization. This in turn may be due, at least partly, to an intensification of the policy of protecting the small-scale sector.

Profitability

Profitability is here defined simply as gross value added less the wage bill, divided by capital. Capital is defined as the undepreciated book value of fixed assets at constant (1980) prices, plus working capital. Profitability is thus gross of interest for which reasonably reliable figures could not be obtained, and also gross of depreciation. It might perhaps be better described as the rate of surplus. Two sets of figures are given for the mean profitability by size in table 12-17. We also calculated profitability by adding imputed

Table 12-16. *Mean Values of Growth Rate of Employment and Sales by Industry*
(percent per year)

Industry	GRE	GRS	GRE3	GRS3
Powerloom	14.7	10.2	9.3	15.3
Soap manufacturing	7.1	7.2	5.8	7.5
Machine tool manufacturing	17.5	14.3	7.3	21.8
Printing	27.4	20.7	6.3	16.0
Shoemaking	15.5	18.3	4.8	17.5
Metal casting	0.7	6.4	0.7	3.9

Note: GRE: growth rate of employment per year for the life of the unit. GRS: growth rate of deflated sales per year for the life of the unit. GRE3: growth rate of employees in the last three years. GRS3: growth rate of deflated sales in the last three years.

wages for non-wage-earning workers to the wage bill. The results of this calculation are not reported, but where it made an obviously substantial difference this is noted in the text.

We first consider the relation of unweighted mean profitability to size. In machine tools there is no relation, except that the smallest group of four firms (group 6–10) has low profitability. In soap also there is no relation. In powerlooms there is a clear tendency for profitability to rise; but in shoes profitability falls with size, except for the smallest group whose profitability is nevertheless not far below the overall average. In printing there is a tendency for the middle size, 11–25 workers, to be most profitable. We tested the significance of the differences in the mean profitability of the smallest size group compared with the means of larger groups. For machine tools and metal casting the *low* profitability of the 6–10 group is significant (5 percent level) as compared with the 51 plus group, but not as compared with intermediate size groups. In printing, the relatively *high* profitability of the 1–5 group is significant (10 percent level) compared with the 51 plus group, but again not as compared with intermediate size groups. In shoes the relatively *low* profitability of the 1–5 group is significant (5 percent level) as compared with the extremely profitable 6–10 group, but not as compared with larger groups.

The overall average profitability or rate of surplus is very high for shoes (128 percent), powerlooms (64 percent), and metal casting (57 percent), high for soap (37 percent), and "normal" for printing and machine tools (29 percent and 20 percent). Here, however, we should note the adjustment for wages of non-wage-earning workers; the profitability of shoes and powerlooms, the industries with most family workers, falls markedly to 58 percent and 52 percent, while remaining high. In shoes, though not powerlooms, the fall in profitability is most marked in the smaller size groups. This is in accordance with what we know of the technology. Shoes but not powerlooms become less family-based as size increases. The other industries are little affected, the rate falling by 1 or 2 percentage points only. It should also be recalled that powerlooms were heavily financed by trade credit, probably at high interest rates, so that the net profitability may be much lower than 52 percent.

We turn to the variability of profitability with size. This is itself very erratic, as can be seen by looking at the coefficients of variation in table 12-17. But there is on average less variability in the larger size classes. The overall variability is very high in shoes, printing, machine tools, and castings, but less in powerlooms and soap.

We next consider the composite mean profit rate. In machine tools and powerlooms, the weighting by capital makes little overall difference. In soap, the profitability is reduced, and in printing and metal casting still more so. This indicates that the more capital-intensive firms are less profitable. Shoes are special. The overall result is dominated by the three large capital-

Table 12-17. *Profit Rate, Coefficient of Variation, and Number of Firms, by Firm Size, India*

Firm size (number of workers)	*Machine tools*	*Powerlooms*	*Printing*	*Soap*	*Shoes*	*Metal casting*
1–5						
Unweighted mean profit rate	—	0.38	0.26	0.42	0.99	0.27
Coefficient of variation	—	0.00	0.66	0.67	1.02	0.00
Number of firms	0	1	7	10	23	1
Composite mean profit rate: mean weighted by capital	—	0.38	0.27	0.26	0.59	0.27
6–10						
Unweighted mean profit rate	0.06	0.45	0.24	0.35	2.36	0.29
Coefficient of variation	0.52	0.39	1.15	0.60	1.17	1.07
Number of firms	4	6	31	19	25	5
Composite mean profit rate: mean weighted by capital	0.06	0.46	0.14	0.26	0.84	0.21
11–15						
Unweighted mean profit rate	0.14	0.50	0.38	0.39	1.24	0.32
Coefficient of variation	0.73	0.84	0.71	0.45	0.92	0.94
Number of firms	9	3	8	6	17	13
Composite mean profit rate: mean weighted by capital	0.10	0.38	0.18	0.35	1.09	0.22
16–20						
Unweighted mean profit rate	0.19	0.67	0.58	0.24	0.79	0.44
Coefficient of variation	0.80	0.63	1.39	0.73	0.70	0.52
Number of firms	11	7	9	4	12	6
Composite mean profit rate: mean weighted by capital	0.15	0.56	0.63	0.19	0.62	0.41
21–25						
Unweighted mean profit rate	0.27	0.66	0.54	—	0.55	0.93
Coefficient of variation	1.02	0.31	0.63	—	0.29	0.00
Number of firms	8	5	2	0	4	1
Composite mean profit rate: mean weighted by capital	0.09	0.65	0.45	—	0.41	0.93

26–30						
Unweighted mean profit rate	0.22	0.35	0.17	0.16	0.47	0.92
Coefficient of variation	1.74	0.48	0.63	0.00	0.74	1.33
Number of firms	10	5	2	1	3	3
Composite mean profit rate: mean weighted by capital	0.14	0.38	0.19	0.16	0.35	1.16
31–40						
Unweighted mean profit rate	0.34	0.76	0.23	0.34	0.65	0.41
Coefficient of variation	0.41	0.64	0.74	0.63	0.66	0.44
Number of firms	6	5	6	2	5	3
Composite mean profit rate: mean weighted by capital	0.32	0.65	0.21	0.32	0.48	0.33
41–50						
Unweighted mean profit rate	0.17	1.05	0.11	—	0.31	0.22
Coefficient of variation	0.45	0.03	0.00	—	0.09	0.00
Number of firms	8	2	1	0	2	1
Composite mean profit rate: mean weighted by capital	0.17	1.04	0.11	—	0.30	0.22
51 or more						
Unweighted mean profit rate	0.19	1.16	0.09	0.37	0.17	0.87
Coefficient of variation	0.54	0.26	0.69	0.54	0.69	1.74
Number of firms	22	3	5	5	3	16
Composite mean profit rate: mean weighted by capital	0.19	1.13	0.08	0.34	0.11	0.28
All						
Unweighted mean profit rate	0.20	0.64	0.29	0.37	1.28	0.57
Coefficient of variation	0.95	0.58	1.27	0.60	1.35	1.63
Number of firms	78	37	71	47	94	49
Composite mean profit rate: mean weighted by capital	0.17	0.67	0.17	0.30	0.12[a]	0.33

— No firms.

a. This is 0.59 without the three firms with 51 plus employees.

intensive firms with low profitability—the result being a dramatic fall in average profitability from 128 percent to 12 percent. If these three are eliminated, however, there is still a large fall to 59 percent, which again indicates that the more capital-intensive firms are less profitable.

We should finally comment on the difficulty of measuring capital, most especially for small sizes in the more traditional sectors. Where capital is very small, consisting of a few hand tools and a little working capital, errors of measurement can be relatively large and make a big difference to the rate of profitability. The fact that land and buildings are often rented is another disturbing factor. These items account for most of the capital used in very small units, and the absence of their value from the denominator (and rent from the numerator) could result in the profitability of the residual capital being much higher than if both had been included. These reservations probably make the most difference to shoes, and they make the very high rates of return to small shoemakers particularly suspect.

Our samples included examples of profitability both rising and falling with size, and there is therefore nothing to be gained by comparison with the overall relations between size and profitability presented in chapter 7.

Except for soap, a comparison of average profitability can be made with the Reserve Bank of India survey, from which we calculated figures for production profits gross of both depreciation and interest, related to the undepreciated value of fixed assets plus inventories—this corresponding to our definition. This can be compared with the figures on composite mean profit rate of table 12-17. For convenience, the results are given in table 12-18.

The closeness of the estimates in the most modern industries—machine tools and printing—is remarkable. But in shoes and powerlooms our estimates are both relatively and absolutely very high, and also relatively high for metal casting. In the case of powerlooms and metal casting, a possible explanation is that the RBI surveys contain a far higher proportion of very small firms, which we have found to be less profitable. The RBI survey for shoes also had a much higher proportion of very small firms, but our survey shows very high profitability for the very small. The possibility that we may have seriously underestimated the amount of capital in small shoe enterprises has already been remarked on. However, very high profitability has

Table 12-18. *Comparison of World Bank and Reserve Bank of India Surveys of Profitability*
(percent)

Survey	*Machine tools*	*Powerlooms*	*Printing*	*Shoes*	*Metal casting*
World Bank	18	67	17	59[a]	27
Reserve Bank of India	20	26	17	31	12

a. This figure excludes the three largest firms which would also have been excluded from the RBI survey.

been observed elsewhere, especially in craft industries. In Africa figures similar to ours for powerlooms and shoes have been observed (see Page 1979, p. 24). In Korea, in industries in which small firms predominated profitability ranged from 42 to 58 percent (see table 7-2 above).[4] It must finally be remembered that these high figures exclude the negative profits of firms that died.

Explaining Growth and Profitability

We next try to explain the profitability and growth of the firms in terms of various entrepreneurial variables and the experience of the labor force along the lines pursued in chapter 11, in which a very similar analysis of technical efficiency was presented. In addition to the above variables, the age of the firm and its initial size were entered as controls. In stepwise regressions age indeed proved to be the most significant variable and the most powerful in terms of accounting for the variance in growth rates, whether of sales or employment. It was much less important for profitability and efficiency (see chapter 11).

For growth, we present the analysis in terms of the growth of sales rather than employment as better reflecting the success of the firm and the aims of the entrepreneur.[5] For profitability, we used the simplest measure (nonwage value added divided by capital) rather than that making allowance for the imputed earnings of unpaid workers.[6]

The two dependent variables used might be expected to be related to each other and to the technical efficiency variables. We therefore present the zero order correlation matrix between them in table 12-19. The analysis excluded metal casting, for which there were too many gaps in the data. There were also no technical efficiency estimates made for powerlooms. It is of course to be expected that technical efficiency and profitability would be related, and this duly shows up. In view of their definitional similarity, it is perhaps a little surprising that they are not still more closely related. The other case in which a stable but not very strong relationship is discernible is between the past growth rate of sales and present technical efficiency. The reason for this is obscure. The real growth rate of sales might have been expected to be positively related to profitability, especially if there was increased use of

4. But these figures relate profits net of depreciation provision to the depreciated value of fixed capital only. It is unclear whether the figure would be reduced or raised by the difference of definition.

5. Employment growth regressions were made but are not presented. Except for printing, the R^2s were higher for employment growth, but this was caused by the greater explanatory power of the age variable. The entrepreneurial variables did not, as one would expect, enter any more significantly than for the sales growth regressions.

6. Regressions were also made for the second measure of profitability. The results were sufficiently similar for it to be tedious to present them.

Table 12-19. *Correlation Matrix of Dependent Variables*

Machine tools, printing, soap, and shoes	*GRS*	*TE*
TE	+0.20	
	+0.13	
	+0.38	
	+0.10	
PFT	+0.28	+0.57
	+0.01	+0.46
	−0.32	+0.16
	−0.11	+0.46

Note: GRS: growth rate of sales. TE: technical efficiency. PFT: profit rate.

capacity. It was, however, negative for soap, albeit positive for machine tools and very small for printing and shoes.

It should be noted that these dependent variables are truly dependent in that feedback to the explanatory variables is in most cases impossible—they cannot affect the entrepreneur's formal education or father's occupation, nor the age or initial size of the firm. It would also be very farfetched to suppose that they could affect the entrepreneur's or his workers' experience.[7]

We turn to the explanation of growth of sales. For machine tools no variable was even remotely significant. For metal casting only initial size was (negatively) significant, with three stars.[8] For printing, age entered negatively with three stars, and graduate education positively with two stars. These together with initial size (negative, with almost one star) explained 25 percent of the variance. For soap, age and graduate education again entered negatively and positively with two stars and almost one star, respectively. Together with experience of the entrepreneur (positive, two stars), 30 percent of the variance was explained. For shoes, age and college education again figured negatively and positively (both one star); together with initial size (negative, one star) 11 percent of the variance was explained. Thus, for three of the industries higher education was clearly significant; it is mysterious that it failed to show for machine tools. Experience of the entrepreneur showed significantly only in the case of soap.

We turn to profitability. No variable was significant for metal casting. In machine tools, age and the experience of workers were significant with two stars, but these together explained only 8 percent of the variance. For printing, college education almost achieved one star and explained 6 percent of

7. The explanatory variables are: (1) experience of entrepreneur; (2) age of firm; (3) initial employment size of firm; (4) education of entrepreneur—illiterate, high school, or college graduate—the base being primary education; (5) employee experience; and (6) occupation of father—a businessman or a production worker. The measurement of the experience of entrepreneurs and workers is explained in chapter 11.

8. Three stars here mean significance at the 1 percent level, two stars at the 5 percent level, and one star at the 10 percent level.

the variance (age was insignificant). In soap, only high school education entered with significance (two stars), and it explains 22 percent of the variance. In shoes, college education was significant with two stars. Together with high school education (nearly one star), 9 percent of the variance was explained. Thus, as in the case of sales growth, with the exception of machine tools, education was clearly a significant factor though its explanatory power was fairly weak. Unlike growth, age was insignificant for profitability except in the case of machine tools.

It is interesting that education above the primary level was significant for growth and profitability, while for technical efficiency it was found in chapter 11 that only literacy was important. The dog that never barked at either growth or profitability was the occupation of fathers, whether businessmen or production workers. The initial size of the firm twice entered significantly and negatively for growth, but never for profitability.

A brief comparison of the above results with other researchers' findings is of interest. In Colombia the most comparable independent variable used was private benefit cost. This was defined as value added divided by the sum of the wage bill and the estimated opportunity cost of capital. For metalworking, both primary and university education were positively significant compared with secondary education. For food processing the same coefficients were negative but not significant at the 10 percent level (Cortes, Berry, and Ishaq 1987, tables 3-9, 3-10). It is difficult to make anything of this.

In Africa it has been noted that entrepreneurs with functional literacy perform significantly better than illiterates. But further education does not appear to contribute significantly to the profitability of the enterprise and may in fact be correlated with poor entrepreneurial performance (Page 1979, p. 31).[9] Harris (1971) found a weak, barely significant positive relationship between education and success in Africa, and he suggested that most of the explanation for the weakness might be that formal education was competitive with learning on the job. It has also been suggested that only the least able of those with middle or higher education in Africa become entrepreneurs. However, our results for India show that education beyond functional literacy was a significant positive factor for growth and profitability. A possible explanation for the difference is that opportunities in the civil service and in large firms relative to the number of those with secondary or higher education are fewer in India than in Africa.

Summary

Our surveys provided some glimpses of the small Indian entrepreneur. All but one of the surveys were done in North India, which partly accounts for

9. The author cites several studies of African entrepreneurship in reaching this conclusion.

the prominence of migrants. The soap industry was almost exclusively run by migrants, and they were also in a majority in the modern industries of machine tools and printing. This prominence echoes that found in many other countries throughout history.

Turning to the social background, caste played an important role in two industries, powerlooms and shoes. In these, and also in soap, the entrepreneur's father was also a proprietor or partner, however small, in about half the cases. In the other three industries, the family background was quite diverse. The education of the entrepreneur was about equally divided between primary and below, middle and high school, and intermediate or higher. But only one-third of the fathers of the present generation had more than primary education.

Most entrepreneurs started in a small way, with less than ten workers; and between a half to two-thirds of the firms in the six industries were started with less than Rs50,000 (about $5,000). However, a few, especially in metal casting but also in machine tools and printing, had an initial investment of over Rs1 million ($100,000). Large initial size and the education of the entrepreneur were strongly related. As everywhere, the initial fund came predominantly from own savings. But nearly one-fifth of the new firms in machine tools, printing, and shoes had got bank loans. This is probably a higher proportion than in most developing countries and may reflect the Indian government's drive to make the banks lend to the small. However, bank loans hardly figured at all in the soap and powerloom industries.

Turning to the growth of firms, those of the small and the young that survive grow fast, faster than larger firms. But shoes and soap, in which primitive methods are used in most or all enterprises, are to some degree exceptional. In these industries a large minority of small new firms grow old and stay small, with less than ten employees.

The gross rate of economic surplus (gross value added minus the wage bill, divided by the book value of assets), which we call "profitability" for short, had no clear relation to size. In some industries it rose, in others it fell; or there was no apparent relation to size at all.

We next tried to explain growth and profitability in terms of a set of variables that included the age of the firm, the experience of workers, and various entrepreneurial characteristics. For growth, the negative influence of age predominated: for obvious reasons, including the absence of non-survivors, this was to be expected. Controlling for age, we found that either college or graduate education was positively significant but not strongly influential in three industries, printing, soap, and shoes; for profitability, age was not significant, but the educational variables were. In machine tools and metal casting neither age nor the education variables were significant for growth; we have no explanation of this. For profitability, age and the experience of workers were positively significant. The occupation of the entrepreneur's father was never significant. It is of some interest to note the significance of education at higher than primary levels for both growth and profitability in three out of four industries, whereas in chapter 11 it was found that only literacy, and not any higher level of education, was significant for technical efficiency (despite a fairly high simple correlation between profitability and technical efficiency). Finally, the experience of workers was the main explanatory variable for technical efficiency, but it was not significant for growth or profitability (except weakly for profitability in the case of machine tools).

Table 12-20. *Percentages of Indian Machine Tool Firms in Present Size Groups (Rows) by Initial Size Group*

Size (number of workers)	*1–5*	*6–10*	*11–15*	*16–20*	*21–25*	*26–30*	*31–40*	*41–50*	*51+*	*Row total*
1–5	0.0	13.3	23.3	26.7	15.3	16.7	0.0	0.0	6.7	38.5
6–10	0.0	0.0	3.8	11.5	11.5	15.4	23.1	15.4	19.2	33.3
11–15	0.0	0.0	0.0	0.0	0.0	0.0	0.0	27.3	72.7	14.1
16–20	0.0	0.0	25.0	0.0	0.0	0.0	0.0	25.0	50.0	5.1
21–25	0.0	0.0	0.0	0.0	0.0	0.0	0.0	0.0	100.0	1.3
26–30	0.0	0.0	0.0	0.0	0.0	0.0	0.0	0.0	0.0	0.0
31–40	0.0	0.0	0.0	0.0	0.0	0.0	0.0	0.0	100.0	2.6
41–50	0.0	0.0	0.0	0.0	0.0	50.0	0.0	0.0	50.0	2.6
Missing	0.0	0.0	0.0	0.0	50.0	0.0	0.0	0.0	50.0	2.6
Column total	0.0	5.1	11.5	14.1	10.3	12.8	7.7	10.3	28.2	100.0

Note: 78 firms in sample.

Table 12-21. *Percentages of Indian Printing Firms in Present Size Groups (Rows) by Initial Size Group*

Size (number of workers)	*1–5*	*6–10*	*11–15*	*16–20*	*21–25*	*26–30*	*31–40*	*41–50*	*51+*	*Row total*
1–5	8.8	41.2	29.4	5.9	5.9	0.0	0.0	8.8	0.0	47.9
6–10	4.2	16.7	33.3	16.7	0.0	4.2	4.2	4.2	16.7	33.8
11–15	0.0	12.5	25.0	0.0	12.5	0.0	12.5	12.5	25.0	11.3
16–20	0.0	0.0	0.0	100.0	0.0	0.0	0.0	0.0	0.0	1.4
31–40	0.0	0.0	0.0	0.0	0.0	0.0	33.3	0.0	66.7	4.2
Missing	100.0	0.0	0.0	0.0	0.0	0.0	0.0	0.0	0.0	1.4
Column total	7.0	26.8	28.2	9.9	4.2	1.4	4.2	7.0	11.3	100.0

Note: 71 firms in sample.

Table 12-22. *Percentages of Indian Powerloom Firms in Present Size Groups (Rows) by Initial Size Group*

Size (number of workers)	*1–5*	*6–10*	*11–15*	*16–20*	*21–25*	*26–30*	*31–40*	*41–50*	*51+*	*Row total*
1–5	20.0	40.0	20.0	0.0	0.0	20.0	0.0	0.0	0.0	27.0
6–10	0.0	14.3	7.1	28.6	14.3	14.3	0.0	21.4	0.0	37.8
11–15	0.0	0.0	0.0	30.0	30.0	40.0	0.0	0.0	0.0	27.0
16–20	0.0	0.0	0.0	33.3	33.3	0.0	0.0	0.0	33.3	8.1
Column total	5.4	16.2	8.1	21.6	16.2	21.6	0.0	8.1	2.7	100.0

Note: 37 firms in sample.

Table 12-23. *Percentages of Indian Shoe Firms in Present Size Groups (Rows) by Initial Size Group*

Size (number of workers)	*1–5*	*6–10*	*11–15*	*16–20*	*21–25*	*26–30*	*31–40*	*41–50*	*51+*	*Missing*	*Row total*
1–5	47.8	23.9	14.9	6.0	0.0	1.5	0.0	0.0	4.5	1.5	71.3
6–10	16.7	25.0	25.0	16.7	0.0	16.7	0.0	0.0	0.0	0.0	12.8
11–15	0.0	0.0	33.3	33.3	0.0	0.0	16.7	0.0	0.0	16.7	6.4
16–20	50.0	50.0	0.0	0.0	0.0	0.0	0.0	0.0	0.0	0.0	2.1
Missing	14.3	42.9	0.0	0.0	14.3	14.3	0.0	14.3	0.0	0.0	7.4
Column total	38.3	24.5	16.0	8.5	1.1	3.2	2.1	0.0	4.3	2.0	100.0

Note: 94 firms in sample.

Table 12-24. *Percentages of Indian Soap Firms in Present Size Groups (Rows) by Initial Size Group*

Size (number of workers)	*1–5*	*6–10*	*11–15*	*16–20*	*21–25*	*26–30*	*31–40*	*41–50*	*51+*	*Missing*	*Row total*
1–5	30.0	40.0	16.7	3.3	6.7	0.0	3.3	0.0	0.0	0.0	60.0
6–10	0.0	44.4	22.2	11.1	0.0	0.0	0.0	0.0	22.2	0.0	18.0
11–15	16.7	33.3	0.0	0.0	0.0	0.0	0.0	0.0	50.0	0.0	12.0
16–20	0.0	0.0	0.0	0.0	0.0	0.0	0.0	0.0	0.0	0.0	0.0
21–25	100.0	0.0	0.0	0.0	0.0	0.0	0.0	0.0	0.0	0.0	2.0
26–30	0.0	0.0	0.0	0.0	0.0	0.0	0.0	0.0	0.0	0.0	0.0
31–40	0.0	0.0	0.0	0.0	0.0	0.0	0.0	0.0	0.0	0.0	0.0
41–50	0.0	0.0	0.0	0.0	0.0	0.0	0.0	100.0	0.0	33.3	2.0
Missing	66.7	0.0	0.0	0.0	0.0	0.0	0.0	0.0	0.0	0.0	6.0
Column total	26.0	36.0	14.0	4.0	4.0	0.0	2.0	2.0	10.0	2.0	100.0

Note: 50 firms in sample.

Table 12-25. *Percentages of Indian Metal Casting Firms in Present Size Groups (Rows) by Initial Size Group*

Size (number of workers)	*1–5*	*6–10*	*11–15*	*16–20*	*21–25*	*26–30*	*31–40*	*41–50*	*51+*	*Missing*	*Row total*
1-5	14.3	14.3	14.3	14.3	0.0	28.5	14.3	0.0	0.0	0.0	15.2
6–10	0.0	0.0	71.4	0.0	14.3	0.0	0.0	0.0	0.0	14.3	15.2
11–15	0.0	0.0	50.0	50.2	0.0	0.0	0.0	0.0	0.0	0.0	4.3
16–20	0.0	0.0	0.0	66.7	0.0	0.0	0.0	33.3	0.0	0.0	6.5
21–25	0.0	0.0	0.0	50.0	0.0	0.0	0.0	0.0	50.0	0.0	4.3
26–30	0.0	2.0	0.0	0.0	0.0	0.0	0.0	0.0	0.0	0.0	2.2
31–40	0.0	0.0	0.0	0.0	0.0	0.0	0.0	0.0	0.0	0.0	0.0
41–50	0.0	0.0	0.0	0.0	0.0	0.0	50.0	0.0	0.0	50.0	4.3
51+	0.0	9.1	9.1	0.0	0.0	0.0	0.0	0.0	72.7	9.1	24.0
Missing	0.0	9.1	36.4	9.1	0.0	9.1	9.1	0.0	27.2	0.0	24.0
Column total	2.2	8.7	26.0	10.9	2.2	6.5	6.5	2.2	30.4	4.4	100.0

Note: 46 firms in sample.

13

Product Markets and the Size of Firms

Product Quality and the Degree of Mechanization: Some Theoretical Considerations

The coexistence in an industry of small and large firms which make use of different levels of capital per unit of labor is usually explained by differences in factor prices. Firms producing a widely consumed common product, for example, laundry soap, still operate in segmented capital and labor markets. Typically the organized capital market makes the price of capital available to large firms cheaper than for small firms; while the organized labor market helps to establish wage levels faced by large firms at a higher level. Thus differences in prices of both factors of production would tend to make the profit maximizing large firm adopt a technology with a higher capital intensity.

Products, even when they are serving a basic standard purpose like washing clothes, are, however, not homogeneous. Differences in the perceived qualities of a product produced by different units could be partly related to packaging, but more substantive differences exist in terms of the physical characteristics of the product. Heterogeneity of products can produce differences in capital intensity by firm size even in the absence of factor price differences.

A firm is in general able to influence sales by varying the quality and price of the product and its advertising budget. An increase in the quality of the product normally requires an increase in costs; otherwise there is no need for anybody to produce a product of low quality. The increased cost could result from more use of labor relative to capital, or the other way around. There is no a priori reason for the higher-quality product to be more capital-intensive, but it will be argued below that technologically, and as an empirical phenomenon, this is likely to be so in many cases.

The optimum combination for a particular firm of product quality and the price which covers the cost of producing that quality has been analyzed by

Dorfman and Steiner (1964, pp. 826–36). If the demand function for a product is

$$q = f(p,x)$$

where q is the sales per unit of time and is a function of price, p, and a quality index x; and if the average cost of production c is a function of q and x: $c = c(q,x)$, then the condition for profit maximization is shown by Dorfman and Steiner to be

$$-f_p = \frac{f_x}{c_x}.$$

They write:

> The left hand side of the equation is the slope of the ordinary demand curve. The right hand side measures essentially the rate at which sales increase in response to increases in average costs incurred in order to increase quality. If the expression on the right hand side of the equation is greater than on the left, the indication is that an increase in quality will increase demand more than enough to compensate for the loss of sales that would result from an increase in price just sufficient to cover the increase in cost. Under such a circumstance quality should be increased.

The analysis suggests that quality variation for a product depends very much on the dispersal of consumers with respect to their sensitivity to quality variation on the one hand (f_x) and to price variation on the other (f_p). In the higher-income part of the market we will expect the first term to be relatively high and the second relatively low. Thus a firm catering to demand in this part of the market will tend to produce more of a higher-quality product, depending on the production cost characteristics of the particular product, that is, the value of c_x.

Two additional points should be noted at this stage. First, consumers who are quality conscious, and perhaps less sensitive than others to price variations, can be expected to respond significantly to advertising. High-quality products meant for the high-income segment of the market will usually have a substantial advertising expense in addition to the higher costs of production. The firm will, in theory, select the optimum package of advertising expense, product quality, and product price so as to maximize its profit. Dorfman and Steiner in their paper have provided a formal solution to this problem of joint optimization of advertising, quality, and price. It is not necessary to reproduce the results in detail here. It is sufficient to note that the amount of advertising expense incurred clearly depends, as does the choice of product quality, on the balancing at the margin of the extra sales due to an increase of advertising expense and the fall in demand in consequence of the increase in price required to cover the extra advertising cost.

The second point is that the argument given above suggests that some firms will concentrate their efforts on the high-income segment, but it does not provide a theory of complete specialization of different types of firms by levels of product quality. If then we do find in some cases, as discussed below, that specialization by quality does occur at the firm level, additional considerations will have to be brought into the analysis.

Product Quality, Capital Intensity, and Specialization

The set of considerations which have to be added to the above general argument to lead to the result that large firms, in some cases, will be more capital-intensive and will specialize in the production of higher-quality products, is related to the technology of the industry in question. We give below the example of the laundry soap industry, in which the modern process of soapmaking, yielding a higher-quality product, uses some sophisticated machines requiring a capital-labor ratio which is higher than the capital-labor ratios observed in less mechanical firms producing a lower quality of soap.

In figure 13-1, M represents the mechanical techniques and *NM* the nonmechanical. The isoquant M is drawn to show that there is a lower limit to the feasible value of the capital-labor ratio, because after a point adding more labor to the process does not save capital significantly. The *NM* isoquant is well-behaved over its entire range but lies below and to the right of the isoquant M. The nonmechanical technique is not dominated by the mechanical technique in the sense of it requiring more of both factors of production to produce a unit of output. Rather, it requires much more labor, which is balanced by a smaller requirement for capital.

Conversely, the unit isoquants and price lines are drawn to show that, with the same factor price ratio facing the firms, those choosing the nonmechanized technique have the lower costs. But firms may still use the mechanized technique because the higher cost of producing the superior product is offset by a higher market price. The total cost and price gap between the superior and inferior products is also likely to be widened because it pays firms producing the superior product to spend more on advertising.

Several conclusions follow if the above is a reasonable stylization of the industry.

- Some firms will be producing with a much higher capital-labor ratio, even with similar factor price ratios facing all firms. The market price of their product will be significantly higher, as will the advertising expense.
- The firms producing for the "superior" segment of the market are likely

Figure 13-1. *Isoquants for Mechanized and Nonmechanized Soapmaking*

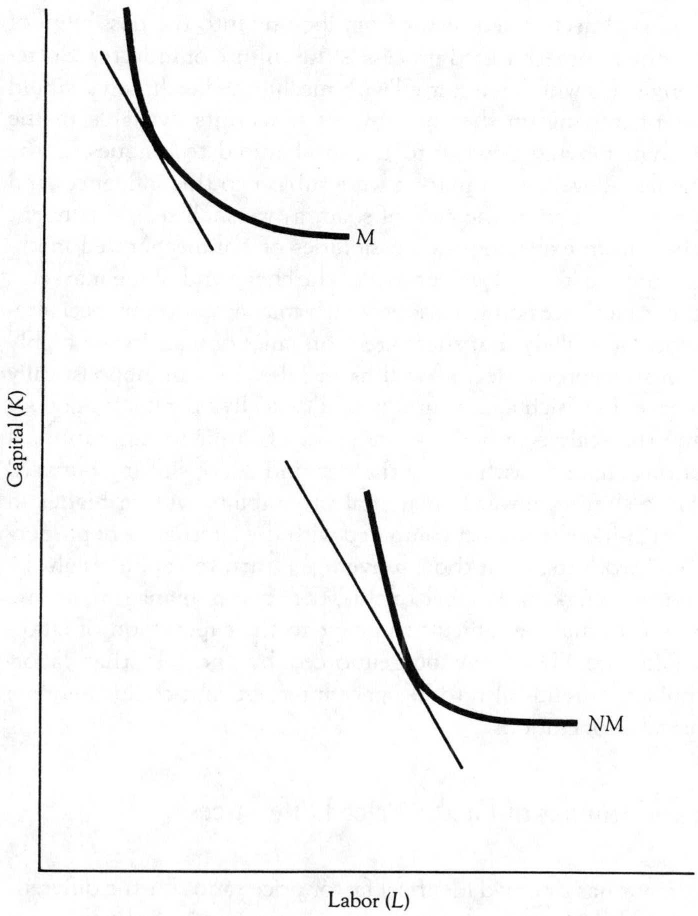

to be large because of the indivisibility of the capital equipment involved in production by the mechanized process. A minimum size of plant is necessary for economical operation with this process.

- The more mechanized technique in the larger firms is appropriate to producing a higher-quality product that can be sold at a relatively high price. It cannot be used to manufacture a product of lower quality which can compete with nonmechanized units catering to the needs of the lower end of the market.

Large firms will then specialize in production for the high-quality market. They can, of course, include on their premises—or within the same organizational structure of the firm—units using the nonmechanized process producing for the lower end of the market. But there are several reasons why such diversification may not take place.

A noneconomic factor which has probably played a role in preventing entrepreneurs in the mechanized sector from looking into the possibility of expansion with the nonmechanized process is the culture of industry. Entrepreneurs and engineers who have started with mechanized techniques would think in terms of moving up the spectrum of blueprints available in the world, rather than moving down into nonmechanized techniques. If the firms that originally developed a market were subject to this influence (and they seem to have been so in the case of soap, as we shall see), this might discourage others from exploring the possibilities of nonmechanized methods with a resultant lag in the development of the cheap end of the market.

There are other more economic reasons which may account for specialization. For instance, it is likely that there are economies of scale in the highly mechanized firms—in production as well as in sales. In a monopolistically competitive market in which a large firm would typically operate, it may not have exhausted the scale economies at its point of profit maximization. If market conditions change—with either the demand curve shifting outward or the cost curve shifting inward—marginal profitability will be higher in expanding the high-income market compared with the alternative of producing lower-quality products, even though average return to capital might be the same in the two markets. Further to this, the style of management in a capital-intensive firm may be difficult to adapt to the supervision of labor-intensive manufacture. This may be reinforced by the fact that labor-intensive manufacture tends also to be space-intensive and could therefore require a dispersal of operations.

The Cause and Impact of Factor-Price Differences

So far the argument has assumed identical factor price ratios for the different types of firms. In other chapters of this book we have examined the evidence about firms of different sizes facing different prices for labor and capital. The traditional explanation of the coexistence of firms producing the same kind of product, but using techniques of different degrees of mechanization, is that this phenomenon is in response to such differences in factor prices.

But firms using different techniques can also coexist because their products, though bearing the same name, supply different markets and are barely competitive with each other. In this case, and if the firms producing the high-quality product are oligopolistic, then their relatively high productivity could be the cause rather than the result of high wages, especially as high capital intensity implies that wages are a small part of total costs. However, it should be noted that, even where this is the case, high wages may still feed back to a choice of yet more capital-intensive techniques for producing the high-quality product.

The Impact of Government Restrictions in the Large-Scale Sector

We have so far been discussing the problem of firms specializing in different parts of the market in the absence of any government interventions. The Indian economic scene, however, has been, at least since the mid-1950s heavily influenced by policies which put restrictions of one kind or another on the large-scale units with a view to promoting the interests of the small-scale sector.

The question arises whether this type of intervention would tend to produce the kind of specialization by the quality of product which has been discussed in this section. On the question of the excise duty, the answer is simple. If, as has been the case in India, the excise duty is on the *quantity* produced by the mechanized units, the marginal profit from increasing quantity is reduced, and firms will find it more profitable at the margin to shift to higher-quality products than otherwise.

Does quantitative restriction on capacity or output of the mechanized firms have a similarly unequivocal effect on the quality of the product? One effect is clear. The restriction will increase the relative price of the higher-quality product, and it will pay the firm to move up in the quality spectrum. Apart from this price effect, demand can be expected to shift more to high-quality products with increase in income over time. Producers in the mechanized sector will be more inclined to take advantage of this market trend when they have a restricted capacity.

An Example from the Soap Industry

Soap production per head in India was for long at very low levels. The large firms specialized on the higher end of the quality range and did not adopt nonmechanized methods. It seems that the lower end of the range remained rather unexploited, at least until after independence and partition, when migrants rapidly developed the nonmechanized sector (at least in North India; see chapter 12).

Two Alternative Technologies

Soapmaking in South Asia makes use of two sharply differentiated types of technologies. One uses electric power; the other is traditional and labor-intensive and does not use power.

In India, government policy has defined the small sector as that which does not use power (irrespective of the size of the unit), and various advantages accrue to this sector, for example, exemption from excise duties and liberal permission to expand capacity. This policy may have contributed to the sharp demarcation between the two techniques. But the experience of Bangladesh, where there is no report of such legislation, suggests that labor-intensive units still do not make use of electric power in the production of soap (Mubin and Forsyth 1982). Evidently, the gap between the two alternative processes of producing soap is very large—and in effect, they lead to products which are separated in the chain of substitutes by a discrete interval. This point is discussed further below with reference to the World Bank survey of the Indian soap industry.

The technological disadvantage of the labor-intensive process results in a minimum involvement of these firms in the production of toilet soap. In the World Bank survey, it was found that 76 percent of the firms surveyed in the small-scale sector had 100 percent of their sales accounted for by laundry soap, and another 10 percent of the firms had 90 percent or more. Only 6 percent of the firms had a substantial proportion (between 50 and 60 percent) of sales contributed by products other than laundry soap.

Product Market Segmentation

The differential in price between the inferior and superior soap ensures their coexistence in the market. It is clear that a smaller proportion of the total expenditure on soap will be spent on the inferior product as we go up the income scale of households, and it may not be consumed at all beyond some level of income. If different classes of consumer distinguished by income level are primarily interested in one or the other of the two types of products, then the price elasticity of substitution between the two types will be low. In this case capital-intensive (large) firms producing the superior product will tend to compete more in terms of the quality than the price of the product.

The ILO study in Bangladesh devoted attention to the nature of markets for the two basic types of soap (no such market survey is available for India). According to this survey, the average retail price of laundry soap made by the capital-intensive process for the sample of consumers was Tk7.60 per pound as against Tk3.49 per pound for the labor-intensive soap. It is clear from the data that the higher-income urban consumers constitute a market segment for soap in the sense discussed above. Labor-intensive soap is a sizable proportion of total consumer expenditures on soap up to an income level of Tk750 for the urban areas, but it drops to insignificant proportions beyond this level.

Price and Quality in the Indian Soap Industry

Detailed data on the type of soap produced were collected from 35 nonpower units and from 5 power sector firms in and around Delhi.[1] Most firms

1. For a full report, see Suri (1983). The date of this survey is May–July 1982.

Table 13-1. *Distribution of Sample Firms by Employment Size and Price of Soap Produced*

	Price per kilogram (rupees)							
Employment size (number of workers)	2.50–3.50	3.50–4.50	4.50–5.50	5.50–6.50	6.50–7.50	7.50–8.50	8.50–9.50	Total
Nonpower sector								
0–5	—	—	5	1	1	—	—	7
6–10	3	2	1	4	—	—	—	10
11–20	—	1	1	2	—	—	—	4
21–50	—	3	1	1	—	—	—	5
51–100	—	—	—	1	3	—	—	4
101 and above	—	—	—	—	3	2	—	5
Total	3	6	8	9	7	2	—	35
Power sector	—	—	—	—	1	—	4	5
Grand total	3	6	8	9	8	2	4	40

— No firms.
Source: World Bank data.

produced more than one brand of soap. Table 13-1 gives the distribution of different brands of soap by size of firm and the price range.[2] The brands produced by the power sector firm are concentrated at the highest price range—and sold for twice as much as some of the lower-priced brands produced by the nonpower units. Nonpower units—particularly the larger ones—in this group were, however, producing some brands which were near the price level of the expensive power sector brands without actually getting into the range. The overlap in the price distribution of brands produced by power and nonpower sector firms is minimal. But it cannot be said that nonpower sector firms produced only very cheap soap. Most brands in firms of less than 50 employment size were in the price range Tk3.50–Tk6.50, while those produced by larger nonpower firms were in the range Tk6.50–Tk8.50.

The material suggests (and it was confined to the field interviews) that in the twenty years since the nonpower soap industry had been established, considerable progress had been made in the quality of soap produced. Nonpower sector firms climb up the quality range within the confines of the labor-intensive technology by using more expensive oil and by producing soap with a lower quantity of fillers. High-quality pure soap is desired by some consumers—although even such soap does not have the advantage of improved lathering properties, which the mechanized method of plodding, milling, and distilling is able to contribute.

An attempt was made to study the relationship between price, quality of product, and firm size with the help of the material collected from the Bank's

2. The inquiry could not cover all the brands of soap produced by each firm. A sample of forty brands of soap was collected from the survey units.

Table 13-2. *Determinants of the Ex-Factory Price of Laundry Soap*

Nonpower firms only ($N = 35$)			
$P = 1.58 +$	$0.08T$		$R^2 = 0.705$
	(78.7)		
$P = 1.93 +$	$0.06T +$	$0.008E$	$R^2 = 0.783$
	(55.0)	(11.6)	
All firms (including power sector firms) ($N = 40$)			
$P = 1.51 +$	$0.08T$		$R^2 = 0.570$
	(50.3)		
$P = 1.57 +$	$0.07T +$	$0.008E$	$R^2 = 0.785$
	(71.7)	(36.9)	

Note: P = Ex-factory price per kilogram, T = TFM (percentage of total fatty matter by weight), and E = employment size of firm. (F rates are given in parentheses.)

follow-up survey. The technique of analysis used is that of hedonic regressions, which express the price of a good as a function of several of its characteristics. This may provide a concise description of the structure of prices across quality specifications and enable us to undertake a more detailed analysis of the determinants of product superiority as revealed by price.

Chemically, the quality of soap can be identified by the total fatty matter (TFM) present in the soap, by the amount of pure alkali present, and by the quality of fillers per unit weight.[3] The sample of soap collected during our special survey was subjected to chemical analysis, and the data were used to see the relationship of these quality indicators to price. The size of firm was included in the hedonic price equations. However, the interpretation of the size variable, if it were significant, is not clear-cut. On the one hand, it could capture the quality of the product, which is not easily quantified by the three chemical characteristics which we measured. On the other hand, it might represent the effect of market power exercised by the big firms through advertising brand names and so forth, as discussed above. Most likely the size effect includes the influence of both sets of factors in unknown proportions.

Table 13-2 reports the relevant equations, for nonpower firms only, and for the entire sample. Of the three quality characteristics used, TFM was the most powerful and was preferred when used by itself. The firm size variable was also significant in the equations for both the samples, but there was a major difference in its contribution to the variance explained. For the nonpower firms the additional contribution was only 8 percent, but for all firms taken together it was nearly 30 percent. In both cases the final equation performed very well with an R^2 of 0.78.

3. The higher the TFM, the better the quality. The amount of pure alkali and fillers reduces the quality of soap.

The explanation of the difference in the strength of the firm size variable in the two equations is that there is a lower simple correlation between employment size and measured quality (TFM) among the nonpower firms taken by themselves. This bears out the points made earlier that mechanical firms produce soap of a quality not fully measured by chemical analysis, at the same time that they exercise greater market power.

The Cost Structure of Laundry Soap in the Two Sectors and Choice of Techniques

For cost data, we have to go back to a study of laundry soap undertaken for the Indian government Bureau of Costs and Prices in 1970. Some relevant figures are reproduced in tables 13-3 and 13-4. The three points brought out by these tables are:

- The capital and wage costs per unit of laundry soap produced are significantly higher in the power sector, so that in the absence of the price difference, production of laundry soap in the mechanized sector would be unprofitable.
- The proportion of nonmaterial costs accounted for by selling expenses are very different in the two sectors—8.6 and 25.9 percent for advertising expenses alone, and 10.5 and 39.7 percent including selling commission.
- Wage costs, which account for 25.8 percent of nonmaterial costs in the small units, amount to only 9.1 percent of such costs for large units.

The first two points are consistent with the theoretical ideas advanced in the first section. The last point bears on a traditional issue in the literature, namely, that higher wages in the large-scale sector are responsible for higher capital intensity. The alternative argument of this chapter has been that differentiation in the product market, and the difference in technology required to produce a commodity of different qualities, may in some industries be responsible for the coexistence of firms of widely different degrees of mechanization. The small share of wages in the large-scale sector adds strength to the argument that the choice of techniques in this sector is not highly sensitive to wage variation.

We have so far been discussing the choice between the nonmechanized and mechanized techniques. However, the sophistication of the mechanized soap industry plant is itself subject to variation, and ancillary activities can be more or less mechanized. The World Bank's survey included three firms in the large-scale sector. Data on wages, capital per worker, and labor productivity for these firms are given in table 13-5. There is remarkable difference in the wage levels of the three mechanized firms—unit I faces wages

Table 13-3. *Cost per Ton of Laundry Soap*
(rupees)

Unit size	*Total cost*	*Raw material cost*	*Total cost minus material cost*	*Conversion cost*		*Other costs (include packing, overhead, depreciation)*	*Advertising and distribution costs*	*Selling commission*	*Excise*
				Wage cost (includes labor cost of packing)	*Fuel and power cost*				
Small	1,839	1,630	209	54	15	116	18	4	Nil
				(25.8)			(8.6)	(1.9)	
Large	2,805	1,829	976	89	60	226	253	135	207
				(9.1)			(25.9)	(13.8)	

Note: Figures in parentheses give the percentages of total nonmaterial costs in each sector.
Source: Government of India Enquiry into Prices of Soap, Bureau of Costs and Prices, 1970.

Table 13-4. *Value Added, Capital and Wage Cost in Laundry Soap, 1970*

Category	Per ton: Value added	Per ton: Value added minus excise	Per ton: Capital cost	Per ton: Wage cost	Cost per thousand rupees of value added: Capital	Cost per thousand rupees of value added: Labor
Five large units[a]						
A	839		873	80	1,040	105
B		638			1,368	125
Seven small units	348	—	509	58	1,463	167

— No excise tax paid.

a. A gives the capital and wage cost per thousand rupees of value added without taking into account the excise tax paid by the large units. B gives the same calculations on the basis of value added after payment of the excise tax.

Source: Government of India Enquiry into Prices of Soap, Bureau of Costs and Prices, 1970.

which are only about 30 percent higher than in the nonmechanized sector.[4] The wage levels are also closely correlated with capitalization and labor productivity, and it therefore seems likely that the wage differences do influence the techniques of production within this sector.

However, the experience of unit I suggests that the *transition* from the nonmechanized to mechanized methods of soapmaking has little to do with wage variations. This was brought out in interviews with the management of unit I during the field survey. Historically the decision to obtain the Mazzoni plant was taken in order that the firm might be able to break into the high-quality market. It was mentioned during the interview that there was a relationship between the quantity produced and the quality of the product. In order to expand production, the firm had to tap marketing channels other than the traditional ones. These marketing methods involved the sale of higher-quality products. While the desire for expansion thus involved switching to higher-quality ranges, two other points were mentioned by the entrepreneur which suggested that mechanization was partly induced by supply constraints. These points, however, had nothing to do with wage levels. One of the constraints was that of space, which has already been mentioned in the theoretical discussion above. It is generally difficult to increase capacity of soap production beyond, say, one ton per day with the traditional process. A second point mentioned was the management problem. As the volume of production increases, the operations of stamping, wrapping, and packing would require a large increase in the labor force if they were carried out with the manual methods of the traditional system. The difficulty of managing too many workers, it was suggested, might induce some degree of mechanization in these processes.

4. We should recall the difference in wage levels noted in chapter 7 between the sample and census sectors of the ASI.

Table 13-5. *Wage Productivity and Capital per Worker in the Power Sector*

			Value of fixed capital (thousands of rupees)				
Unit	*Unskilled wage (rupees per month)*	*Number employed (all shifts worked)*	*Land and building*	*Plant, machinery, and fixtures*	*Total*	*Value of fixed capital per worker (thousands of rupees)*	*Production per worker (tons per year)*
I[a]	285	80	213	3,018	3,231	40.4	12.83
II	561	553	11,943	14,968	26,911	48.7	25.34
III	965	170	5,405	6,096	11,501	67.7	58.16

a. Unit I had a disproportionately low investment in land and building because it started as a unit producing labor-intensive soap and had only recently acquired a Mazzoni plant. The layout of its factory had not yet been designed for large-scale production, as was the case of the other two units.

Source: World Bank data.

An Example from the Textile Industry

Another example of the continued coexistence of mechanized and non-mechanized techniques based on differences in product quality, and serving markets of different income levels, is provided by the textile industry in India. In chapter 4, we discussed the relative profitabilities of handlooms, powerlooms, and mills. While the three techniques can be ordered in terms of their capital-labor ratios for a homogeneous product, and wage differentials *could* explain their coexistence, the switchover interest rate for the survival of handlooms was seen to be very high, if the World Bank's survey data are to be considered representative. Yet the handlooms have successfully competed with mills ever since the establishment of factory industry in India.

In the period before World War II when protection of the handlooms was not an element of government policy, the handlooms never lost ground absolutely. On the contrary, about a third of the increase in domestic production was contributed by handlooms (mill output grew fastest, for powerlooms became important only after World War II). Before World War I the increase in handloom production took place in free competition with foreign imports as well as domestic factory production (see Mazumdar 1984a, part I, section 2; and Little 1982b). After independence, the mills lost heavily in cloth production because of government fiscal policy and other restrictions. But although the new powerloom sector was the most dynamic, handlooms continued to expand absolutely in spite of the fact that the effective incidence of the fiscal measures was no higher for powerlooms than handlooms (Mazumdar 1984a, p. 16).

An important reason for the steady but slow growth of the handloom sector is the partial specialization of the three sectors in different types of cloth. A rough indication of the quality of cloth produced is given by the count group of yarn which is used by the three sectors. Table 13-6 gives the

Table 13-6. *Percentage Distribution of the Total Quantity of Yarn Consumed in Different Sectors by Count Group, 1942–79*

Count	*1942*		*1979*		
group	*Mills*	*Handloom*	*Mills*	*Powerloom*	*Handloom*
1s–10s	8.1	20.0	10.7	8.7	24.1
11s–20s	49.6	34.4	34.1	18.2	33.2
21s–30s	25.7	19.6	37.3	21.3	13.8
31s–40s	11.6	14.2	18.2	32.4	15.5
40s	5.0	11.8	4.5	19.4	13.4

Note: Civil deliveries of yarn in hank form are considered to be consumed by the handloom sector; the rest by the powerloom.

Source: Report of the Fact-Finding Committee on the Textile Industry, 1942, Government of India; *Handbook of Statistics,* 1979, Indian Cotton Manufacturers' Federation, 1980.

percentage distribution of yarn by count groups used by the three sectors for 1979 and for 1942. In 1979, compared with the mills, the handlooms wove a significantly larger proportion of cloth from coarse yarn of less than 10s, as well as from fine yarn of more than 40s. This pattern is seen also in the figures for 1942. During the 40-year period, mills have clearly "gone finer," with a reallocation of 15 percentage points from the 11s–12s count group to the 21s–30s and 31s–40s count groups. Handlooms, conversely, had roughly the same distribution by count groups at both dates, with relative specialization in coarse and very fine cloth.

The fact that handlooms specialize in the production of cloth for poor people is verified by data collected on consumption in the seventeenth round of the National Sample Survey. The figures presented in table 13-7 show the importance of the lowest expenditure class in the consumption of the handloom sector. The other specialization by handlooms in fine cloth (using yarn of more than 40s count) is a characteristic of the clothing industry throughout the world, which requires both labor and skill intensity for the production of superior garments. (It should be remembered that handloom saris require weaving and no tailoring.)

The increase in the percentage of yarn woven in the higher count groups by the factory sector underestimates the trend in this sector toward a better quality of product. The strength of the cloth produced depends not only on the count of yarn used but also on the closeness of the weave. The mills in the last two decades have been weaving cloth with a larger number of picks per meter. Equally important has been the very substantial increase in mechanical preshrinking, dyeing, and printing of cloth within the mill sector—and the proportion of automatic looms installed (Mazumdar 1984a, part III, section 1). The factory industry has thus been trying to create a niche for itself by producing more durable, mechanically processed cloth—no doubt induced partly by government restrictions and partly by competition from less mechanized producers.

An Example from the Machine Tool Industry

The machine tool industry provides an interesting contrast to the patterns of technical choice and product quality encountered in textiles and soap manu-

Table 13-7. *Handloom Clothing as a Percentage of the Total of Mill-Made, Handloom, and Khadi Clothing*

	Monthly household expenditure class (rupees)					
Locality	*0–25*	*25–50*	*50–100*	*100–150*	*150–300*	*300 plus*
Rural	20.8	10.8	15.9	15.7	10.3	12.5
Urban	57.1	15.6	16.2	12.0	14.1	12.0

Source: National Sample Survey, seventeenth round, 1961–62.

facturing. Our analysis of machine tool manufacturing in chapter 10 revealed that there was no discrete break in techniques of production between small and large-scale enterprises engaged in the production of workshop machine tools such as lathes, drill presses, and milling machines. Rather, these enterprises exhibited a continuous range of factor substitution possibilities along a single production surface.

The surveys did, however, reveal a tendency for large firms (those with more than 100 employees) to be more capital-intensive and to employ somewhat newer capital equipment than firms in smaller size categories. But within-group variation for the large-scale firms was substantial, and the differences between capital intensity among size groups were frequently not statistically significant at conventional levels. Smaller-scale firms, however, were more skill-intensive, and the scope for capital–skilled labor substitution was revealed to be substantial. Under these circumstances, the relationship between product quality and the technique of production is not so clear-cut as with soap and textile manufacturing.

As a supplement to our basic survey, we commissioned a special survey of prices and characteristics of the most popular product of the mechanical engineering firms surveyed, machine shop lathes. Fifty-seven observations covering product prices and specifications were obtained from firms within the machine tool sample. Engineering principles and a general knowledge of the machine tool market provided guidance as to the machine characteristics to include in the survey. In this section we use the technique of hedonic regression to explore the relationship between product quality, price, and firm size.

The Impact of Restrictions on Large-Scale Enterprises

In chapter 4 we noted that the capacity of large-scale enterprises producing machine tools was restricted to the levels prevailing when SSI reservation was introduced. As all machine shop tools, including lathes, milling machines, drill presses, and grinders, are restricted to production by the small-enterprise sector, larger firms cannot increase their output of these items. Responses to our surveys indicated that, given the restrictions on expansion, many of the larger enterprises had attempted to upgrade the type of machine produced, both to take advantage of the export market, which is unrestricted, and to secure a market niche which was less open to competition from small-scale producers.

In order to move into the higher-quality end of the market, however, larger firms would have to improve on the characteristics which determine higher product prices. The machine tool market is not characterized by unrelated segments. Purchasers trade off price against product characteristics which determine the durability and flexibility of the machine. These characteristics are related in turn to design and care in manufacture, but unlike the case of soap or textile manufacturing are not technologically infeasible for firms in the small-enterprise sector to achieve. Thus, despite the capacity restrictions, small and large-scale firms in machine tool manufacturing continue to compete in the same market.

The Observed Structure of Prices

Table 13-8 presents the results of a hedonic regression of price per lathe on the product characteristics which our engineering surveys indicated were the major determinants of choice by purchasers. The regression performs rather well. It explains about 70 percent of the variation in machine prices. In most cases signs of the independent variables conform to a priori expectations. The hedonic regression is a reduced form of the structural equations operating on the supply and demand sides of the market. Variations in a particular product characteristic affect both cost (and hence the supply price) and willingness to pay. It is therefore generally impossible to distinguish between a characteristic's effects on supply and demand prices. It is possible, however, to describe in a general way, using our understanding of the production

Table 13-8. *Hedonic Determinants of Price per Lathe*

Variable	*Coefficient*
Length of bed (millimeters)	−2.09
	(2.15)
Width of bed (millimeters)	36.55[a]
	(2.56)
Height of center (millimeters)	21.38
	(18.29)
Admit between centers (millimeters)	6.06[a]
	(2.56)
Horsepower of motor	−617.33
	(841.11)
Gear drive	1,249.99[a]
(1 = present)	(594.26)
Bed preparation	2,611.51[a]
(1 = hand-scraped)	(614.62)
Availability of test	2,241.77[a]
chart and specifications	(639.18)
(1 = present)	
Availability of spares	−490.11
(1 = present)	(606.32)
Constant	−5,019.85
	2,710.76
R^2	0.699
F(9,56)	14.47
Observations	66

Note: Standard error given in parentheses.

a. Significant at the 0.95 level or above.

process and the equipment market, the role of each of the independent variables in the regression.

Four variables represent the major capacity specifications of the machine. The length of the bed and its width determine the potential size of work which can be accommodated by the lathe. In addition these two specifications determine the size and difficulty of forming the cast body of the machine. The height of the chuck and the admit between centers determine the maximum diameter and length of the work which can be turned. It is possible to trade off, to a degree, the overall size of the lathe, as determined by length and width, with height of center and admit, which permits greater use of the potential working area. Greater height of center and admit require more careful machining and assembly, however, which thus adds to production cost and skill requirements.

Of the four size-related variables two, one in each pair, are significant determinants of the product price (at the 0.95 level). These are width of bed and of admit between centers. Length of bed, which determines the maximum range of admit, and height of center do not significantly affect variations in product price within our sample.

The remaining characteristics in the regression primarily represent variations in quality given the overall size of machine established by the four preceding variables. First, we include the horsepower of the motor and the type of drive assembly. The combination of motive power and gearing determines the ability of the lathe to turn the work and the degree of vibration encountered. Gear drives, which are represented in the regression by a dichotomous variable, permit greater turning power for a given horsepower of motor at lower levels of vibration. Thus, for finer work gear drives are preferred to belt drives, which have a lower production cost. In the regression the variable for gear drive is significantly and positively related to price while that for horsepower is not.

Second, two other machine characteristics were found to affect product price significantly. The first was the presence of hand scraping to determine bed level. This technique, which produces greater uniformity in the bed at the cost of increased skill requirements, has a strong positive association with product price. The presence of a test chart and accompanying specifications was also strongly and positively associated with price. This variable to some extent reflects consistency in design and ability to meet standard specifications which are important for precision use of the machine. Interestingly, the availability of spares, which was chosen as an indicator of the extent of after-sales service, was not significantly related to price.

Firm Size and Machine Quality

Table 13-9 provides a cross-tabulation of machine prices and characteristics by employment size of firm. Machine price rises with firm size up to the level

Table 13-9. *Cross-Tabulation of Lathe Characteristics by Employment Size of Firm*

	Firm size (N = number of workers)			
Characteristic	N < *50*	*50* < N < *100*	*100* < N < *200*	*200* < N
Price	11,682.9	14,666.6	16,500.0	15,166.6
	(3,179.5)	(3,500.0)	(1,732.0)	(2,565.8)
Length of bed (millimeters)	1,692.3	1,773.8	1,802.0	1,742.0
	(319.3)	(389.2)	(236.9)	(281.2)
Width of bed (millimeters)	269.9	271.3	295.7	290.0
	(35.8)	(36.4)	(31.5)	(34.7)
Height of centers (millimeters)	211.4	215.1	219.0	216.0
	(29.2)	(28.1)	(32.0)	(32.9)
Admit between centers (millimeters)	920.4	991.1	908.25	811.3
	(227.4)	(276.2)	(183.6)	(198.4)
Horsepower of motor	2.27	1.89	2.52	2.41
	(0.22)	(0.31)	(0.29)	(0.42)
Gear drive (1 = present)	0.24	0.45	0.75	0.66
	(2.75)	(1.52)	(1.32)	(0.99)
Hand-scraped bed	0.00	0.45	0.75	0.66
	(0.00)	(1.52)	(1.32)	(0.99)
Test chart	0.07	0.66	1.00	1.00
	(1.67)	(0.99)	(0.00)	(0.00)

Note: Standard error given in parentheses.

of 100–99 employees, after which it falls. The only significant difference between groups occurs, however, between small firms (less than 50 employees) and all larger sizes. Machine prices do not differ significantly between groups within the larger size range.

Machine specifications show a similar pattern of increase up to the size 100–99 employees, followed by a fall in each of the capacity dimensions in the largest size range. In most cases, however, differences between means of the various size classes are not significant. Only one variable, width, shows a significant difference between the smallest and other size groups in which the medium-large size class exceeds that for the small size class at the 0.90 level. There is thus little variation in machine capacity associated with size of firm. Both small and large enterprises manufacture lathes of similar size specifications.

Differences between firms arise most strikingly in the quality related characteristics that are significantly related to price variations, gear drive, bed preparation, and availability of test chart and specifications. Hand scraping of the bed is done almost exclusively by firms with more than 50 employees, while availability of test data and specifications occurs in firms of more than 100 workers. Presence of gear drive is observed with greatest frequency in size category 100–99, which shows a significantly higher incidence than the smaller size categories.

Apparently, the observed differences in product prices are primarily due to differences in such characteristics as gearing, bed preparation, and design specifications. These are attributes which to a large extent determine superior quality as opposed to increased capacity of the individual machine. The quality related characteristics differ between small and larger firms, with firms below 50 employees exhibiting relatively fewer quality related variables. They appear to peak, however, in the medium-large size group and then to diminish at the upper end of the size range.

Some Welfare Implications of the Choice of Techniques

A tentative conclusion of this chapter is that in some industries the degree of mechanization could be more significantly related to product quality than to factor price differentials. For such industries the case for the subsidization of the less mechanized process will depend not just on the social opportunity cost of factors of production (relative to their market prices) but also on the welfare consequences of promoting the consumption of goods of a particular quality.

Coarse handloom cloth is mostly bought by the lowest income classes in India. With a significant rise in real income, the market for this type of product will probably tend to disappear. But so long as per capita income growth stagnates, it is possible to argue for subsidization of handlooms on distributional grounds. However, our account of Indian textile policies in chapter 4 shows that evaluation of results is complicated by the unintended encouragement of powerlooms and by the fact that synthetic fibers have not been given a chance to compete for the low-income market.

The case of laundry soap is different. Although hard data do not exist, the prevalent opinion in the industry is that much of the growth of the labor-intensive sector has been caused by new demand from a growing class of consumer that previously used very little soap. Demand for labor-intensive soap probably would have grown anyway because of rising urban incomes in the post-independence era. But the establishment of the industry in North India was helped by the migration of entrepreneurs after partition from Sind (Pakistan), where the industry had taken root. The industry in Delhi seems to have been started by these migrants before the restrictions on large-scale soap production, but it was clearly helped in its growth by this government policy.

Protection against technologically advanced domestic factory production is analogous to the infant industry argument for protection against foreign competition. The nonmechanized soap industry is a case in which protected infants have made some progress, for there was ample evidence of product improvement despite the fact that improvement by using power was ruled out. While this may be held to have justified such protection, it is doubtful whether it should become a permanent feature.

There are good distributive reasons for retaining an excise tax on superior soap. But it should be ad valorem, not specific, for the latter gives an artificial inducement to large-scale producers to move further up the quality range. Given such an excise tax, is there any ground for retaining the restrictions on productive capacity in the power-using sector? Our earlier discussion of the history and technology of the industry suggests that it is unlikely that the power-using sector would seriously invade the low-cost territory, though the latter might eventually shrink as incomes rise. The main reason to fear such an invasion would be the resultant decline in the demand for low-wage labor. This would certainly be the direct effect in the soap industry itself. But we have not evaluated the claim of the technologically advanced units that the production of their material inputs is far more labor-using than that of the nonpower sector.

The machine tool industry offers a marked contrast to the two preceding examples. There is no sharp break in technique of production associated with small and large firms; rather, there appears to be continuous substitution along a range of factor intensities. This continuity of technical choice is accompanied by a similar continuity in product quality. Small and large firms compete in essentially the same market for machine shop lathes and produce equipment with similar dimensional characteristics. Superior product quality and price are determined primarily by superior design characteristics, which entail a greater degree of technological mastery in production. These characteristics are exhibited primarily by firms in the size range 100–99 employees, which are superior to both smaller- and larger-scale competitors.

It is difficult to advance any argument on distributional grounds in favor of the capacity restrictions on large firms in this producers' goods industry. Lower-cost machines embody lower quality characteristics which can be objectively evaluated by purchasers. The fact that firms across the full size range of the industry produce machines of varying quality indicates that there is no potential benefit to be gained from restricting the output of any size class of producer.

14

Labor Markets in India and Elsewhere

We saw in chapter 7 that a feature of the economic scene in India was the wide difference in labor earnings in firms of different size groups. It is important to take a close look at these differentials and the reasons for them. The first section of this chapter brings together the evidence for recent years. The various sources give a convincing picture of a large differential. The question immediately arises: to what extent is this evidence of distortions in the labor market? As discussed in the second section, formal institutional impact on wages in the large-scale factory sector is of recent origin and became firmly established only in the post-independence era. Furthermore, the historical evidence shows that wages in large-scale factories were established at a relatively high level in periods preceding the growth of unions or government labor legislation. In the third section, some theories which might account for the existence of wage differences between small and large units are reviewed. Finally, the fourth section presents comparative evidence from a few other countries.

Evidence on the Wage Differential

The Central Statistical Organization Data

The Central Statistical Organization (cso) has provided an overall picture of wage differentials between the small and large sectors by states (India 1979). The major source of the information is the Annual Survey of Industry, which has been collecting information on "average annual wages" (that is, earnings) for workers for manufacturing units along with other data. Wages include all payments made to workers in cash, such as basic wages, dearness allowances, overtime payments, bonuses, and so forth, but do not include the money values of benefits in kind (India 1979, table 9, p. 60).[1]

1. The money value of benefits was not, however, separated for workers and salaried workers

Table 14-1. *Average Earnings in the Sample Sector of the ASI as a Percentage of Average Earnings in the Census Sector, India, 1958–75*

Period	*Bihar*	*Gujarat*	*Maharashtra*	*Tamil Nadu*	*Uttar Pradesh*	*West Bengal*	*All India*
1958–59	35	48	62	67	78	64	59
1964–65	32	50	61	47	59	60	45
1968–69	33	49	47	47	61	56	50
1970–71	30	49	48	49	53	51	49
1974–75	34	42	52	39	48	49	45

The ASI, covering all registered factories, is divided into two parts. The "census sector" (so called because information is obtained from all units in this sector) comprises factories employing 50 or more workers if they are using power and also factories employing 100 or more workers if they are not using power. The rest of the factories are covered on a sample basis—and this sector is called the "sample sector." Since information for the two sectors of the ASI is published separately, we have an important source for the study of wages as well as other economic variables for small and large factories.

Table 14-1 gives the earnings differentials for various states between the two sectors at different dates between 1958 and 1975. Differences in skill composition of the workers are not controlled for in this table, but the differentials seem too wide to be explained by that factor alone (it should be emphasized that salaried workers are *not* included in the data). As is discussed more fully in the next section, the influence of government labor legislation has had a more pronounced effect on wage determination in the post-independence period. The legislation extends to establishments covered by the Factory Act, comprising units in both the sample and census sectors of the ASI. If in fact wages under the Factory Act had been uniformly raised in all units under its purview, it would have squeezed the wage differentials between small and large factories during this period. Table 14-1, however, shows that in three of the six states covered—in Tamil Nadu, Uttar Pradesh, and West Bengal—the differential has in fact widened, while the other states show no marked trend.

Average wages in the unregistered sector, except for urban Gujarat, are even lower than in the sample sector (see table 14-2). But the differential between the urban unregistered and the sample sectors is smaller than that between the sample and census sectors. For example, in 1974–75 for all of India, while average wages in the sample sector were only 45 percent of those in the census sector, earnings in the urban unregistered sector were 80 percent of sample sector wages—and this pattern holds good for the individ-

for all years. Estimates had to be made on rough assumptions for this portion of earnings accruing to workers. For details, see pp. 63–64 of the report.

Table 14-2. *Average Earnings in Different Sectors of Employment, Selected Indian States, 1974–75*

		Annual wages (rupees)						
Period, source	*Sector*	*Bihar*	*Gujarat*	*Maharashtra*	*Tamil Nadu*	*Uttar Pradesh*	*West Bengal*	*All India*
					Registered manufacturing			
1974, ASI	Census	5,180	4,105	5,130	4,289	3,431	4,872	4,288
	Sample	1,763	1,732	2,669	1,666	1,636	2,402	1,913
		(34.03)	(42.19)	(52.03)	(38.84)	(47.68)	(49.30)	(44.61)
					Unregistered small-scale manufacturing			
July 1974–June 1975, NSS 29th round[a]	Urban	1,489	2,033	1,778	1,230	1,528	1,560	1,551
		(28.75)	(49.52)	(34.66)	(28.68)	(44.54)	(32.02)	(36.17)
	Rural	1,319	1,526	968	583	1,101	1,059	822
		(25.46)	(37.17)	(18.89)	(13.59)	(32.09)	(21.74)	(19.17)

Note: Figures in parentheses indicate percentages of wages in the ASI census sector.

a. NSS is National Sample Survey.

ual states reported in table 14-2. Average money wages in the rural unregistered sector were considerably lower, but benefits in kind are likely to be more important in rural than urban areas.

We conclude that the Factory Act, and the associated labor legislation, does not distinguish the high-wage sector from the low-wage one. Explanations for the wage differentials have to be sought for elsewhere.

The World Bank Surveys

Our next set of evidence on the wage differential between small and large units comes from the data collected in the World Bank's surveys. The average annual wage of unskilled workers in the four industries surveyed in North India is given in table 14-3. The units, although ranging in size, were all small, employing less than 150 workers. The powerloom units were located in a small provincial town in North India (Mau) where the wage level is significantly lower than in Delhi, the location of most of the other industry surveys. There is evidence of an increase in unskilled wages by size group within the small-scale sector surveyed, but it is not marked except for powerlooms. An idea of the difference in wages between these small units and the large factories can be obtained by looking at the data on earnings of the "lower-paid operatives" in cotton textile mills, which is regularly reported in the *Indian Labour Journal.* At the date of the surveys (1979), the annual earnings of these operatives in Delhi were Rs5,896, including only basic wages and dearness allowance. Bonuses and other benefits could add another 20 percent to this figure. In Kanpur, the nearest large city to Mau (about 150 miles away), the basic wage and dearness allowance of the low-grade textile workers amounted to Rs5,700 (*Indian Labour Journal* 1981, pp. 1749–1750). These figures are two to three times those shown in table 14-3 and support the findings from the cso material that wages in large units are probably as much as twice those in small establishments, even after controlling for skill difference.

Table 14-3. *Average Annual Wage of Unskilled Workers in World Bank Surveys, 1979–80*
(rupees)

Firm size (number of workers)	*Machine tools*	*Powerlooms*	*Printing*	*Shoes*	*Soap*
1–10	2,172	—	2,364	—	2,664
11–25	2,316	1,140	2,436	2,256	2,832
26–50	2,700	1,428	2,604	3,000	2,964
51 plus	2,544	2,244	2,292	—	3,240

— No observation.

The Bombay Study

The most detailed and accurate analysis of the wage differential between small and large units in India comes from a research project undertaken by the World Bank in collaboration with the Economics Department of Bombay University.[2] It involved a special survey of about 5,000 male workers in Bombay City. It was decided to cover three sectors of the market: the single workers and the casuals;[3] the workers in the small-scale sector; and the workers in factories. Owner/workers and family workers were excluded, though they constituted an important proportion of the city's "informal" sector.

The sample frame was provided by lists of establishments in the city—for factories, the list maintained by the chief inspector of factories and, for small establishments, the registers maintained by the municipal ward offices. The sample number of workers to be interviewed was predetermined for each of the three sectors of the labor market, and the total for each sector was distributed among establishments belonging to different industrial groups so as to ensure that the sample of establishments reflected the industrial distribution of the wage earners for each sector of the market. For each of the factories selected, the quota of workers to be interviewed was selected at random from the employee rolls, while for the small-scale sector, whose average employment was 2.9 workers per establishment, all the wage earners in each of the sample establishments were surveyed. The size of the sample was about 2,700 workers in factories and 2,000 workers in small-scale units.

Casual workers are not attached to any establishments. It was, therefore, decided to survey 11,000 casual workers randomly selected from those who gathered for work in the well-known market places for casual workers, but taking care to ensure that the industrial composition of the sample reflected the industrial distribution of the population of single workers in the city, as derived from census data.

The sample was analyzed to study the difference in male earnings between the three sectors distinguished—casual, small-scale, and factory—and between relevant subsectors within them. The earnings of such workers are affected by their personal attributes as well as by their sector of employment. To investigate wage distortions we need to control for the skills of the workers.

The form of analysis used is multiple classification analysis: "a method of displaying the results of the analysis of variance especially when there are no significant interaction effects" (*Statistical Package for the Social Sciences* 1975).

2. For further details, see Deshpande (1979) and Mazumdar (1984b).

3. The Bombay labor market includes a large number of workers employed on daily contract, as and when work becomes available, much like old-style dock labor in industrial countries.

Table 14-4. *Determinants of the Log of Earnings of Bombay Manual Workers, 1978–79*
(multiple classification analysis; grand mean = 5.73)

		Unadjusted		*Adjusted for independents*		*Adjusted for independents plus covariates*	
Variable and category	N (1)	*Deviation* (2)	*Eta* (3)	*Deviation* (4)	*Beta* (5)	*Deviation* (6)	*Beta* (7)
Occupational sector (number of workers in factory)							
10–99	498	0.11	—	0.10	—	0.06	—
100–499	452	0.43	—	0.40	—	0.35	—
500–999	174	0.58	—	0.50	—	0.41	—
1,000 plus	1,339	0.54	—	0.53	—	0.41	—
Small	1,580	−0.35	—	−0.34	—	−0.26	—
Casual	900	−0.57	—	−0.54	—	−0.44	—
			0.77		0.73		0.58
English							
No English	3,878	−0.07	—	−0.03	—	−0.02	—
Some English	1,065	0.25	—	0.10	—	0.07	—
			0.22		0.09		0.06

Training							
Trained	1,233	0.28	—	0.09	—	0.08	—
Untrained	3,710	−0.09	—	−0.03	—	−0.03	—
			0.27		0.09		0.08
Education							
Illiterate	1,022	−0.10	—	−0.04	—	−0.10	—
Literate, no schooling	53	−0.34	—	0.08	—	−0.01	—
Grades 1–3	507	−0.03	—	−0.01	—	−0.05	—
Grades 4–6	1,319	−0.03	—	0.02	—	0.00	—
Grades 7–10	1,029	−0.02	—	−0.02	—	0.02	—
Grades 11–12	943	0.17	—	0.02	—	0.10	—
Diploma and degree	70	0.33	—	0.18	—	0.05	—
			0.17		0.05		0.12
Occupation							
Skilled blue collar	1,031	0.20	—	0.07	—	0.07	—
Unskilled blue collar	3,912	−0.05	—	−0.02	—	−0.02	—
			0.18		0.06		0.06
Multiple R^2					0.62		0.69

— Not applicable.

This form is particularly useful when the explanatory variables are in categoric terms, so that they can be entered into the analysis only as dummy variables. The details can best be explained by reference to table 14-4, which gives the results of the analysis of the log of earnings[4] in terms of five explanatory categories and two continuous variables, age and $(\text{age})^2$. The grand mean of 5.73 at the top of the table refers to the natural log of earnings of the entire sample. The headings at the left of the table are the explanatory variables, and within each group the categories which have been distinguished. Column 1 gives the number of observations falling under each category. Column 2 gives the gross deviation of the mean earnings of the particular category from the grand mean. Column 4 displays the "net" effect of each variable on earnings—that is, after the effect of the other variables is controlled for. These variables which are entered in categoric terms are called "independents." The statistics—eta and beta—correspond to simple and partial correlation coefficients. Eta (column 3) gives the proportion of the variance in earnings explained by each package of the subcategories used, for example, size and education.[5] Beta is the proportional reduction in the residual variation after the effect of the other variables has been taken into account. The beta statistics (columns 5 and 7) thus give an idea of the ranking of each explanatory variable in explaining the monthly earnings of workers. Column 7 gives the set of deviations of earnings from the grand mean for each category when we control for the other explanatory variables that are entered as continuous rather than categoric terms (called covariates). These are age and $(\text{age})^2$. It is seen that the gross deviations for the various categories are reduced when other factors are controlled for in columns 5 and 7.

The model performed well, explaining 69 percent of the variance. The program gives the significance of the variables included (not shown). All were significant at the 1 percent level. The significance of the interaction among the independent variables was also tested. Only two of the ten possible two-way interactions were significant—that between size and training and that between training and occupation. Even in these two cases the F values were low—2.26 and 2.30, respectively—compared with the explanatory variables taken singly.

This analysis of earnings shows the remarkable importance of the sector of work as the major determinant of earnings for manual workers in the Bombay labor market. This is revealed by the ranking given by the beta values as well as the spread of earnings between the subcategories within each variable. Thus, taking education as the next most important variable in the

4. Earnings are defined as total monthly earnings in cash *plus* income in kind (food, accommodation) in the main employment.

5. Note that the proportions will add to unity only if there is no interaction at all among the explanatory variables.

determination of earnings, the net spread in the log of earnings between illiterates and those with grades 11–12 as read from column 6 is 0.20. Conversely, the spread of the log of earnings between casual workers and those working in large factories is no less than 0.84. This implies that, controlling for other factors, earnings of workers in the largest factories are 2.3 times those of casual workers. It is also interesting to note that within the factory sector itself, earnings of manual workers are different after controlling for other factors. In particular, those working in factories employing 10–99 workers earn considerably less than those working in larger factories although they earn substantially more than casual workers or those in small establishments. The critical point within the factory sector seems to come at the lower end of the range 100–499 workers. The net differences in earnings in the three size classes above 100 workers are not very large.

The importance of the net effect of the size of establishment on earnings within the factory sector reinforces the point noted earlier about the large wage difference between the sample and census sectors of the ASI. The small-large wage difference exists in a significant way even within the sector covered by wage and labor legislation in India.

We have already noted that the two-way interaction between size and education was not significant. This was independently verified by looking at the percentage distribution of workers in the three sectors by educational levels: no striking difference was observed. Evidently, educational credentialism does not play any role in recruitment to the high-wage sector of the manual labor market in Bombay, as has sometimes been suggested in other developing countries.

Institutional Influence and Size-Related Wage Differentials

The basic question from the point of view of this book is how far this large differential between small and large units in Indian manufacturing is due to overt institutional influence, for example, trade unions or government wage legislation. This is germane to answering the question: "Is the wage differential evidence of substantial distortions in the labor market?"

In this section we approach this question from a historical perspective. When did an institutional influence on wages in the large-scale factory sector become important, and what is the evidence on wage differentials between small and large units before this period? We depend for our evidence on the Bombay labor market, on which substantial historical research has been done by one of the authors (Mazumdar 1973).

The Period before World War I

Factory industry in India had its beginning early in the second half of the nineteenth century and was concentrated in cotton textiles, jute, and coal.

There was no union or government influence on wage settlements. Wages were fixed by individual contracts between employers and workers. They seem to have varied a lot from mill to mill in the same city. This was perceived by some as a problem encouraging disputes. But the employers' organizations, such as the Bombay Mill Owners' Association, explicitly rejected as impractical the idea of imposing standardization (Kulkarni 1979).

A curious aspect of reports on the factory labor market from this period is the chaotic nature of the remarks made about it—complaints about "veritably a labor famine" existing side by side with testimonies about the adequate supply of labor (for a fuller description, see Mazumdar 1973). Managers in the industrial concerns were British and had to depend on intermediaries or jobbers for recruitment of labor. This system has been blamed for the periodic shortage of labor; but it is a controversial point, to say the least. During those early years, the high rate of absenteeism was a chronic complaint. The rural origin of the labor force led to instability of supply, and throughout the period and later, a persistent theme in the evidence given by employers to various commissions of inquiry on labor was that instability—in the sense of a lack of attachment to industrial work—was the major labor problem facing them. Yet the evidence suggests that employers nevertheless managed to develop a stable core of workers, which coexisted, however, with a much more unstable mass of labor—often hired on daily contract (Mazumdar 1973). The wage level in the industry was set high enough to attract stable labor with a higher supply price. Thus, it appeared that wages were higher than the alternative earnings of labor in sectors of the market where there was no need to retain stable labor.

As early as 1892 the collector of customs informed the Royal Commission on Labour, in response to the criticism of the chief inspector of factories, about the constancy of factory wages: "The point which Mr. Moon seems to me to miss is that though today wages have not risen they have continued greatly in excess of the wages the worker seems to have been earning or can earn in any other employment" (Report of the Royal Commission on Labour 1892, vol. 36, p. 128). Sixteen years later, the Indian Factory Commission confirmed this statement: "The wages of operatives in textile factories vary from place to place, but are everywhere considerably higher than those earned by the same class of men in other employment."

Perhaps the best measure of the level of factory wages is the ratio of earnings of factory labor to the agricultural wage in the region from which the bulk of factory labor was recruited. Material assembled by Mazumdar shows that, for the entire period 1900–37, this ratio was remarkably stable, varying between 2.0 and 2.5 (Mazumdar 1973, app., p. 495).[6] The cost of living difference between city and country could not be nearly as large. Thus, another study by Mazumdar (1958) showed that in 1954–55, when

6. There was a short period in the early 1930s when the ratio went above 3.0 as factory wages fell more slowly than prices.

reasonably firm data were available, the budget of a typical agricultural worker cost 45 percent more in the city, while the budget of an urban industrial worker cost 32 percent less in the country. It might be objected that average earnings of factory labor include the effect of skill formation in the industry. But as against this, the daily agricultural wage rate overestimates the earnings of agricultural laborers, who typically get employment for relatively fewer days in the month than factory workers.

Further strong evidence that textile wages were significantly higher than the supply price of labor is provided by the peculiar system of labor utilization in the textile industry, known as the badli system. The badlis constituted a casual wing of the labor force—labor that was employed on a day-to-day basis much as in the docks—alongside a permanent core of workers paid on a monthly basis. Every morning they gathered in front of the mills and were hand picked as needed by the jobbers responsible for recruitment. There was no guarantee of a day's work. Indeed, there is evidence of considerable underemployment among the badlis. Some studies put the employment secured by a badli during a month at between fifteen and eighteen days (Shah 1941, p. 106; also see Bombay Textile Research Association 1963). So even if badlis were paid at the same rate as the regular workers (which they generally were not), their monthly earnings could have been as little as half those of the full-time workers. This would be a measure of the gap in the established wage level in the textile factories and the supply price of urban labor.

It is also clear that the system of badli labor—and the associated underemployment of badlis—predates any institutional influence on the textile labor market. Evidence given to the Royal Commission on Labour in 1892 suggested that the badli labor force (or "the floating residuum of mill hands," as they were called) came to about 25,000 at a time when the average full-time employment in the industry in the city was 65,000. Neither unions nor the government was much in the picture of wage determination at this time. Thus, the maintenance of a relatively high wage in the textile factories cannot be traced to institutional factors.

We discuss in the next principal section how wages in the factory sector might have been established at a relatively high level before 1914 despite the evidence of abundant supply of labor. Some rent may have accrued to stable labor, which was in scarce supply. However, the elasticity of supply of stable labor could not have been very low because in spite of the very high rate of growth of textile employment, there is no evidence of any upward movement of real wages (Mukerji 1959, p. 92, table 6).

The 1914–47 Period

World War I produced a period of inflation, which was followed by the postwar depression and a second bout of inflation during World War II.

These large fluctuations of prices resulted in stronger efforts by Indian factory labor to protect its real standard of living.

Trade unions and collective bargaining were *not* the mechanism whereby factory labor tried to protect its interests. The first workers' organizations were more in the nature of welfare groups. It was not until 1925 that the first textile labor union was founded. But by March 1927 this union claimed a membership of only 6 percent of total mill employment. Besides, there were more than a few unions on the scene, often based on the interests of different political groups or parties. The leaders of the unions were middle-class politicians involved in a variety of activities. But in spite of the absence of trade union organization, strikes by textile labor became frequent and often effective. A twelve-day industrywide strike in Bombay City in January 1919 involving 150,000 workers was followed by a month-long stoppage the following year. The strikes of 1924 and 1925 bore close resemblance to these earlier industrial actions (Kulkarni 1979).

The strikes were set off by the large increase in the cost of living. The employers had a more permanent organization to represent them in the form of the Bombay Mill Owners' Association, whose recommendations to member mills, although not binding, had considerable influence. The government also came increasingly into the picture as a peacemaker and extended its influence through recommendations made by specially appointed commissions.

This type of institutional influence did have a significant effect on the wage level. A cost of living allowance was granted by the Mill Owners' Association immediately after World War I. The association's subsequent attempt to cut wages in 1925 by 11.5 percent was abandoned when the government offered to abolish the excise duty on cloth. The depression and combined fall in prices led many individual mills to start cutting wages in 1933 and 1934. This provoked a prolonged strike of three months in 1934, but although it fizzled out in the end, the Mill Owners' Association was induced to recommend a consolidated minimum wage for textile workers. In the early stages of World War II, the association granted a war bonus of about 16 percent to the workers.

The net effect of these influences seems to have been, according to one calculation, a doubling of the real wages of textile labor between 1919 and the mid-1930s; in spite of subsequent fluctuations in real wages, this level was maintained at the end of World War II (Mukerji 1959). As against this, the urban-rural wage difference increased by only 50 percent in this period (Mazumdar 1973). In any event, it is clear that textile workers in Bombay City had emerged as labor aristocrats with high wages relative to other types of factory labor. Thus in 1957 the minimum wage (including bonuses and allowances) of a textile worker was Rs124 per month, while the average monthly earnings of unskilled labor in the factory sector of the city were Rs103 (Kulkarni 1979, p. 363; also see table 6-5 in the present book). In 1936 there was one official survey of earnings in a nonfactory sector—retail trade (Bombay Labour Office 1936, app. B). This showed that about one-third of the workers in retail trade had earnings below the unskilled wage in factories, and perhaps half were below the level of low-skill workers in the textile industry.

A point of some importance emerges from the discussion in this subsection. It is remarkable how the textile workers, in spite of their lack of organization, were on some occasions able to bring about a general stoppage of the mills for prolonged periods and achieve a significant increase of real wages. We have referred earlier to the existence of a large mass of badli labor in the textile industry—already in excess supply and with earnings below the full-time wage. It might have been expected that the badlis would provide a reserve army of labor, which would have made successful strikes difficult to achieve in the absence of powerful trade union organization. Evidently the permanent workers in the industry were in some sense a preselected and established body of workers who could not be easily replaced by badli labor. This nonhomogeneous character of labor available to the textile industry will be of importance in our search for explanations of the small-large wage differential in the next major section.

Post-Independence State Intervention in Wage Setting

The state took a much more active role in wage determination after independence in 1947. The Minimum Wage Act came to provide a floor to wages in agrobased industries, road construction, transport, and other industries dominated by small or decentralized firms. But the size of the minimum wage was left to the judgment of advisory boards constituted from time to time. The recommended wages are generally at low levels, and the degree of enforcement is also somewhat uncertain.

The impact on wage determination in the large-scale sector seems to have been much more significant. In the late 1950s and the 1960s industrywide wage boards were set up consisting of representatives of employers, employees, and the government. Their recommendations on wage scales were not mandatory, but if accepted by the government they had a significant effect on the wage levels that were finally contracted. Another type of body—tribunals set up under the Industrial Disputes Act of 1947—can make recommendations in the case of disputes which are binding on the parties, although the government is free to accept the awards with modifications and the parties can also appeal to the Labour Appellate Tribunal set up under the 1947 act.

The state has thus been very much in the business of wage determination since independence, and its importance has increased over the years. We can judge its impact by comparing the evidence from a few surveys of small-scale industries in the early 1950s with the material presented in the first section of this chapter for more recent years. These surveys were commissioned by the Planning Commission in selected cities for all small-scale units taken together. Wage data from three of the surveys are presented in table 14-5 together with comparable data on wages from the factory sector.

If we look back at the analysis of wage differentials given for Bombay City for 1978–79 in table 14-4, it is clear that the differential has become much larger. But the important point to note—which has been mentioned earlier also—is that the differential is significantly wider for larger factories employing more than 100 workers. Factories employing 10–99 workers have wage levels about 58 percent higher than small units, and those employing more than 1,000 have wages 144 percent higher. The earnings data for Bombay City given in table 14-5—which refers to average earnings in factories of all sizes—show that the factory sector had wage levels about 65 percent higher than small units. Without knowing the distribution of employment by size of factories in 1955 in Bombay City, we cannot strictly say how the size-weighted differential at this date compared with that in 1978–79. If we take the weights given in the 1978–79 sample, the wage differential was 80 percent higher in all factories compared with the small-scale, if the adjusted figures from column 6 of table 14-4 are used, and 119 percent higher if the unadjusted figures of column 2 are used.

While it is impossible to estimate precisely the lift given to factory wages by institutional factors in the post-independence period, it is clear that wages in larger factories have been affected most. The CSO study of wage trends in the sample and census sectors of factory industries confirms this. The indexes of

Table 14-5. *Comparison of Annual Earnings of Unskilled Workers in Small-Scale and Factory Units in Selected Cities*

City	*Small-scale*	*Factory workers (different dates)*	*Factory workers (adjusted)*[a]
Bombay	678[b]	1,244[c]	1,112
Delhi	780[d]	1,243[e]	1,110
Calcutta	856[f]	1,194[g]	1,382

a. These figures have been adjusted by the cost of living index for the respective cities to correspond to the different dates of the surveys of small-scale enterprises.

b. Lakdawala and Sandesara (1960, p. 96, table 5-9; earnings of unskilled workers only, 1955).

c. Earnings of "labourers not elsewhere specified." "Report on Family Living Survey among Industrial Workers, 1958–59, Bombay." Labour Bureau, Government of India, table 2-3, p. 9.

d. Dhar (1960, p. 67, table 8-3; average earnings of all time-rated workers, 1954).

e. Average income per paid employment of all employees. "Report on Family Living Survey among Industrial Workers, 1958–59, Delhi." Labour Bureau, Government of India, table 2-2, p. 7.

f. Reserve Bank of India (p. 336, table 130; average earnings of all workers in small units employing less than 20 workers, 1960–61).

g. Average income per paid employment of engineering workers. "Report on Family Living Survey among Industrial Workers, 1958–59, Howrah." Labour Bureau, Government of India, table 2-3, p. 10.

real wages prepared by the CSO at 1961 prices have the following values for the two sectors over a fifteen-year period between the early 1960s and mid-1970s (India 1979, p. 13, table 3-2):

Years	*Census sector*	*Sample sector*
1960–61	100.9	99.0
1961	100.0	100.0
1974–76	111.1	85.9

Statewide data analyzed by the CSO show that this pattern is observed in most of the individual states.

The question might be asked, "If the government wage-setting machinery has, as indicated above, jurisdiction over all establishments covered by the Factory Act, how has the large size differential within this sector been widened?" The answer to this puzzle is that industrial tribunals and wage boards incorporated the principle of "capacity to pay" of the firm in question in recommending wage levels. To quote a person who had the opportunity of observing the government labor legislation at close range for a number of years: "There are many industries in this country in which we find large and modern units existing side by side with small and inefficient units. Though there might not be much difference between large and small units as regards the quality or intensity of work expected of labor, their capacity to pay would vary greatly. Consequently, wage differentials based on size have traditionally developed, and have been maintained by tribunals and wage boards" (Subramanian 1979, p. 137).

In Search of the Causes of the Wage Differential

It should be apparent from the above review that while institutional factors increased wages, particularly in the larger factories, they cannot fully explain the wage difference between the small- and large-scale sectors in Indian manufacturing. Three theories discussed in the literature are of relevance in our search for an explanation of these differentials. These are:

- The different supply prices of temporary and permanent migrants and the different mixes in which they are used in the small and large sectors.
- The wage-efficiency relationship.
- Internal labor markets in the large-scale sector.

The three hypotheses are really complementary, and they might apply sequentially in the course of the formation of a labor force in the factory sector of a developing country. We will discuss the first group as probably contributing to the establishment of a high-wage sector in the first place. The Bombay labor market research project also produced empirical material consistent with this theory.

Temporary and Permanent Migrants

Two distinct types of migrants are attracted from rural to urban areas. There are migrants who come for short periods of their working lives, usually without their families; they include seasonal migrants, those working to meet a specific income goal, and those wanting to survey the urban labor market without committing themselves to any long-term career there. Then there are migrants who spend most of their working lives in the towns, even if they retain some links with their rural areas for holidays or retirement. They are mostly settled in town with their families. The supply price of the permanent (family) migrants is, for several reasons, higher than that of the temporary (lone) migrants.[7] First, the loss of income in the family farm due to the absence of an individual may be considerably less than the total farm income, because other family members are able to substitute for his labor on the farm. This is particularly true when the absence is during the slack periods in agriculture. Second, the earner-dependent ratio for a family is significantly lower in the urban sector of India (as in many other developing countries) because of the more limited role of women and children when compared with rural market activity. Third, the cost of living in town for a family is higher not only because of higher commodity prices, but also because the person who has migrated with his family incurs the cost of finding for himself protection against old age, unemployment, and ill health—protection ordinarily provided by the social security system in a developed country and which the rural family would provide for the individual migrant.

Given this difference in the supply price of individual and family migrants, if demand for labor in the urban market were undifferentiated, little family migration would take place as long as the operation of the agricultural economy allowed for a plentiful supply of individual migrants. But if employers perceived a strong link between stability and efficiency of labor in particular types of firms, the wage level would be set high enough to attract family migrants. For those types of firms or sectors in which the link between stability and labor productivity is weak, wages would be lower, nearer to the supply price of individual migrants.

Organized sector firms using expensive machinery are likely to value stable labor more than firms operating at the lower end of the market. A positive relationship between firm size and the wage level would then exist insofar as larger firms use more sophisticated technology. This will be particularly so when the labor force requiring new skills for modern large-scale industry is being developed and the cost of learning new work routines and skills is high. Once a stabilized labor force has been formed, it is not replaceable by the "floating mass," as was noted in the history of textile labor in Bombay.

7. For the argument that follows, it is not necessary to assume that all permanent migrants are family migrants. It is sufficient that the marginal permanent migrant is a family migrant.

The Wage-Efficiency Relationship

Wages high enough to attract permanent rural-urban migrants define the lower limit of the wage differential relative to sectors in which nonstable labor predominates. It discourages return migration to the rural areas and thus deals with one of the major causes of instability of labor in markets attracting a large volume of temporary migrants. But migrants, although committed to permanent urban work, may move to other firms offering better rewards in the same urban labor market. The wage required to attract stable family migrants is not necessarily high enough for an optimum rate of turnover for the firm in question. Some firms may set wages to create a firm-specific labor force with a very low rate of turnover. But the wage which minimizes labor cost per unit of output may be still higher if higher wages increase working efficiency. With a firm-specific labor force, the "least cost" wage is increased for several reasons. First, the firm is dealing with relatively few workers separated from the rest of the work force, so that the benefits of a wage increase are not shared among a large number, as would happen if they worked for a number of employers over a period of time. Second, the employer-employee relationship takes on some of the characteristics of an implicit contract, with the understanding that the employee would achieve a certain level of efficiency and that the employer would not pass on short-term fluctuations in demand by cutting wages. Third, management costs are smaller the smaller is the work force, and hence there is an incentive for employers to increase wage rates rather than hire extra workers as long as efficiency responds to wage increases.

In early expositions of the wage-efficiency relationship, the physical effects of better nutrition associated with higher wages were stressed (Leibenstein 1957). This hypothesis had limited empirical support and probably little validity for urban workers, whose wage levels are relatively high. In more recent work the effect of high wages on worker morale or incentives has been stressed. It is, however, clear that if employee contracts specified that the agreed wage would be tied to particular productivity levels, then the firm would be equating the wage offered to the supply price of superior labor. In practice, however, employment contracts cannot and do not specify the worker's productivity level at all precisely. The reasons for incomplete contracts could be uncertainty about contingencies which may affect the worker's duties; the difficulty of specifying an objective measure of the individual employee's contribution to the productivity of the firm; and the high cost of operating a screening mechanism when an individual worker's potential productivity is not objectively known before a period of employment. With incomplete information, no general market for superior labor available to all firms exists. Each firm is then faced with an inelastic supply of potentially productive workers who will be responsive to the incentives of high wages. Up to a point, an increase in the wage will increase the proportion of such workers in the enterprise and lead to a more than proportionate increase in labor efficiency (Maleanson 1981, pp. 848–66; see also literature there cited).

Here we have one reason why the wage-efficiency relationship could be expected to operate more strongly in large firms. With an inelastic supply of potentially productive workers to the individual firm, the wage offered is higher the larger the requirement for such workers. Furthermore, insofar as large firms use more sophisticated technology, their need for firm-specific skills in their labor force is relatively high.

Internal Labor Markets

A third group of ideas tending to support the hypothesis of a positive relationship between the wage level and the size of the firm is the formation of internal labor markets in large firms. The characteristic of internal labor markets is that jobs within the enterprise are arranged in lines of progression. New workers are recruited principally to fill jobs at the bottom of the ladder, while vacancies at higher levels are filled as much as possible through promotion. This arrangement reduces training costs because skills absorbed in a particular job can contribute to training required for the job at the next level. The possibility of internal promotion also increases the incentive to learn. In addition, costs associated with screening recruits are reduced because the ability and performance of existing employees are already known to the employer, and the costs associated with hiring new workers are lower the lower is the skill level at which the recruitment is made.

Internal labor markets tend to develop an elaborate set of rules and procedures for evaluation of performance and promotions. Such arrangements and the opportunity for on-the-job training tend to provide incentives for a stable labor force for the firm.

Well-developed internal labor markets are possible only in firms of a large employment size. Workers in such firms will show a strong positive relationship between experience and earnings. The prediction of this model is that the lifetime earnings of a typical worker will be higher in large firms, but it does not suggest any reason why the entry wage of an unskilled worker will be any higher than in small firms. Indeed, if workers were concerned with expected earnings over a long period, entry wages would (in a competitive model) be predicted to be lower in large firms. If then we find that entry wages are significantly higher in larger firms,[8] other considerations, such as the wage-efficiency relationship discussed above, have to be brought into the argument.

8. The evidence from the Bombay Labour Market Survey is analyzed in Bowden and Mazumdar (1982). On the Japanese experience in age-earning profiles by firm size, which is particularly striking, see Economic Research Institute (1959, chap. 14, p. 16).

Empirical Evidence

The empirical implication of all three groups of theories is that labor in the large-scale sector will be more stable in terms of its length of service in the same firm. In addition, the theory of wages based on permanent migrants suggests that the unorganized sector will have a larger proportion of temporary migrants and single-member families. Data from the Bombay Labour Market Survey are consistent with both these implications.

THE INCIDENCE OF TEMPORARY MIGRATION. Table 14-6 gives the distribution of male workers by the duration of the current job. The larger proportion of workers in the small-scale sector with a short duration of work in the current job may be caused by one or more of three factors: faster growth of employment in the sector; greater interfirm mobility within the sector; and a higher incidence of return migration.

The first reason carries little weight. Joshi and Joshi (1976) have calculated the growth of employment in the organized sector of Bombay City between 1961 and 1971 from the returns of the Directorate of Industry, and they have contrasted it with the growth in employment for the entire city as given in the population censuses of the two dates. After correcting for the change in the definition of a worker in the two censuses, growth in organized sector employment in Bombay City lagged behind total employment by about 1 percent per year. But if we include the satellite town of Thana with Greater Bombay, the lag would be *half* this annual rate. Such a small difference in the growth rate of employment in the organized and unorganized sectors cannot explain much of the observed difference in length of service.

Turning to interfirm mobility, the Bombay survey collected data on the number of job changes. The small-scale sector showed a somewhat higher level of turnover. But generally, the degree of mobility was surprisingly low in both sectors. Further, although migrants showed more interfirm mobility than natives, the difference was not large—for the two sectors taken together, 40 percent of the migrants and 50 percent of the natives reported having no change of jobs at all in the urban market. This proportion drops with age, but even for the age group 35–40 the percentages of nonmovers were as follows: migrants: 38 in factories, 35 in small-scale sector; natives: 44 in factories, 30 in small-scale sector.

Table 14-7 gives the *average* number of job changes by duration of residence for the migrants in the two sectors. Migrants on the average might have had one other job after being in Bombay about five years in the small-scale sector and ten years in the factory sector. These data on the low turnover rate of workers remaining in the urban labor market, together with the evidence in table 14-6 that in the small-scale sector a large proportion of the sample holds jobs only for short periods, suggest that there is a high incidence of return migration from this sector.

Table 14-6. *Percentage Distribution of Male Workers by Years Worked in Current Job*

	Years worked in current job									
Sector	*Less than 1*	*1–2*	*3–4*	*5–6*	*7–9*	*10–14*	*15–19*	*20–29*	*30 plus*	*Mean*
Casual	14.1	35.0	22.3	8.2	10.4	5.7	1.9	1.9	0.4	4.2
Small	19.3	27.7	15.3	10.6	8.0	9.1	4.7	4.2	1.0	5.4
Factory										
Employing 1–99	4.6	8.6	14.3	13.7	17.1	19.9	10.8	9.2	1.8	9.7
Employing 100–499	0.7	2.9	11.5	20.4	16.2	25.4	11.9	9.5	1.5	10.6
Employing 500 plus	0.7	4.1	6.3	11.8	10.7	18.2	16.9	23.2	8.1	15.4

Source: Bombay Labour Market Survey, 1978–79 (Mazumdar 1984b).

Table 14-7. *Migrants: Mean Number of Jobs Changed in Urban Market by Duration of Residence and Sector of Current Job*

Duration of residence (years)	*Sector*	
	Small-scale	*Factory*
Less than 1	0.21	0.19
1–3	0.49	0.39
3–5	0.85	0.43
5–7	1.01	0.51
7–9	1.17	0.91
9–14	1.40	1.01
14–19	1.54	1.20
19–29	2.69	1.29
Above 30	1.85	1.05

Source: Bombay Labour Market Survey, 1978–79 (Mazumdar 1984b).

We know from the work of demographers that return migration is significant in Bombay City. Zachariah (1968), using information from the 1951 and 1961 censuses, found that in that decade, out-migration from Bombay was about 500,000 persons, of whom three-quarters were returning migrants. Roughly half the influx of population into Bombay during the decade was offset through out-migration. As a percentage of the stock of migrant population in 1961, the rate of out-migration was about 20 percent. Contrary to some prevalent models, the peak rate of out-migration was not among the older, postretirement age groups. The highest rate of out-migration was observed in the 30–35 age group.

The 1971 census out-migration data have not yet been fully analyzed, but calculations by Joshi and Joshi (1976) show that return migration continued in the decade 1961–71. However, along with a reduced rate of migration relative to natural increase, the volume and rate of out-migration seem to have fallen. Out-migration from 1961 to 1971 probably offset one-third rather than half of gross in-migration as in the previous decade.

Table 14-8 brings together data on the percentage distribution of migrants for the small-scale and factory sectors in our sample. Taken in conjunction with evidence presented earlier of a much shorter duration of jobs in the small-scale sector and only slightly lower incidence of interfirm mobility among workers in this sector, the data strongly suggest that the incidence of return migration is much higher in the small-scale sector.

Differences in Family Size. The hypothesis that the lower wage in the small-scale sector is due to the lower supply price of lone migrants (many of whom are temporary) is supported by the data from the Bombay survey. The proportion of single-member families was 66 percent among male migrants in the small-scale sector, as against 38 percent for migrants observed in the factories. The average family size was 2.1 for the former and 3.4 for the latter.

In the nature of things, the evidence on family size cannot be a conclusive

Table 14-8. *Percentage Distribution of Male Workers by Years of Residence in Bombay*

Sector	*Native*	*Less than 1*	*1–2*	*3–5*	*6–10*	*11–20*	*20 plus*	*Mean years of residence (migrants only)*
Casual	20.9	4.9	14.8	21.2	20.7	12.4	5.1	12.5
Small	17.8	4.8	9.4	17.0	18.6	19.9	12.3	11.0
Factory								
Employing 1–99	19.1	0.6	1.6	7.2	15.9	35.7	19.9	16.0
Employing 100–499	16.7	0.0	0.4	5.6	16.4	40.0	20.9	16.1
Employing 500 plus	21.8	0.0	0.3	2.7	7.0	28.3	39.9	22.0

Source: Bombay Labour Market Survey, 1978–79 (Mazumdar 1984b).

proof of the hypothesis but can only be consistent with it. This is because there is no way in which we can be sure which way the causal link works. If wages are determined exogenously in the factory sector, then it is possible to argue that high income induces factory workers to settle in the city more easily with their family. The presumption that the higher supply price of family migrants causes factory wages to be higher is based on a theoretical deduction plus the observation that factory wages were relatively high in periods during which there was little evidence of any institutional determination of wages.

Summary

We conclude that there is some support for the theory that the wide differential of wages observed in India between small and large firms had its origin in the different supply prices of temporary and permanent migrants and their different demand prices in the two sectors. But as discussed above, other influences tend to widen the differential once a stable labor force has been created in the factory sector. We also saw in the second section of this chapter that institutional factors became stronger after independence. They could have widened the size of the differential, particularly for the large factories. However, the influence of institutions, even in recent years, should not be exaggerated. As discussed, the enforcement of wage awards is channeled through the Factory Act, and if the influence of labor legislation were pervasive, size-related wage differentials within this sector would have been reduced. In fact, the evidence is that the differentials have moved in the opposite direction. As can be seen from tables 14-5 and 14-7, small factories employing less than 50 or 100 workers have a more unstable labor force than larger factories, and in this way resemble establishments in the small-scale sector. Although all factories are legally under the purview of wage boards and industrial tribunals, the awards of these bodies have taken into account the economic differences between small and large factories.[9] Thus they have reinforced—and indeed somewhat widened—wage differentials which had been established by economic forces in the years when institutions were not so important.

A point of some importance emerges if the significance of autonomous economic forces in explaining the observed wage differentials is recognized. Profit-maximizing firms will not voluntarily pay high wages unless they expect such a wage policy to lead to lower labor cost in the long run. Thus, the difference in the efficiency wage of labor would be substantially less than the difference in wages per worker as between small and large firms. It follows that the problem created by the large wage differential is best viewed only

9. This is not to say that in specific applications some awards may have used proxies for capacity to pay (like gross turnover) which are institutional rather than economic reasons for a differential.

partly as one of "distortions" in the neoclassical allocative efficiency sense and partly as one of the distribution of income. All other things being equal, large-scale units tend to promote a labor aristocracy—offering a small volume of employment at a relatively high wage. Policies to promote small-scale industries can then be justified on distributional grounds—as an effort to change the shape of the earnings distribution in the manufacturing sector, with more employment being created at lower levels of wages.

Comparative Experience from Other Countries

In many developing countries, it has been found that wages, after skill is controlled for, rise with the size of the firm. The steepness of the rise varies, and there are differences in the impact of institutional factors. Apart from this, the economic factors discussed in the last section lead one to expect that the differential would be large in cases when a plentiful supply of circulatory migrants keeps up with the demand in the small-scale and casual labor markets. At the other end of the scale, the importance of large-scale firms making use of capital-intensive modern technology would increase the size-related wage differential.

Indonesia and Malaysia

The importance of rural-urban circulatory migrants varies with the nature of the economy, particularly the rural economic structure. A contrast is provided by the examples of Indonesia and Malaysia. There is very little institutional influence on wages—at least in the private sector—in these two countries. But the difference in migration patterns is large. In Indonesia circular migrants play a major role in the urban labor market. In a survey of fourteen villages in West Java in 1973, Hugo (1977, p. 59) found that no less than two-thirds of his sample of migrants were nonpermanent (that is, did not meet the census criteria of being absent for six months or more).[10] By contrast, urban Malaysia has much less incidence of temporary or return migration, as is shown in a World Bank research project (Mazumdar 1981, pp. 202–04, 221–23). The importance of stage migration—that is, migrants stopping at smaller intermediate urban areas before reaching the bigger cities—suggests a more permanent urban drift and a different pattern of adjustment to the urban economy.

The size-related wage differential in the two economies is consistent with our hypothesis. In an intensive survey of three industries—weaving, cigarettes, and kerrek—in Jogjakarta (Central Java), Manning found that the hourly earnings of the major semiskilled operator in the industry concerned

10. For further details on this and other points on wages in Indonesia, see Lluch and Mazumdar (1983, pp. 91–106).

were Rp39, Rp56, and Rp80 for firms of employment size of less than 100, 100–499, and 500 or more, respectively (Manning 1979, pp. 240–42). Even larger wage differentials have been noted in the one official survey that exists, for 1974 for all manufacturing in Java. Average wages and salaries per day for all operatives in large industries (defined as those employing more than 100 workers) were two-and-one-half times the level in cottage industries (less than 5 workers), and twice the level in small industries (5–20 workers) (World Bank 1979, p. 12). Thus, the Indonesian wage pattern in manufacturing is very much like the Indian one discussed in this chapter.

The Malaysian story seems to be different. An analysis similar to the one performed on the 1978–79 Bombay sample was undertaken by Mazumdar for a sample of three cities in urban Malaysia. Size of firm was a significant determinant of the earnings of unskilled labor, but the level in enterprises of 1–9 workers was only 32 percent lower than in enterprises of 100 or more workers (Mazumdar 1981, p. 162, table 9-6).

The migration-related hypothesis explaining wage differentials in India, Indonesia, and Malaysia almost seems to fit too well. Further probing will no doubt reveal that wage differentials are influenced by other factors as well, such as have been discussed in the preceding section. An interesting point from Manning's study is the close association between size of firm, degree of mechanization, and the wage differential. Highly mechanized firms (with a capital-labor ratio of more than Rp1.5 million) pay two-and-a-half times the wages in nonmechanized firms, while foreign-owned firms stand apart in paying four times as much (Manning 1979, p. 242). The influence of technology on wage levels is hard to separate from that of size per se. Labor instability (as measured by turnover or absenteeism) was much higher for nonmechanized firms, but within the mechanized sector, differences of wage level for different groups were positive without corresponding differences in turnover rates—adding support to the argument that influences other than the migration hypothesis also affect wage differentials by size of firm.

The wide wage gap observed in Indonesia between large and small firms has little to do with direct intervention in the labor market, whether by government or by trade unions. In this sense, size-related wage differentials without labor market distortion can be said to exist in Indonesia in a more pronounced way than in India. However, distortions in capital markets favoring large firms and foreign enterprises (in the form of negative real interest rates, tax holidays, and so forth) might indirectly encourage such firms to adopt a high-wage policy. The contrasting Malaysian case is a warning against generalization—even for economies in the same geographical area.

Colombia

The labor market was studied in the research on small-scale enterprise development in Colombia as part of the World Bank project. At the aggregate level, wage differentials by size category seem to be high. The average earnings including fringe benefits in large industry are over twice that in small and medium-size establishments (Cortes, Berry, and Ishaq 1987, chap. 2).[11] The difference in average earnings, although marked, is less than the difference in labor productivity. But much of the difference is accounted for by the difference in skill mix and by the fact that the differential for skilled and white collar workers increases much more than for blue collar workers. "Blue collar wages vary only moderately by size up to the plants of 100–199 workers. There is a marked increase between that category and the plants of 200 or more workers, suggesting a sort of wage cliff (Cortes, Berry, and Ishaq 1987). It will be recalled that there was a similar wage cliff observed for India at the employment size of 100 or more workers.

However, earnings function studies, like the one done for Bombay City, suggest that the size-related differential is not nearly as important for Colombian cities as it was seen to be for Bombay. Mohan (1981, p. 75) estimated an earnings function for a sample of male workers in Bogotá which produced the result, after education, experience, and a number of other factors were controlled for, that for every doubling of firm size, earnings increased by 2.3 percent. The study unfortunately does not address the problem of the wage cliff because the firm-size variable was introduced as a continuous one. Cortes, Berry, and Ishaq (1987, chap. 2) concluded that the adjusted differential between large (more than 100 workers) and small firms (less than 50) would be on the order of 35–55 percent; the higher figure includes fringe benefits.

Like several other Latin American countries, the Colombian labor market is distinguished by the importance of fringe benefits (social security, severance pay, and so forth) that increase with the size of the firm. Even within the set of small-scale metalworking firms surveyed for the World Bank project, fringe benefits were 31 percent of basic wages in firms of 1–10 workers and 50 percent in firms of 61 or more workers (Cortes, Berry, and Ishaq 1987, chap. 2). We can conclude that the size-related wage differential in Colombia, although not spectacular, is to a large extent institutionally determined and is a labor market distortion favoring small firms.

The Developed Countries

Examples have been given in this chapter of several developing countries in which size-related wage differentials are very large. How does the experience of developed countries fit into the picture?

There is enough evidence to show that wage differences between establishments are positively correlated with size. In one study of U.S. manufacturing

11. The cutoff point used in this study is an employment size of 100 workers. The original source of the data is Departamento Administrativo Nacional de Estadística.

in the late 1960s, it was found that the largest establishments paid 25–30 percent more than the smallest, and the differential is increased if fringe benefits are included (Lester 1967, pp. 51–67).[12] Also, the differential has existed since the 1920s. Similar differentials have been noted in the United Kingdom. But the magnitude of the differential is much smaller than in many developing countries, as is apparent from the examples given in this chapter. Furthermore, in developed countries only a part of interestablishment pay differentials is explained by size, which shows that the association between size and the wage level is not very close (Wilkinson 1973).[13]

Several reasons can be cited why we would expect size-related wage differentials to be smaller in developed countries. First, the role of temporary rural-urban migration in providing a flow of cheap labor is nonexistent in these economies. Moreover, it is doubtful whether, at the developed country level of technological development, there would be many manufacturing establishments which could compete in the market using the type of non-stable labor which temporary migrants provide. Second, with economic development, labor employed in the manufacturing sector would tend to become more homogeneous in basic skills, and there would be less incentive for expanding firms to retain a special body of workers with higher wages. The labor market, in other words, tends to get more integrated in terms of general skills. This would seem to be confirmed by the fact that in Korea around 1970 the wage differential by size was very large, but as the labor market tightened and real wages increased throughout the decade the differential was reduced for both men and women and for both production and administrative personnel (Park 1981). Third, institutional wage bargaining in developed countries tries to achieve uniform standards for workers in different establishments, whether by industrywide contracts or in comparison with unionized firms. The contrast with the awards of industrial tribunals in India, which perpetuate and indeed widen size-related wage differentials, should be recalled in this connection.[14]

If, in fact, institutional influences in developed countries tend on balance to iron out size-related wage differentials, economic factors must be operative to maintain the moderate differentials which have been observed. It has been argued above that some major factors operating in developing countries are small in the developed country context. But others are significant. The

12. This section draws on the discussion in Wood (1978, p. 167–78, which quotes Lester) on "intra-industry relativities."

13. This study found that half the variance in wages was explained by size.

14. Even if small firms in developed countries may not be unionized to a great extent, it has been argued that "the industry wage rates negotiated by the representatives of managements and unions tend to be treated as a social minimum, below which wage offers are infrequently encountered." In Britain, wage councils set statutory rates in industries in which collective bargaining is poorly established by paying special attention to negotiated rates in unionized sectors. (Pencavel 1974, pp. 196–98.)

internal labor market hypothesis discussed earlier probably exerts an upward pressure on wages in large organizations in developed, as in developing, countries. Wood (1978, p. 169, citing Ingham 1970) has suggested an additional hypothesis particularly valid for high-wage economies: the preference of workers for employment in small enterprises. The reasons singled out by Wood are the impersonal nature of work in large establishments and the longer distance (and cost) of travel.

Conclusion: Is India a Special Case?

To sum up: the evidence does not suggest that India is a special case as far as the maintenance of a large wage differential between small and large firms is concerned. The differential is probably larger in most developing countries than in developed countries. Unlike the situation in developed countries, in India institutional factors have tended to widen the differential somewhat, pushing India up in the ranking of developing countries by the size of the differential. But still it ranks below Indonesia, where the differential is apparently maintained by purely economic factors. Finally, there is no trend yet in India for the differential to narrow over time, as happened in Korea. On the contrary, institutional factors have been partly responsible for widening the differential in the 1960s and 1970s.

15

Capital Markets in India and Elsewhere

The factor intensity of operations, which was surveyed in chapter 7, is governed, in part at least, by the working of factor markets for both capital and labor. The main reason why it has been believed that small enterprises deserve special support lies in the supposed malfunctioning of these markets.

So far as capital markets, the subject of this chapter, are concerned, it is thought, first, that the supposed labor intensity of operations is explained partly by the inaccessibility or high cost of capital; and, second, that such inaccessibility or high cost is itself a market imperfection which should be removed or reduced (though, if the first hypothesis is valid, this would increase the capital intensity of the small firms but not necessarily that of the whole aggregate of firms).

Some A Priori Generalizations

It has been increasingly realized in recent years that it is almost impossible to imagine perfect capital markets. This would imply that every potential competitive lender was able to divide potential borrowers into a set of groups, each of which contained only members who had an equal probability of repaying any loan with given terms. Even when lenders are competitive, it is closer to reality to assume that any substantial loan that is transacted involves unique elements. The lender has to assess the probability of repayment (and of partial repayment), which ideally requires knowledge of many attributes of the borrower, including where relevant his or her business or project, these of course varying from person to person. Furthermore, two lenders having the same description of a borrower may assess the probabilities differently and so estimate a different present value of the loan. Finally, the expected present value of a loan will not be the only determinant of whether the lender is willing if some lenders are more risk averse than others.

The expected yield of a portfolio of loans (of given terms) will rise with the expenditure on acquiring knowledge of borrowers and what they are likely to do with the money, so that the proportion of bad debts is reduced. There is clearly an optimum expenditure on investigation. But this optimum will vary with the size of loan, since it is likely to cost as much to increase the present value of a hundred small loans by *x* percent as it is to increase the value of a hundred large loans by *x* percent. Indeed, it may cost more, since the small loans go to small operators who may not keep accounts and who often have less experience and ascertainable track records. With very small loans, investigation is virtually excluded so that, except in very highly moral and prudent communities, lending is usually confined to those who anyway have good knowledge of the borrower, which is therefore costless—friends, relatives, and business contacts.

Requiring collateral is another way of increasing the present value of a loan. Collateral reduces the risk of nonrepayment, but some administrative cost will always be involved in arranging collateral and in foreclosing if need be. More important, many would-be borrowers do not have title to suitably secure collateral, that is, more or less immovable property which the bank or other lender can seize (or movable property which the bank can hold, for example, jewelry). Other would-be borrowers may be more or less reluctant to commit collateral. In the case of a corporation the lender may insist on the personal guarantees of directors; but this is not of much concern to us since the vast majority of SSEs are partnerships or sole proprietorships with unlimited liability. Finally, raising the average rate of interest may raise the estimated value of a portfolio. But, other things being equal, the probability of default itself rises with increasing interest rates, since high interest rates decrease the ability and willingness of a borrower to repay. Moreover, they may increase the proportion of dishonest or overoptimistic borrowers and of risky projects (known as "adverse selection"). Consequently, there may be a maximum rate of interest, such that any higher rate would *reduce* the expected present value of the portfolio (see Stiglitz and Weiss 1981).[1]

In an honest community, the risks of nonrepayment can be better assessed (since honesty is not always the best policy for borrowers). Further, if project risks could be objectively assessed, then even the most risky projects could be financed by loans. The bank or other lender would need to charge very high rates of interest for such loans, which however need be no burden on the borrower if principal and interest are repaid only in the event of success and if failure carries no serious stigma. This is in effect assuming away the problems of moral hazard and adverse selection. In practice, a bank cannot easily distinguish the degree of risk and would in any case seldom be permitted to

1. But it should be noted that there seems to be no direct empirical evidence of adverse selection, though credit rationing, which it explains, is the rule. In certain circumstances, higher interest rates could result in benign selection.

vary interest rates to the degree required. Equity participation becomes a desirable alternative, though not one that is feasible for banks in the case of very small enterprises.

In a dishonest community with poor collateral, the proportion of bad risks is high. Since it is difficult and expensive to separate good from bad, lenders could seldom vary interest rates from loan to loan so as to accommodate differences in risk. For some sectors, there may then be no potential portfolio for which the yield-maximizing rate of interest would be high enough to cover the cost, which includes the cost of borrowing as well as the administrative costs, and provision for expected defaults. Then there is no market at all. Certainly there will be some potential borrowers, with good uses for borrowed money, who are crowded out by the dishonest and reckless. Even in the absence of this extreme case, the rate of interest to good borrowers will be raised by the presence of the bad, and some good projects will therefore not be undertaken.

It is also possible that good projects will be foiled even in the most honest community, if all but lenders have rational expectations. It is widely thought that bankers are unduly cautious, though this view could well be prejudice. The reverse is also possible, bad projects being undertaken from unjustifiable optimism. Even in cases where loans, and the projects for which they are intended, are all justifiable, optimum arrangements between lender and borrower will not be possible where the two sides have different (rational) expectations (see Virmani 1982).[2]

Imperfections in the process of intermediation—channeling the savings of some to the investments of others—are inevitable, since the future is uncertain and people are unreliable. Imperfection does not imply that improvement is possible, though of course it may be. From a policy point of view, it becomes important to understand the reasons for imperfection. It has been attributed in the past to a lack of development of formal lending institutions, to bankers' pessimism or risk aversion (these resulting in high interest rates or lack of access to credit), and to monopoly. The consequence has been that governments have put ceilings on interest rates and tried to suppress informal lending institutions, while also often subsidizing banks, as well as giving them special incentives, and also directives, to lend in particular directions. New-style institutions specializing in longer-term industrial lending were intended to overcome the pessimism of traditional bankers. But high interest rates for some, as well as lack of access for others, may just as well be signs of a healthy capital market as of a badly operating one.

The above analysis suggests that reducing the risks of lending, while permitting lenders to compensate for risk so far as they can, are the fundamen-

2. Such expectations may be rational on both sides, given different information, or differing ability to process it, or both.

tal means of improvement. But before elaborating on this, a survey of the actual state of capital markets, with special reference to the manner in which they operate for small business in developing countries, is in order.

The State of Capital Markets in Developing Countries (Excluding India) in Relation to Small-Scale Enterprises

Our information comes both from borrowers and from lending institutions. We first consider small firm surveys outside India, which is considered separately.

Borrower Surveys

The evidence from Africa up to 1977 has been well summarized in Page (1979, pp. 21–24).[3] Initial investment is overwhelmingly from own savings or those of relatives or friends. Neither banks nor moneylenders nor trade creditors play a significant role.[4] Start-up capital is often tiny. For instance, in Sierra Leone a mean of about $70 (in 1974–75) is reported (Liedholm and Chuta 1977, table 8), and a similar figure has been given for rural industry in Kenya (Child n.d.). Further investments are financed to a very high degree from retained earnings, though institutions play a somewhat larger role. When, rarely, small enterprises obtain loans, the real rate of interest is very low, government often fixing maximum nominal rates. Moneylender rates are mostly irrelevant. There is no evidence, here or elsewhere, concerning charges on loans from friends or relatives. The opportunity cost of these, and of the use of own savings, is very hard to estimate. Page (1979) suggests a real rate of 15–25 percent. Most surveys stress complaints of a lack of working capital. But this is hard to interpret, as it does not always imply that more and profitable sales could be made if only the unit could borrow to buy more materials (or perhaps give more credit). Moreover, even if inadequate stocks of materials or work in progress is a constraint, it may be a failure in material markets rather than in capital markets. We have already noted that complaints of material shortages are very general.

Evidence from other parts of the world suggests less divergence from the above sketch than one might expect from the fact that capital markets are least developed in many parts of Africa. Ninety percent of the start-up finance for units with less than 25 employees in Jamaica came from own savings, with friends and relatives providing another 6 percent. About 13

3. Small refers to less than 50 or sometimes less than 30 workers.

4. Ghana is an exception, according to Steel (1977, table 4-3), where 10.8 percent of firms with less than 30 workers are recorded as having used bank credit to establish themselves.

percent had applied for loans to commercial banks or government organizations, of which only about 20 percent were successful. The more educated, and those in the towns, applied more often and were more successful. Lack of collateral was the main reason for refusal, and often also a reason for not applying (as everywhere). The average initial investment was $800, varying from $70 to $17,000 (Fisseha and Davies 1981). In Haiti, the picture is similar. Banks provided less than 3 percent of start-up finance, which ranged from $100 to $6,000 for enterprises employing less than 50 workers, but 8 percent of finance for subsequent investment. Ninety-four percent of the units had never applied for a loan. Half of the few applications failed for lack of collateral. Most loans from whatever source were predominantly used for working capital, which was perceived as the most serious problem; but again, the significance of this is not altogether clear (Haggblade, Defay, and Pitman 1979).

The situation is not very different in more developed Colombia. The main source of start-up finance has been personal savings. This often took the form of severance pay, especially for firms started by production workers and with initial employment of less than 8 workers. Severance pay was often supplemented by other personal savings and family loans and by suppliers' credit. Institutional credit hardly figured (Cortes, Berry, and Ishaq 1987, chap. 4). This was despite the fact that initial investment, while judged to be "generally low" by the authors, was high by the standards of small new enterprises in Africa or the Caribbean (see above) or Asia (see below). Thus in a sample of 68 metalworking firms of a mean size of 35 workers, the average start-up size was 14 workers with an investment of $70,000. Of the firms that started with 7 or fewer workers (60 percent of the total), the average start-up investment was $13,500. The 16 firms started by mechanics had an initial average investment of $9,000, reckoned to be within the reach of skilled workers after five to ten years. Finally, it is worth remarking that even in the most industrialized countries start-ups are predominantly financed directly by personal savings.

Institutional finance played a bigger but still minor role as firms grew larger and older. In the two sample sectors, metalworking and food processing capital used was about 75 percent self-financed. In the latter industry, this proportion fell markedly with both size and age—for instance, in the size range 1–7 workers the self-finance rate was 94 percent. In metalworking the proportion of self-finance also fell with size, but less markedly. In both sectors less than half the firms used any institutional finance, the paperwork required and sheer dislike of indebtedness being the main reasons for not applying; and lack of collateral, as ever, the main reason for refusals. Education and social position were influential in applying for and obtaining loans (Cortes, Berry, and Ishaq 1987, chap. 4).

The Colombian study attempted to relate the use of different forms of outside finance—public banks, private banks, and moneylenders—to the economic surplus generated.[5] While there was some association between private bank credit and economic surplus (despite the relative costliness of such finance), the main line of causation may well have led from success to bank finance (Cortes, Berry, and Ishaq 1987).

Moving to Asia, we have some establishment survey information for Korea, Malaysia, the Philippines, and Thailand. In Korea in 1973, 9 percent of units in the size group 5–49 workers cited financial institutions as the main source of start-up finance, more than elsewhere. Thirty-two percent started with less than $2,500, and 82 percent with less than $25,000. The mean employment was about 13.[6] One-quarter looked to outside sources for working capital, mainly to banks, but also to suppliers and friends and relatives. The larger the size, the larger the share of institutions. About one-third of investment was financed from outside sources, much less for the smallest units (Ho 1980, p. 75).

In an all-size sample in the Philippines, it was found that the modal size of start-ups had been 6,000–20,000 pesos (1978 prices)—that is, $800–$2,700—and that 8 percent had borrowed from financial institutions for the purpose. But after start-up, the use of institutional credit for both fixed and working capital increased (Anderson and Khambata 1981, pp. 110–11).

We have no start-up figures for Malaysia, but of the firms with new investments in size group 1–50 workers, 18 percent got some associated bank credit, and 20 percent obtained credit from banks for working capital. However, more than half of working capital came from suppliers until the size class of over 100 workers was reached, where banks became predominant. The proportion of establishments with bank credit rose sharply with size (but not the proportion of bank credit to total liabilities among those served). Entrepreneurial characteristics (experience, qualifications, other business interests, and knowledge of available assistance) were more important in explaining credit use than establishment characteristics (size and age), though these were of some significance (Bruch 1982).

In Thailand (1977) financial institutions, almost wholly commercial banks,

5. The authors use the term "private benefit cost," which consists of value added divided by the sum of the wage bill (including an estimate of the entrepreneur's opportunity cost) and capital costs (including an estimate of the opportunity cost of own funds).

6. In comparison with Colombia, it must be noted that firms with 1–4 workers initially were excluded from the Korean inquiry (Ho 1980), while many were of this size initially in the Colombian sample. If they had been included, considerably higher proportions would have been found with initial investment of less than $2,500. Although the Korean figures are at 1972 prices (on average) and those for Colombia at 1977 prices, it seems safe to conclude that initial investment costs in the Colombian sample were much higher than in the Korean sample. Finally, the figures cited are for only part of the metalworking sector in Colombia and for all manufacturing in Korea. However, table 5-1 in Ho (1980) gives figures for metal products and other industries as well as for all manufacturing. Sixty-eight percent of the firms in the metal products sector started with an investment of under $25,000, and the mean employment size of new firms was about 13—so that this sector is fairly representative of manufacturing as a whole.

provided about 2 percent of start-up finance for firms in the size group 0–9 workers, and 10 percent for those in size group 10–49. Organized markets played only a slightly larger role in the provision of working capital to going small firms. The Small Industry Finance Office played a negligible role, although it was set up for the purpose. Few had ever heard of it, and only a few of them applied. In general, as elsewhere, the larger the firm the greater the role of institutions. Again as elsewhere, firms were more concerned with availability and speed than with interest rates, and lack of collateral was the main reason for not getting loans from the banks (Tambunlertchai and Loohawenchit 1981, pp. 221–26).

Lender Surveys

One of the major pieces of evidence comes from a World Bank review of ten loan projects aimed at assisting small and medium-size enterprises, through either commercial banks or other official promotional institutions, mainly by loans though also by technical assistance.[7]

The most important thing to note is that, except in the case of Sri Lanka, and perhaps Jamaica, few loans had been made to small enterprises employing less than 50 workers and probably none to the very small (less than 10 workers).[8] But the upper limit of eligibility was defined in terms of asset size, ranging from $63,000 in Bangladesh to over $1.75 million in Mexico, so that one cannot be precise about employment size. Within the limits loans were skewed to the medium-size and large. In Bangladesh only about 7 percent of commitments were for small loans below $4,000, disappointing the intention of the project to reach down to "cottage-shop" industry. Medium-size loans averaged $36,000. In the Philippines, only 3.3 percent of the value of subprojects financed were those of establishments with assets of less than $14,000 (100,000 pesos), and 69 percent were in the range $14,000–$70,000. In Mexico, 60 percent of the number of loans went to small firms, but these were not very small, being defined as those with assets less than $310,000, and so on. Only Sri Lanka differed much from the other countries involved (Bangladesh, Cameroon, Colombia, Jamaica, Kenya, Mexico, Morocco, Philippines, and Portugal), the modal loan falling in the range $500–$2,500. In general, the bias toward the larger was much greater than the World Bank intended or expected but amply confirms the natural tendency of commercial banks and other specialized lending institutions to want to lend where costs are lower and risks less.

7. It seems to be correct to describe this under lender surveys, although the consultants did make superficial surveys of a small sample of subprojects. The summary report of the review is in Levitsky (1983).

8. In the case of Jamaica, the consultant suggests that the formal criterion of small may have been met, while the firm or the entrepreneur's activities were far from small. See Helm (1981).

In some countries a high proportion of the loans made were for start-ups, more so where the intermediaries were development finance companies rather than commercial banks. In some of these, however, notably Bangladesh, Jamaica, and Morocco, although the enterprise was new, the entrepreneur was an old hand and often seems to have been wealthy and already in control of several enterprises. It is clear from many surveys of small firms that the clientele of the lenders was very far from being representative of new firms in general. The latter usually start very small, are rarely started by established businessmen, and seldom get institutional loans.

In other countries where a high proportion of loans went to new enterprises, for example, Cameroon and Kenya, the default rate was very high. It was also high for new projects financed by the Development Bank of the Philippines (DBP) and very high in Jamaica for loans not channeled through commercial banks. However, default rates are no guide to what is possible, if the money is slushed to "borrowers" with no intention of repaying, as has happened to official funds in a number of countries. On balance, funds were channeled with less default through commercial banks, though the guarantee given in a number of cases did not stop the banks requiring 100 percent collateral or more.

Nominal interest rates on development finance corporation subloans were in the range of 10 to 15 percent for most countries and usually lower than commercial rates. They were no higher and sometimes lower for smaller enterprises. Since inflation rates were high and varied, the real rates were often low and sometimes negative.

Another major study already referred to, largely from the official and lenders' point of view, is that of Anderson and Khambata on the Philippines. Saito and Villanueva (1980) provide further information on the costs of Philippine banks. The main points from these studies are as follows:

- Commercial banks mainly lend short and to large enterprises. They are not permitted to charge more to small enterprises.[9] Nor, on average, do they act as a conduit for government funds (from the Industrial Guarantee Loan Fund [IGLF], a rediscounting and guarantee facility), finding the spread of 5 percent inadequate, though a few banks—those with the firmest appraisal and surveillance—find such operations profitable.
- The most successful user of IGLF funds was the Private Development Corporation of the Philippines, which concentrated on the larger loans, averaging $50,000.
- The 35 decentralized private development banks were unsuccessful in handling IGLF funds and incurred high default rates. Yet they had been remarkably successful in using their own resources for very small loans,

9. This, as well as the 5 percent spread referred to below, has apparently been changed. Thus Levitsky (1983) reports, "In the Philippines, the commercial banks received an 8 percent spread for subloans made to small enterprises and 6 percent for those to medium enterprises" (p. 47).

below the $7,000 lower limit for the IGLF. These small loans were collateral- and character-based and depended on local knowledge partly acquired through small business deposits. The larger IGLF loans probably required a more project-oriented approach, which was not in the style of these small banks. Saito and Villanueva (1980) attribute their success to a particular style of management. Private development banks are generally managed by the major stockholder, who tends to be a prominent and often wealthy person from the local area.

- Apart from the IGLF, the DBP makes public loans, about 10 percent of its industrial lending going to small and cottage industry (it also makes some resources available to the private development banks.) These operations are still not profitable, although arrears have declined so that large industry loans are subsidizing the small (Anderson and Khambata 1981, para. 2-31). But while arrears may have declined, the rate remains very high. Thus Levitsky (1983, p. 74) reports that about 57 percent of DBP's SSE portfolio was affected by arrears of over three months in 1981.

The record of arrears and default for many institutions in many developing countries is discouraging. But it is important also to point to the fact that it is possible in some conditions for very small loans to be made to poor people and to very small enterprises. Arrears and default on a large scale usually arise from dishonesty, or recklessness bordering on dishonesty, rather than from miscalculation or bad guesses about the future. A good capital market and reasonably cheap credit presuppose an honest community. In the case of larger loans the lender may have recourse to the law, and this helps to keep the borrower honest: but for very small loans this becomes hopelessly expensive. For these to be possible, either informal methods of securing payment or a moral community which puts pressure on potential defaulters are required. This is, of course, why friends and relatives play so large a role. But there are also success stories of wider cooperative communities.[10]

Indian Capital Markets

Borrower Surveys

In 1977 the Reserve Bank of India conducted a large sample survey of small-scale units assisted by commercial banks under the Reserve Bank's Credit Guarantee Scheme. Since the guarantee scheme covers almost all small-scale industrial units, and since banks are expected to cover all loans to SSEs, the survey can be regarded as a survey of those industrial SSEs which have borrowed from the commercial banks. At the time, the definition of small

10. See, for instance, the account of the Federación de Cajas de Crédito in Kahnert (n.d.).

was that the original investment in plant and machinery should not exceed Rs1 million (Rs1.5 million for certain ancillary units).[11] Eighty-six percent of the estimated population of enterprises was engaged in manufacturing, the rest mainly in "job work" (presumably servicing and repairs, often also classified as manufacturing); it should be noted that other guarantee schemes, operated by the Credit Guarantee Corporation of India since 1971, cover those working in agriculture and husbandry, transport operators, retailers, and self-employed professionals. Very small manufacturing business (with less than Rs150,000 of original investment in plant and machinery) could in principle be covered under these other schemes, but probably almost none were, since the terms for the banks under the older Reserve Bank scheme, started as long ago as 1960, were better.[12]

The RBI survey sample numbered 12,356 units, out of an estimated population of 233,218 assisted units. It is interesting to try to compare the total of assisted units (there was no lower limit to size by any measure) with the total number of units in the same size range. Manufacturing establishments with employment of less than 20 numbered 4.55 million according to the 1971 census. By 1976 this might have grown to 7 million (projecting past growth rates between 1961 and 1971) (see Mazumdar 1983, part I, table 12). The number of manufacturing units in the RBI survey population with employment of less than 21 may be estimated at 177,000 (see Reserve Bank of India 1979, tables 1-2-6 and 1-2-7).[13] Thus about 2.5 percent were assisted by the commercial banks. Even if the growth in the number of manufacturing establishments had slowed almost to a standstill since 1971, so that the total was only 5 million, the proportion assisted would still have been only 3.5 percent. If one considered only the nonhousehold sector, the proportion would be considerably higher, since the banks lend little to the more traditional sectors.

The proportion of all manufacturing workers in the assisted units was considerably greater. The average number of employees per unit in the estimated population was about 13. Since the proprietor, partners, and unpaid family workers seem to have been excluded, we add a guess of 1.5 workers, which results in an estimate of employment of 3.33 million. Of these, some 2.86 million were in manufacturing. Total manufacturing employment in 1971 is estimated to have been 22 million. In 1976, this might have risen to 24 million, so that assisted units had roughly 13 percent of manufacturing employment. It is worth noting that over 70 percent of manufacturing employment was in the 12 percent of units with over 20 employ-

11. The limits have since been raised to Rs 2 million and Rs2.5 million.

12. The credit guarantee schemes have since 1981 been brought together in one institution, the Deposit Insurance and Credit Guarantee Corporation, which has initiated a new scheme of its own.

13. We ignore the discrepancy between the limits of 20 in the census and 21 in the survey.

ees (Reserve Bank of India 1979, table 3-4). Although we have no comparable figures for other countries, it seems that this degree of commercial bank penetration, which has been growing rapidly since 1960, and especially since 1970, is greater than in most developing countries. It is perhaps even more unusual for commercial banks to reach down to the tiniest units. Twenty-nine percent of the estimated population of assisted small units had no employees, 12 percent had net assets less than Rs1,000, and 37 percent had net assets less than Rs10,000. Reaching down to the very small takes very little indeed in terms of loanable funds. Units with less than Rs100,000 of total net assets accounted for 66 percent of the estimated population of assisted SSE units, and 5 percent of their total use of formal credit (banks and financial institutions); those with net total assets of Rs100,000–Rs1 million accounted for 29 percent of the population and 38 percent of its total use of formal credit, and those with net assets over Rs1 million were 5 percent of the population and used 57 percent of the total formal credit (Reserve Bank of India 1979, vol. II, table 1-2). Since the mean employment of units, with Rs200,000–Rs500,000 of fixed assets (corresponding to about Rs500,000–Rs1.25 million of total net assets), was 45, it can be deduced that about half the finance for small units went to those that were of medium size as measured by employment (more than 50 employees), and a very small proportion to units with less than 20 workers.

The above account of the penetration of the commercial banks ignores many other financial institutions assisting SSEs, of which the most important are the state financial corporations.[14] We do not know any details, but the RBI survey shows that of those units assisted by the commercial banks, credit from other financial institutions was only 12 percent of that of the banks and was concentrated on the larger of the SSEs. The work of Sandesara (1982), discussed below, supports this latter point.

Some details of the changes in the pattern of liabilities and assets as one moves from the very small to the not so small (among the assisted SSE units) are given in table 15-1. The main features of interest are as follows:

- On the asset side, the proportion of inventories in total assets is constant up to total net assets of Rs100,000 (with one anomaly). From Rs100,000 to Rs2 million there is a somewhat higher plateau; the size class of over Rs2 million holds exceptionally high inventories, while receiving rather than giving trade credit.
- On the liability side, if one ignores the two smallest size classes with net assets of less than Rs10,000, the proportion of short-term borrowing from the banks rises strongly and monotonically with size, in accordance with expectations. It is hard to explain the very high proportion

14. For a full account of the many institutions which directly or indirectly provide finance and other support for the small-scale sector, see Ojha (1982).

Table 15-1. *Assets and Liabilities of Assisted Units Classified According to Value of Total Net Assets, 1976–77*
(percent)

Assets and liabilities	*Total net assets (thousands of rupees)*										*All units*
	Up to 1	*1–9*	*10–19*	*20–49*	*50–99*	*100–199*	*200–499*	*500–999*	*1,000–1,999*	*2,000 or more*	
Assets											
Net fixed	40	55	70	53	45	40	40	39	40	32	38
Inventories	32	32	19	30	32	38	38	39	38	59	44
Short-term assets	27	13	12	16	22	21	22	23	22	9	18
Loans and investments less trade dues, and so forth	9	2	3	7	13	13	14	13	12	−4	8
Cash	8	6	5	5	5	5	4	5	4	6	5
Other (including tax advances)	10	5	4	4	4	3	4	5	6	7	5
Capital and liabilities											
Capital, resources, and provisions	36	67	66	60	54	54	46	41	38	32	40
Long-term borrowing	10	12	15	14	14	15	17	18	19	12	16
Banks	10	11	13	12	10	8	6	6	8	5	7
Financial institutions	—	neg.	1	1	1	3	5	5	5	3	4
Other	—	1	1	1	3	4	5	6	6	3	5
Short-term borrowing	54	21	18	25	32	32	37	41	44	56	44
Banks	52	19	14	18	19	21	22	23	24	37	27
Other (mainly deposits)	2	2	4	7	13	11	15	18	20	19	17

—Not available.
neg. Negligible.
Source: Reserve Bank of India (1979).

of bank borrowing in the tiny size with assets of less than Rs1,000, even remembering that the sample is one of assisted units only. It cannot be accounted for by the presence of many virtually bankrupt firms with substantial fixed assets, for the mean size of fixed assets is tiny. It is however probable that fixed capital and total assets in use (and possibly equity) are much higher than the balance sheet figures, because either fixed capital is rented or though it is owned is not counted (for example, where working and living space are the same). But although some small units may be more substantial than appears from the figures, it is doubtful whether this is an adequate explanation. However, an adequate explanation is not important insofar as these tiny units account for an insignificant proportion of the finance extended to small units.

- The proportion of long-term borrowing is scarcely related to size. Given that the units are all assisted units, this is much less surprising than it would otherwise have been.

Sandesara (1982) has surveyed small and medium-size enterprises in a number of particular subindustries in three cities to investigate whether assistance seemed to improve performance. He used two sets of firms, "sample" being those assisted for at least three years by a state financial corporation and "control" being those not so assisted. Because of difficulties in finding enough controls in the same industry, the control firms include some that have been assisted by other organizations (and even by a state financial corporation for a short period). The sample enterprises proved to be on average close to the upper end of the range of the small. They were nearly three times the size of the control units (about Rs2 million of total assets on average against about Rs700,000). Equity was a little over one-quarter of total liabilities, much the same for sample and control, but less than for the RBI sample, for which it was about two-fifths. Short borrowing was about five times long borrowing (as also in the RBI sample). Net fixed assets as a proportion of the total varied greatly from industry to industry, but the average was not very different from that in the RBI survey—32 percent against 38 percent.

The above surveys of firms were concerned with those assisted. Assisted firms form a biased sample which cannot represent all enterprises, the vast majority of which are very small and have no links to any capital market institution. We saw that only about 3 percent had probably received commercial bank loans by 1977. We do not know how many others receive any form of governmental assistance, cooperative loans, or informal sector loans other than from friends or relatives. Things may have improved since the West Bengal survey of 1965–66, cited by Bose (1978), which disclosed that only a minute fraction of SSEs had ever heard of, let alone used, any of the rather numerous official sources of financial and other assistance. A still earlier 1960–61 survey of small engineering firms (with less than 20 workers) in Howrah (Jadavpur University 1964) showed that the sources of borrowing for those that did borrow were predominantly friends, relations, and dealers. Banks and government agencies hardly figured. But about a quarter of the medium-size borrowing firms (20–200 workers) borrowed from the commercial banks.

Our own surveys, although not yielding much capital market information (and no reliable information on the cost of loans), do suggest such an improvement. In a pilot survey of 58 miscellaneous small enterprises (with less than 50 workers) in Bombay, most of which were worth less than Rs1 million, Timberg (1980) reported that almost all got some raw material allocations under the Small Scale Industries Act, and well over half had received bank loans. Most of them were therefore probably benefiting from the RBI guarantee scheme and the general governmental pressure on the commercial banks to lend to SSEs.

Some results from our later surveys concerning start-up finance are given in chapter 12. Except in the case of powerlooms, for most firms over 80 percent of initial finance came from own personal savings. Friends and relations also figured, but the amounts were small compared with personal savings. In the case of powerlooms, trade credit dominated. Powerlooms are evidently still an example of the putting-out system, whereby the producer is financed by the merchant. Bankers still scarcely lent to small units in the more traditional industries, powerlooms and soap. But they were of some importance in the modern industries, machine tools and printing, in which 17–19 percent of operations had obtained some initial credit from banks, possibly reflecting the higher educational level of the entrepreneurs. They were, however, of similar importance in the traditional industry of shoes.

Lending Institutions

Although SSE loan guarantee schemes have existed since 1960, the RBI survey must also have been showing the results of the bank expansion following nationalization in 1969 and the increased pressure on the banks to lend to small firms.

The bank expansion has been studied by Rangarajan (1980). We report briefly only on its apparent effects on SSEs. According to Rangarajan, bank credit to SSEs grew from Rs2,860 million in June 1969 to Rs12,200 million in June 1976, and the number of recipients grew from 57,000 to 301,000, an impressive rate of increase.[15] The share of SSEs in total bank credit rose from

15. But according to the RBI survey, bank credit outstanding to the estimated population of assisted SSEs was Rs13,190 million (somewhat more than Rangarajan's figure), while the estimated number of assisted was 233,000, much less than Rangarajan's figure. We do not know the reasons for these discrepancies.

8 to 12 percent from June 1969 to April 1975 but stabilized around 11 percent until March 1978.[16] Some later figures are available from the RBI. In June 1980 outstanding credit to SSEs was Rs25,340 million and 11.9 percent of total bank credit. But by June 1981, it had jumped to Rs34,060 million and 13.2 percent of total bank credit (Ojha 1982, p. 939). This source also gives a figure of 794,000 for the number of units assisted at the end of December 1980, representing an increase of 2.6 times in 4.5 years (if Rangarajan's figure for June 1976 is accepted).

There was an overall ceiling for interest rates on short-term advances of 19.5 percent (in 1981–82). But priority sectors had lower ceilings. The rate for small-scale enterprises varied from 10.25 to 14.5 percent depending on the region and size of loan. Besides this, banks are required to lend 1 percent of their total loans to low-income borrowers at a rate of 4 percent (the Differential Rate of Interest Scheme).

When nominal rates are controlled and changed only occasionally, and the rate of inflation varies sharply, real interest rates also vary sharply and have sometimes become negative. In 1981–82 the inflation rate (of wholesale prices) was about 8 percent, so that real rates ranged from about 2 to 12 percent. Thus the Indian authorities insist on lower interest rates for the riskier loans.[17] The ill effects of this unnatural inversion are counteracted by forcing the banks to devote a certain proportion of their advances to the priority sectors (which include SSI), and by partly compensating the banks for the increased risk through the credit guarantee scheme. It is too early to say how effective the guarantee scheme will be, and at what cost, for it has become large only in recent years.

In the light of much evidence from other countries, it would not have been surprising to find default rates growing alarmingly as a result of the rapid expansion of lending to small enterprises. Under the guarantee scheme, banks are supposed to report defaults independently of claims made. The default rate was less than 5 percent of amounts loaned by June 1977 (Rangarajan 1980, p. 129). However, Rangarajan states that there is some evidence that banks do not report fully. Also when there is a rapid growth of loans, and especially if there are any grace periods extended, the number of loans that are too young for default to be formally possible is above normal. As against this, reported defaults may be collectible (there is no evidence of how far the guarantee scheme has reduced collateral requirements in India—though elsewhere guarantees have not had this effect) (Levitsky 1983, p. 40).

16. The share of agriculture, the other main priority sector, rose more impressively from about 5 percent in 1969 to nearly 11 percent in 1978.

17. This mistake was not made in 400 B.C. Then 15 percent was the lawful rate for the less risky loans (much the same as now), but loans for trade could be charged at 60 percent. See Kangle (1965). Our attention was drawn to this work by T. N. Srinivasan.

But finally the latest evidence from Rangarajan (1980, p. 190) suggests a significant deterioration. Further evidence comes from the RBI. According to provisional data, about 3 percent of the small-scale industrial units financed by commercial banks at the end of December 1980 were "sick," and the "sick" units accounted for about 10 percent of the advances.[18]

A study of the small-scale enterprise component of the World Bank's Calcutta Urban Development Project throws some light on lending to very small, mainly traditional enterprises such as tailoring. These loans were newly made under the differential interest rate scheme referred to above (which enjoys the RBI guarantee). This and some other incentives secured commercial bank coöperation. The bankers were also assisted by voluntary agencies and the Calcutta Metropolitan Development Agency, which identified and prescreened potential borrowers. Notable success in terms of income and employment is claimed. But for reasons already mentioned there is no reliable estimate of arrears, and the real cost of extending such loans is not known. The lesson is drawn that prescreening of borrowers, some follow-up activities including perhaps loan repayment collection by grass-roots private voluntary agencies, and possibly the staff of elected local bodies, can greatly reduce the cost of such schemes to the financial institutions involved; their local knowledge greatly helps in selecting the more creditworthy borrowers (Kahnert 1985). It is also probable that risks were reduced by the fact that the lending was mostly to traditional established trades.

A few further clues as to the operations and possibilities of commercial banks can be found in a study of one bank in India which has in various ways been something of an exception (Thingalaya 1978; also see Bhatt and Roe 1979, chap. 3). The Syndicate Bank, with headquarters in a small town, and founded in the 1920s, always specialized in both borrowing from and lending to poor people in very small amounts. This was clearly profitable, for its deposits grew at a rate of 33 percent per year from 1926 to its nationalization in 1969. In 1968, prior to nationalization, deposits of less than Rs1,500 accounted for 50 percent of deposits and 90 percent of deposit accounts. Thirty percent of its advances were to the subsequently defined priority sectors of agriculture (12 percent), small-scale industrial enterprises (11 percent), and other small business and professional operations (7 percent), as against 8 percent for the commercial banking system as a whole. Although administrative costs and the previously extremely low default cost have risen since nationalization, the Syndicate Bank's costs of lending to small-scale enterprises have remained much lower than for the only other Indian bank

18. A unit is sick if "it has incurred cash loss for one year and if in the judgment of the financing bank it is likely to continue to incur cash loss in the current year as well as the following year and if its total liabilities by way of loans, etc., far exceed its total worth as measured by paid-up capital and free reserves" (Ojha 1982, p. 947).

for which data could be obtained, and indeed much lower than for any developing country bank for which data could be obtained (Thingalaya 1978, app. B, statements IV and IX).[19] In explaining this, Bhatt and Roe (1979, pp. 16–23) put emphasis on the recruitment and promotion policies, high school graduates with a local background being preferred, and on decentralized and informal decisionmaking. This leads to relatively low salaries and a capability of handling more, and of course smaller, accounts than in the banking system as a whole, yet with lower default. All this suggests that there was, and probably still is, a vacuum which could be profitably filled if somewhat innovative methods were adopted; if this is so, the puzzle is why more institutions were not sucked in.

It might be said that the Syndicate's success lay in its being much more akin to the informal market than the commercial banks. The informal or unregulated market in India includes not only friends, relatives, and professional moneylenders, but also a number of large financial institutions and a host of smaller ones which comprise the sector inappropriately known as "indigenous" banking (as if commercial banking were not indigenous!). This has been recently studied by Timberg and Aiyar (1980). "It is clear that these systems provide a large amount of finance—at least 50 percent of that provided by the commercial banks. They serve sectors of the economy or needs that are not served by the banks—especially support wholesale trade—and at rates only slightly higher than those charged by the banks. Their transactions costs are normally lower than the banks, and this results in some increase in the total amount of intermediation" (Timberg and Aiyar 1980, preface, p. i). By and large, these unregulated lenders do not advance a higher proportion of their loans to small-scale industrial enterprises than the commercial banks, the Shikapuris of Bombay and South India being an exception with an estimated 16 percent of loans for SSEs. They normally lend only to the upper part of the range of SSEs, say those with assets between Rs100,000 and Rs1 million. They lend almost entirely short term for working capital, and, in contrast to the commercial banks, without collateral. They lend on reputation and are generally unconcerned with the use to which the money is put. Despite higher interest charges than those of commercial banks, other costs in terms of time spent in loan application are avoided. For these reasons, and for lack of uncommitted collateral, or willingness to commit it, it is probable that in some areas and trades a high proportion of enterprises occasionally resort to this part of the informal market for short-term credit needs (Timberg and Aiyar 1980, table 20).[20]

19. But some resemblance to the private development banks in the Philippines is evident (see the second section of this chapter).

20. However, this section of the informal market appears to be in difficulty as the commercial banks have withdrawn the financing on which the Shikapuris traditionally relied (Timberg and Aiyar 1980, p. 39).

Conclusions

Differences in degrees of uncertainty of repayment of loans, and the reasons for such differences, go far to explain the state of capital markets and the problems which small enterprises often have in getting credit, especially from institutions. A viable bank, or other lender who expects to be repaid, has to assess the risk. The risk will depend to varying degrees on what the money will be used for (the project), on the enterprise, on the entrepreneur, and on the collateral and the cost and difficulty of having resort to the collateral.

Commercial bankers seldom concern themselves with projects. Only if a project is large in relation to the size of the enterprise will it significantly affect risk. Moreover, if a banker has much faith in the sagacity of the entrepreneur, he will not see any point in second guessing him. The cost of project appraisal soon makes it absurd as the size of the loan diminishes.[21] New small enterprises are themselves projects, and the collateral is almost necessarily less than 100 percent. Profit seeking institutions will therefore finance them only if the entrepreneur can himself be assessed. If a commercial bank is used as a funnel for government money, with partial guarantees and given interest spread, it may be induced to lend for projects that need appraisal, supervision, and monitoring. It is then used more as a development bank and takes on activities outside its normal experience; results have been mixed, at best.

Public development banks are often given social aims beyond money making, and regard for these aims requires project appraisal. Many examples show that this implies either that such development banks can serve only medium-size and large industry or that their small-enterprise operations must be subsidized. This also goes for commercial banks which undertake development banking activities.

The nearer the bank, or other financial institution, comes to being "informal," the more likely is it that it can make small loans successfully. The key is local knowledge, which facilitates low-cost appraisal of entrepreneurs and their enterprises. Commercial banks with a well-developed network of branches have an advantage over development banks in this respect. But a financial institution can seldom rival those who have business relations with the enterprise and may extend trade credit; and of course it cannot rival friends and relations. From the borrower's point of view, the informal financial sector often has advantages of speed, approachability, minimal paperwork, and no collateral requirements.

Enterprise surveys suggest that resort to moneylenders is not typical of small enterprises in most countries including India. This is prima facie not true of Korea, however, in which professional moneylenders form a consider-

21. In Jamaica, the Small Enterprise Development Company apparently spent more than twice the amount of small loans on their evaluation. See Wilson (1981).

able part of a large, well-developed informal financial sector to which enterprises of all sizes often turn for short-term working capital needs, willingly paying interest rates about double those of the organized, controlled, and mainly public, banking institutions (Park 1976, pp. 55–59).

But the relative importance of moneylenders in Korea may be as much a matter of definition as a real difference. India also has a large unregulated "indigenous" banking sector that includes quite big establishments with a degree of organization that distinguishes them from the small, isolated moneylenders. Unlike the latter it lends mainly to substantial traders and at interest rates not much above the regulated rates charged by the scheduled commercial banks.

India differs from other developing countries in that greater efforts have been made, especially in recent years, to channel institutional lending, mainly from the banks, to small enterprises. It is important, however, to remember the Indian meaning of "small," which is now defined as having an undepreciated value of plant and equipment of Rs2 million–Rs2.5 million (about $200–$250,000). About half the finance for the "small" enterprises goes to enterprises which would be classified in most developing countries as medium-size in terms of employment (more than 50 employees). The very small, say those with less than 20 employees, still receive a much lower proportion of bank finance than their proportionate contribution to employment or value added. We have no figures from the lending side as to the number of new firms assisted.

As a result of the above efforts, and also of the policy of reserving products for small enterprises, which must have increased the demand, the volume of bank lending to small-scale industry and the number of borrowers have increased dramatically. About 13 percent of bank advances now go to small enterprises, after being negligible around 1960. There are signs that the amount and number of bad loans may be becoming a serious problem. Nevertheless, the record so far compares very favorably with most developing countries, in which default has often become almost the rule.

The picture seen through the small borrower's eyes, especially new firms, still gives an impression which is not very different from other countries. Despite the greater penetration of the banks, the proportion of small firms that get bank loans is very small. As in all countries, including the developed, household savings dominate as the source of finance for start-ups. Banks rarely finance new ventures, and where they do the entrepreneur is often already an experienced business man. The World Bank's inquiry into development finance corporation subloans suggested that where many start-ups were financed, either the entrepreneur was an experienced person of some substance or the default rate was unsustainable.

Where very small loans, not necessarily for productive purposes, are made without serious default, it seems that close informal community knowledge is needed, something which is not impossible for banks if there is sufficient decentralization of decisionmaking (witness the Syndicate Bank in India or the decentralized private development banks in the Philippines), but is not easy either. Of course, as enterprises grow and acquire a track record the involvement of the banks rapidly increases. This is as true of India as anywhere else.

There is nothing necessarily wrong with a scenario in which both inexperienced entrepreneurs starting up in a small way and also small established enterprises are unable to get institutional credit. Easy access to such credit for new enterprises can result in adverse selection whether interest rates are high or low, and then there may be too many founders who will never succeed. And for established small firms the costs of administering small loans, and of selection (for lending blind is most unlikely to be socially desirable), must always be considered. In short it is possible to try too hard to serve small firms with institutional credit.

We started this chapter by posing two questions. The first was whether the high cost of capital, or its inaccessibility, explained the relative labor intensity of small firms (though we have seen that such relative labor intensity is far from being the rule). What is the cost of capital to small firms? It probably varies a great deal. Let us start with new firms. We have seen that new small ventures are overwhelmingly financed from personal savings. The opportunity cost of personal savings is at most times in most countries very low, as measured, say, by the rate of interest on deposits. Ignoring risk, the average cost of capital is thus low. But the average cost is relevant to whether an enterprise is started at all[22] and not to the choice of technique, which is determined by the marginal cost. The marginal cost may be virtually infinite, or anyway as high as moneylender's rates. It could be much lower if own savings or those of friends or relatives are not exhausted. But if they are not exhausted, the implicit rate of interest will still be higher than that on deposits, since the risk element is inescapable in such cases. We may conclude that the marginal cost of capital, although variable, is generally high for new firms and that this must make for labor intensity. If the firm grows, thus establishing a successful record, its access to public or commercial finance at lower rates of interest will surely improve. There can thus be little doubt that the functioning of credit markets tends to make small and especially new firms more labor-intensive than larger ones.

The second question was whether the imperfection of capital markets was a reason for special measures favoring small firms. Capital markets can never be perfect because lenders cannot costlessly and accurately assess the probabilities of payment. The smaller and less experienced the borrower, the

22. One might speculate that the rapid relative decline of small enterprises in Korea was partly caused by the very high rates of interest on deposits which ruled for some time. But greatly improved job opportunities and rising real wages may have been more important reasons.

greater will be the imperfection. It is in the nature of events and persons that some loans for what would have been good ventures will never be made, and this applies especially to small loans. Imperfection does not imply the possibility of improvement and is not therefore a sufficient reason for intervention. The problem is to find measures with benefits that outweigh costs. The search is on and should continue. But the question whether society will benefit from a reduction in capital market imperfections, and in what sectors of the economy, can admit of no general answer.

16

Summary and Conclusions

Our primary goal in this book has been to place small-enterprise development and policy in India in an international perspective in an effort to draw some lessons relevant to industrial policy in developing economies. This chapter summarizes our principal findings and sets out our major conclusions relevant to policy.

Summary

In the developing economies, we usually take "small" to mean less than 50 workers, when size is measured by employment. Within the small, we distinguish household manufacturing and workshops with up to 5 employees. Together we refer to them as cottage shops. India has its own definitions, to which we call attention when necessary.

Household manufacturing, which is confined to a few industries and is most prevalent in rural areas, has declined relatively in all the economies examined (Colombia, India, Korea, Malaysia, Philippines, Taiwan). Workshops have also declined in the economies that have increased industrial employment most rapidly (Colombia, Korea, Malaysia, Taiwan). They may have declined even in India, but this is uncertain. Yet cottage shop manufacturing still accounts for more than half of all manufacturing employment in the poorest countries (India, Indonesia, Philippines, and most African countries).

There has also been a relative fall of employment in small and medium-size factories (in the range 5–100 workers) in the more rapidly industrializing economies (Colombia, Korea, Malaysia, Singapore, Taiwan). Korea is especially notable in that from 1963 to 1975 over 1 million persons were absorbed into manufacturing employment, 86 percent into establishments with over 100 workers. The smallest size group—5–9 workers—showed no change in employment. The story in India, where manufacturing has grown relatively slowly, is the opposite. Manufacturing establishments in the range of 10–100 workers have risen since 1960 relative to the large, especially those with over 500 workers. To some extent this is a redressal of an anomaly, for it was widely observed in the 1950s that India had peculiarly few establishments in this range, employment being concentrated in households and workshops and in very large establishments.

In the industrial countries "small" takes on a different meaning. There has been a relative decline of manufacturing establishments employing less than 500 workers for a long time, and the smaller the size class the greater the decline. But the eclipse of the small was especially pronounced from 1945–70 except in Japan, where the percentage of very small establishments (employing less than 21 workers) in the total actually rose slightly between 1960 and 1975. The fall in the United Kingdom was greater than anywhere else, the absolute number of small establishments (less than 200 workers) more than halving from 1935–68. Although the decline was general, the position reached around 1970 varied widely from country to country. Establishments with less than 200 workers accounted for between one-third and two-thirds of manufacturing employment, with no apparent relation to national income.

Since 1970 the decline seems to have been arrested and even reversed, at least in the United Kingdom and the United States. At the same time there has been a marked change in the climate of opinion in Europe. After a long period of encouraging the formation of very large businesses, there has in recent years been a spate of legislation in favor of small firms, which are now widely believed to be essential to the continuing genesis of a desirable industrial size structure.

The relative decline of small-scale enterprises in most developing countries, including many not mentioned here, has been accelerated by their industrialization policies. Protection, investment incentives, credit control, and the promotion of industry in the public sector have all discriminated against the small. Big firms have been better placed to obtain import permits for machinery, components, and materials. Investment incentives have discriminated, both de jure and de facto, against the small. Credit controls and interest rate ceilings have made banks favor large firms even more than is their natural disposition.

Another factor was important for the highly export-oriented economies of East Asia. Big firms have economies of scale in export marketing, and it is generally true that small units export a much lower fraction of their output than the former. This largely explains the very rapid relative decline of small and small-to-medium-size firms in these countries, a decline which has been in no way inconsistent with a very labor-demanding development.

It is true that in most countries there have been some countervailing measures, such as the establishment of credit institutions for the small, industrial estates and extension schemes. But these have only scratched the surface and have very little impact compared with the general economic policies of the government. India is, however, a special case. For some twenty years after World War II, Indian policy was guided by a philosophy which, itself a compromise between a modernizing and a Gandhian outlook, saw large-scale, import-substituting heavy industry as the keystone of the development arch, while traditional small-scale industry would, with few investment or development expenditures, meet the everyday demands of the people. Thus much small-scale, especially handicraft, industry was protected and preserved. This policy contributed to the dearth of small-to-medium-size modern enterprises. But in the late 1960s, investment in heavy industry became less dominant, and policies favoring the small were strengthened, principally by reserving an increasing number of products (now over 800) for the small, while restricting capacity expansion of the large, and also by promoting a quite dramatic increase in bank lending to small and medium-size firms. The ceiling definition of the small has also been repeatedly changed. It started as less than 50 workers (or 100 without power) and became in 1966 a limit in terms of the original value of plant and machinery only, which now stands at Rs3.5 million (Rs4.5 million for auxiliary units), or about $310,000. This corresponds on average to about $775,000 for total capital employed and admits many units with over 50 workers. These changes have thus widened the range of units which can benefit from the policies favoring small enterprises. The upshot has been a rapid growth of small factory establishments (less than 50 workers) especially in urban areas, while medium-size units (50–100 workers) have grown fastest in rural areas. It was found that there has been a marked change in favor of factories with less than 100 employees in most of the industries for which comparability over many years could be established.

Some industries in which the concentration of employment in large units remains very high, for example, iron and steel, jute, cotton textiles, and sugar, are also among the largest employers. Thus, Indian manufacturing is still very large-establishment-oriented compared with that of other countries. Manufacturing establishments in India with over 500 workers account for a higher share of total factory employment in units of over 20 workers than in the United States and far higher than in Japan, Korea, or Taiwan, despite the rapid growth of large-scale industry in these countries in the past twenty years. Similarly, the medium range of 50–500 workers accounts for well over half of total factory employment in these countries, while in India it accounts for less than a third. A more detailed analysis of all nonhousehold manufacturing in six North Indian states (Bihar, Haryana, Madhya Pradesh, Punjab, Uttar Pradesh, and West Bengal), using the population census of 1971, confirms the above point. In all six states, medium-size establishments account for a considerably lower proportion of manufacturing employment than is the case in Korea, Taiwan, and the United States. But the analysis also shows that the size structure varies widely from state to state despite the supposedly uniform influence of central government policies. Punjab stands out as small-establishment-oriented, and West Bengal as large-establishment-oriented.

In all the states a high proportion (58–76 percent) of employment in units of less than ten workers is accounted for by six (two-digit) industries—food products (20, 21), clothing (26), wood products (27), metal products (34), and repairs (97). The question arises whether the large-size-orientation of West Bengal and the small-size-orientation of Punjab are accounted for by a different industry mix. The answer is yes for West Bengal and no for Punjab. On the one hand, West Bengal is the home of some large-scale industries, especially steel and jute, and this can be ascribed to natural comparative advantage. On the other hand, Punjab is small-establishment-oriented within (two-digit) industries. This is harder to explain. One possibility is that rapid agricultural growth has stimulated small-scale agro-related subindustries, largely located in urban areas. (Nonhousehold manufacturing employment in Punjab, like that in West Bengal and Madhya Pradesh, is three-quarters urban, against about half in Bihar, Haryana, and Uttar Pradesh.)

We next examine the results of Indian industrial policy toward the textile industry, which is by far the most important of those industries profoundly affected by the government's protection and promotion of traditional and other small-scale industries. In the 1950s the government banned the installation of new looms by the mills, except for replacement or export production, and it later reserved synthetic cloth production for powerloom and handloom operations, which also paid no excise duty (except, in theory, if more than five powerlooms were operated).[1] The government's intention was mainly to promote handloom production, and it was envisaged in the second plan that only 10 percent of increased cloth production would come from powerlooms. The outcome was different. From 1956–81 mill cotton cloth production fell from 4.9 to 3.1 billion meters. Small-sector output rose by 3.3 billion meters, but three-quarters of this increase came from powerlooms. Powerloom output rose by about 8 percent per year, and handloom output by about 2 percent. There is no mystery to this. Judging by our samples at least, powerlooms are much more profitable than handlooms. They have the further advantage that handlooms are not very suitable for weaving synthetic cloth. It is not even certain that handlooms have, on balance, benefited at all from the government's policies, for their production was also rising slowly throughout the first half of this century, being "protected" as now by low wages and, to some extent, specialization in certain markets. Mill wages are now two to two and a half times as high as those paid by both powerloom and handloom operators.

1. The looms used by small-scale powerloom operators are very often bought from mills. The new looms used by them are of rather lower quality and cheaper than those used by the mills, but there is no essential technological difference.

The question arises whether powerloom production is socially more beneficial than mill production. A crude cost-benefit analysis—but the best we could do given the availability of figures—suggests the opposite. Ignoring distributional considerations, a switch from powerlooms to mills appears to have a high social yield, at any plausible shadow wage rate. The best one can say for the powerlooms is that they improve the distribution of income in that millworkers cannot be numbered among the poor in India.

Government policy toward sugar has had equally dubious results. The "intermediate" product khandsari has been encouraged in similar ways to powerloom production, and it shows up badly as compared with mill production. Effects on the traditional gur industry have probably been detrimental, but we have seen no cost-benefit analysis of this industry, which appears to have been neglected in sugar policy discussions.

The policy of product reservation for small-scale industry has had some deleterious effects in light engineering. It has resulted in reductions in the linkages between small and large enterprises, and it has limited the development of specialized subcontracting firms. The structure of the industries that developed subsequent to the capacity restrictions on large firms has inhibited the adoption of new technology of both products and processes.

We turn to an examination of how capital-labor ratios, and capital and labor productivity, vary with establishment size. One of the main arguments of those who believe that "small is beautiful" is that small enterprises are labor-intensive and that labor intensity of production is much to be desired in countries where labor is plentiful and capital scarce. But the value of what is produced cannot be neglected. Labor intensity alone cannot be a criterion of social desirability. If capital were the only scarce factor, high capital productivity would be the desideratum, and this may or may not go together with labor intensity. In fact, labor is not a free good, and more complicated criteria for deciding whether one firm size is better than another are needed, for example, total factor productivity or social cost-benefit analysis.

A first step, however, is to see whether the hypothesized relationships—the larger the size, the greater is the capital intensity and the lower the capital productivity (and the higher the labor productivity)—hold good. The short answer is that they hold good for manufacturing in the aggregate (though there are exceptions, the smallest size class not always being the most labor-intensive or having the highest capital productivity). But this of itself is of no interest so far as policy is concerned. They may, on the one hand, arise in large measure because some industries use little capital in small establishments and have high capital productivity and low labor productivity, and vice versa, with little variation between units of different size. If this were so, then the best way to increase the labor intensity of the economy would be to increase demand for the products of industries that are typically labor-intensive and have high capital productivity. The size of establishment, and supply-side intervention in favor of the small, would be irrelevant. If, on the other hand, *within* an industry small units are more labor-intensive and have higher capital productivity than large units, and their products are competitive with those of the large units, then supply-side intervention in favor of the small would be appropriate. It is therefore necessary to disaggregate, that is, to examine how capital intensity and capital and labor productivity vary with size within industries and subindustries that are sufficiently narrowly defined for the products of the different establishments to be competitive with each other.

When manufacturing is broken down into its different component industries, the hypothesized relationships between size, as measured by employment, capital intensity, and the factor productivities, frequently fail. The greater the disaggregation, that is, the more narrowly defined are the industries, the more frequently do they fail. This is shown most clearly by examination of the Korean census, in which the relevant figures are published by size group for industries at the four- and even five-digit level. In virtually no industry was there any regular (monotonic) behavior of rising capital intensity, rising labor productivity, and falling capital productivity with size. It was further found that among 213 subindustries labor intensity peaked with about equal frequency over all size classes in the range 50–500 workers. Interindustry variations are far greater than the unreliable intraindustry variation by size group. The latter seldom vary by a factor of more than three, while the range of the former is more than a hundred.

Capital productivity in Korea peaked in the range 50–500 workers in two-thirds of the industries. We found further evidence in Colombia, India, Japan, and Taiwan that capital productivity tends to be highest in medium-size establishments.

Total factor productivity is a better measure of efficiency in the use of factors of production than capital productivity alone. Results based on the Korean census show that in about half the industries examined this measure peaked in the range of 50–500 workers. In only about 5 percent of industries did it peak in the smallest size class. Moreover, the few industries in which small establishments are most efficient are also small industries in terms of employment.

Our own surveys of product-defined industries in India generally support the above results. For capital intensity, the main feature was that variations within size groups were so great that apparent variations between size groups were rarely significant: and in none of the five industries examined was there a monotonic progression of average capital intensity with size. Only in shoes was there a large significant jump in capital intensity between size groups of less than 50 workers and those with over 100 workers. This corresponded to a difference in technique, influenced by government policies defining the small units, rather than to size as such. In soap, in contrast, a policy-protected establishment could be of any size provided it used no power, and there was no relation between size and capital intensity. If one is looking for labor intensity as such, it is to technology that one should look, not size as measured by the number of workers.

For capital productivity the same high variability within size groups was observed, and even the average results were very mixed. For printing the average tended to fall with size; in machine tools to peak around 25–50 workers; and in soap and metal casting there was no significant relation. In shoes there was a sharp significant fall in the class of over 100 workers, corresponding to the sharp increase in capital intensity; and the smallest size group also had relatively low capital productivity. Technical efficiency (and total factor productivity) did not vary systematically or significantly with firm size except in the machine tool industry, in which the inefficiency of the small size group (5–9 workers) was significant compared with that of larger establishments.

Apart from those relating to technical efficiency, the most interesting results arising from an econometric estimation of production functions were that (1) there was no significant evidence of increasing returns to scale across the full size range in any of the five industries for which tests could be made—that is, printing, machine tools, soap, shoes, and metal casting; (2) there was a high partial elasticity of substitution (about 3.0) between skilled and unskilled labor in the four industries for which estimates could be made (printing, machine tools, shoes, and metal casting); (3) there was a high but somewhat lower elasticity of substitution (about 2.0) between capital and skilled labor in the same four industries; and (4) the elasticity of substitution between capital and unskilled labor was considerably lower—about 1.0 in machine tools and metal casting, and about 0.5 or less in printing and shoes.

In all the above analysis size has been defined by the number of workers. We have found that capital intensity and productivity rarely rise and fall consistently in the expected manner as size increases. In particular, we have found that very often both capital and total factor productivity peak in the medium range of 50–200 workers. Here an important caveat is in order. We have found some evidence that if size were defined in terms of some measure of capital used, then the results obtained might be much more reliable, and small firms thus defined might show up much better. It is notable that the Indian definition of "small" for policy purposes is in terms of capital, which does not exclude some firms that would be medium-size in terms of employment. Unfortunately most census and other tabulations of data, including the Indian, use employment size, and we have not therefore been able to verify widely the validity of the caveat here entered.

We next turn to the founding of our sample firms and their subsequent record as measured by growth of employment and sales and by profitability; and to an attempt to explain the differences between firms in these respects as well as the differences in technical efficiency.

So far as founding is concerned, a very high proportion started small with less than 20 workers, but there was a handful of larger starts especially in machine tools and metal casting (initial investment is mentioned later). The educational level of founders was about equally split among primary and below, middle and high school, and intermediate or higher. This however varied by industry; most entrepreneurs in powerlooms and shoes had little education; middle levels predominated in machine tools, metal casting, and soap; and higher levels in printing. The fathers of the founders had had much lower levels of education, and the social background of the founder was thus quite mixed. But in the caste-dominated industries, powerlooms and shoes, where the educational level was also low, the fathers of founders themselves had frequently been small proprietors or partners. In the other industries, apart from soap, inheritance was negligible. But perhaps the most interesting point is the importance of migrants (the surveys were made in North India except for metal casting). The soap industry was almost exclusively run by migrants, and they were also in a majority in the modern industries, machine tools and printing. This prominence of migrants echoes that found in many countries throughout history.

In considering growth, it must be emphasized that only survivors could be surveyed. Many inquiries bear witness to high death rates among new small firms. Bearing this in mind we found very high growth rates of employment over the lifetime of the firms (15 percent or more) in printing, machine tools, shoes, and powerlooms; while soap employment grew at 7 percent and metal casting at about 1 percent. The growth of sales was similar, but generally rather lower. In the last three years (1977–80), however, the employment growth rate had slowed to about 5–9 percent, except for metal casting, which remained at about 1 percent. But sales growth became higher than employment growth in every industry. We found no reliable explanation of this recent rise in productivity.

Size is closely related to growth, the young survivors growing much faster than older firms; but shoes and soap, in which primitive methods are used in the smaller units, are to some degree exceptional. In these industries a large minority of new small firms stay small.

We tried to explain the growth (of sales) in terms of a set of variables including the age of the firm, the experience of workers, and various entrepreneurial characteristics. The negative influence of age dominated the explanation except in machine tools and castings. Controlling for age, it was found that either college or graduate education, though not strongly influential, was positively significant in printing, soap, and shoes; but no educational variable was significant for machine tools or metal casting.

"Profitability" (or the rate of economic surplus) was defined as value added less the wage bill, divided by capital as measured by the original value of fixed capital in constant prices plus working capital. Even after making allowance for unpaid workers, it was found to be very high—about 50 percent—in the more traditional or older industries (shoes, powerlooms, and metal casting) and more normal—20–37 percent—in the more recently established or more modern industries (machine tools, printing, and soap). The relation of size to profitability was erratic, and the coefficients of variation within size groups were very high, so that there was little significance. However, in shoes, machine tools, and metal casting the smallest size group (1–5 workers) did have significantly low profitability, while in printing this same smallest size group had significantly high profitability. College or high school education was a positive factor for soap, shoes, and printing, but not for machine tools or castings.

For technical efficiency, although similar to profitability in definition, rather different explanatory variables were most significant. The firm-specific experience of workers was the most important variable. Literacy of the entrepreneur was also positive, but there was no evidence of middle or higher levels of education being of any value. Only in machine tools was the size of the firm a positively significant influence; this could reflect economies of scale in that industry, although no direct evidence of such economies was found.

One dog that did not bark at any of the dependent variables was the father's occupation. More surprisingly the experience of the entrepreneur himself in the same occupation was not significant—it may be that older entrepreneurs, while they have firm knowledge in certain matters, become outdated in others.

Thus far we have considered establishment size, and the use of factors of production, without reference to the nature of the markets in which units buy and sell. Yet such markets, together with the technologies that are known and available, largely determine the scale of operations and the techniques chosen.

We first consider product markets. It was seen above that little by way of increased employment could be expected from supply-side encouragement of labor-intensive units (which were, at that point in the argument, also assumed to be small) unless their products were good substitutes for the actual or potential products of large, capital-intensive units.[2] The industries surveyed were chosen with this in mind. The outputs of small and large were at least potentially competitive. Only soap was a partial exception. The small units produce almost wholly a low-quality laundry soap, and they could not produce the more refined soaps or detergents made by the large firms. This arises partly because of the manner in which the government protects the small, by making it impossible to use power without losing their fiscal advantage. Equally, the large firms could not produce the low-price, low-quality soap, given the wages they pay. In soap, high quality and high labor productivity are technologically linked. This fact, together with the oligopolistic structure of the power-using sector, suggests that high wages might be more

2. A subcontractor to a large firm may be considered to be in competition insofar as the latter could produce the item which it subcontracts.

the effect of capital-intensive techniques necessary for high-quality production than the cause of them.

In other industries also, the manner of protection has probably reduced though not eliminated competition by tending to segregate the markets of large and small: an instance is that the output mix of the cotton mills has moved toward the finer counts of cloth. This, however, does not alter the fact that powerloom output has massively displaced that of the mills (albeit, as we have seen, with dubious social advantage).

It has been argued that small enterprises tend to produce things which are "appropriate" for the poor. This is obviously irrelevant for printing, metal casting, and machine tools. The small soapmakers did not answer the relevant question, but other evidence shows that their product is used mainly by the poor. Powerloom output is bought almost entirely by the poor (defined as having an income of less than Rs300 per month), but the mills can and do produce similar cloth, though many, handicapped by the high wage level, have become "sick" and are subsidized. Handloom output is bought by all income groups, but mostly by the poor. Small shoemakers do not appear to relate especially to any particular income group.

No information on subcontracting emerged from the survey material. It is known that in some locations many small shoemakers sell to large firms, for example, Bata, but almost none in our sample did so. Metal casting is another industry in which subcontracting might have been expected, but again hardly any firms in our sample were selling to large engineering firms. Finally, there were few small-scale exporters either directly or indirectly. There were none in soap or powerlooms, and very few in shoes. In machine tools and metal casting there were a few, and more than a few in printing.

It is widely presumed that small enterprises in developing countries are more efficient users of labor and capital than large enterprises because wages are lower and interest rates higher than in large enterprises, both being closer to the true scarcity or "social" prices appropriate to capital-scarce, labor-abundant economies. The validity of this argument depends on three assumptions:

- It must be true that wages and the cost of capital vary with size in the supposed manner. It is of course true that on the average a small firm pays less for labor, and probably more for capital, but it may not be true that there is any steady progression. There may for instance be little difference between small and large within quite wide size bands.
- It must be true that the wage and capital cost differentials which exist do not arise for good social reasons. In short, they must be "distortions."
- Firms must choose techniques appropriate to the different factor price ratios they face. Our survey results, and others, suggest that there is plenty of technological room for factor substitution. But we have also found that industry-specific factor proportions and productivities do

not vary consistently with size (at least as measured by the number of workers).

The above points make it necessary to probe closely into labor and capital markets if the presumption in favor of small enterprises is to be validated or refuted. We turn first to labor markets and consider the progression of wages with size in India, where the average wages in factories with over 50 workers (or 100 without power) are rather more than double those of factories with fewer workers. This does not control for the quality of workers nor tell one whether there is a steady progression or a jump somewhere.

Research into the labor market in Bombay showed that workers in small and medium-size factories in the range 10–99 workers were paid, after controlling for education and training, considerably more than casual workers (some 65 percent more) and that there was another jump between the ranges 10–99 workers and over 100. Workers in the largest factories were paid 2.3 times casual workers, after allowing for skill differences.

With one exception, our surveys covering enterprises in the range 1–150 workers showed only a small progression or none for unskilled workers. (The data set used for this observation excluded a few large firms in the shoe industry, for which a different questionnaire was used.) The exception was powerlooms, in which the size group of over 50 workers paid double the smallest size group. Weavers in mills in a neighboring city were paid double again!

Apart from this exception, the evidence suggests that there is a wage cliff in the range 100–200 workers, with some association of size and wages within the small-to-medium-size range, but not a very strong one. This could go some way to explain both the lack of any monotonic relationship of capital intensity and size in our samples, as well as the fact that in shoes there was a large significant jump in capital intensity in firms with over 100 workers (but for which we have no wage data). There is supporting evidence of a wage cliff in Colombia, where it was found that blue collar wages vary only moderately by size up to plants of around 200 workers, beyond which there was a marked increase.

Large wage differentials by size of enterprise are not unique to India. In Indonesia they are just as large. But in Colombia and Malaysia they are considerably smaller. Differentials also exist in the developed countries, but they are small compared with those in any developing country.

It cannot be lightly assumed that the wage differentials which exist after allowing for skill differences must be distortions that are caused by union activity or government intervention. As against this, large wage differentials existed in the Bombay textile industry before there were any strong unions (though there were successful strikes) and when there was no government intervention. The most plausible explanation is that the mills found it profitable to attract permanent migrants, who, coming with their families, had a considerably higher supply price than transient migrants, who worked mainly for small enterprises or casually. Still higher wages in the largest firms may also be explained by the development of internal labor markets. Further industrial development results in a more settled and homogeneous urban labor force, and differentials narrow. This explanation is consistent with the equally large differentials prevailing in Indonesia and the smaller ones in countries such as Colombia and Malaysia and in the OECD (Organisation for Economic Co-operation and Development) countries.

However, in India, differentials widened after independence. This can be attributed to growing union strength and to governmental intervention in wage setting, which served particularly to raise wages in the largest factories. The upshot is that a considerable part of the very large differentials that exist between big and very small factories and workshops must be reckoned to be a distortion, while much of the still considerable differentials between medium-size factories and the very small is probably not a distortion.

The cost of finance to small firms is hard to determine, and it is important to distinguish between the average and the marginal cost. The great majority of new private firms start very small and rely entirely on personal savings, mostly their own but eked out by loans from relatives or friends: there are exceptions, but these mostly occur in the more modern industries and when the founder is already a substantial businessman. The average cost of personal savings is often very low: the only practical alternative may well be bank deposits, which generally bear a low, and often even negative, real interest rate. The average cost will determine whether an enterprise is set up. But the marginal cost should determine the choice of technique. This can vary from being very high (moneylenders' rates, though moneylenders are rarely resorted to for the initial investment fund) to being quite low if the sources of personal savings are not exhausted by the project that the founder envisages. Such variability may be a partial explanation of the very high variability in capital-labor ratios which we have found among small firms, though unfortunately our surveys yielded no reliable information on either the average or marginal cost of credit, and hence no information on relative factor prices.

As new firms grow and establish a track record the presence of institutional lenders, mainly the commercial banks, rapidly increases. For firms that obtain bank loans, it can be presumed that the interest rates paid are the marginal cost[3] and that this is below the cost of informal credit, and that therefore there is a tendency for the cost of credit to fall. In some countries the cost of such bank loans to small and medium-size firms is higher than that paid by larger firms, but in others it is lower. India is in the latter

3. However, allowance should be made for other conditions that the bank may impose, such as a minimum balance on current account.

category, since interest rates are controlled and banks are required to extend a minimum proportion of advances to small enterprises or other priority sectors. There has been a dramatic increase in bank lending to small firms in the past decade, though the proportion of small enterprises with bank loans is still quite low. There is thus no monotonic rise in the cost of capital, anyway not for all firms.

It is an open question as to how far, if at all, institutions should be created to lend to the small or existing institutions should be subsidized or directed, or both, to lend to the small—especially the very small. This doubt arises both from the nature of capital markets and from the characteristics of small firms and their role in the economy.

Capital markets are inevitably imperfect in the sense that any system whereby all investments that should be financed, given the most objective possible estimate of probable outcomes, will be financed, and others rejected, is a fantasy. The reasons are that projects are risky, that risk is assessed and evaluated differently by borrowers and lenders, that borrowers may be reckless or dishonest, and that it is costly for lenders to acquire information about borrowers and their projects. The smaller the borrower the more risky it is to lend to him (there is plenty of evidence of high death rates among small firms and high rates of default on the part of small borrowers). Furthermore, the cost per dollar lent of acquiring information about the borrower and his project mounts rapidly as the size of the loan falls. It is no wonder that banks insist on a high degree of collateral. But collateral only reduces the problem, for it cannot altogether eliminate the cost and the risk: and in any case small and especially new firms are often unable or unwilling to proffer full collateral.

There is nothing obviously wrong with a scenario in which both inexperienced entrepreneurs starting up in a small way and small established entrepreneurs are unable to get institutional credit. Easy access to such credit for new enterprises may cause, and in many cases certainly has caused, adverse selection whether interest rates are high or low. Then there will be too many founders who founder and sink. And for established small firms the high costs of administering small loans and of selection must be taken into account. The *lack* of institutional credit can be seen as a filter which does more to eliminate those who are dishonest, incompetent, or sluggish than it does to prevent potential climbers from setting foot on the mountain of success. The problem can thus be looked at in a more dynamic way. In the OECD countries the role of new small firms is seen as that of a seedbed. This image may well be appropriate also for developing countries. However, there is no apparent shortage of new starts in the countries that we have looked at, and even in India institutional credit for new start-ups is rare: and if there were any shortage the first step should surely be to remove artificial obstacles to the obtaining of materials, premises, and power supplies—obstacles that are common in developing countries—rather than to provide credit. Institutional credit may then be seen mainly as facilitating the expansion of those that have passed the seedling stage and are able to show promise of further achievements.

But what, it may be said, about the myriad of tiny enterprises that come and go, or remain small, and that still provide the bulk of employment in some of the least developed countries? The fact that there is such a myriad existing without institutional credit is perhaps a sufficient answer. But we have also seen that some institutions of a relatively informal kind do exist, which are able, without subsidization, to lend profitably in very small amounts. These should certainly not be discouraged. Further study of and experiments with such lending are needed.

Conclusions Relevant to Policy

The following conclusions should be useful for policy:

- Employment size is a very poor indicator of capital intensity and productivity, and of total factor productivity or technical efficiency, for narrowly defined industries; and for policy purposes it is essential to look at narrowly defined industries.
- There is considerable evidence that many small enterprises, with less than 10 workers, are often not the most labor-intensive and their capital productivity and technical efficiency are very rarely the highest in the whole size range of establishments in the industry.
- When measured by employment size, firms in the medium-size range of 50–200 workers have the highest capital productivity, and total factor productivity, in most industries in all the countries examined. Also, while the variability of such economic characteristics as factor employment and productivity, and profitability, is very great within all size classes, it is less in the larger size groups than among the small, even ignoring the high incidence of death among the latter.
- The very small (less than 10 workers) should not be looked to for their efficient employment of factors of production. But at the same time they should not be discriminated against. They are there, and they still provide the bulk of employment in the lower-income developing countries. Many of the industrialization policies of developing countries, which on other grounds should anyway be changed, in fact discriminate against them. Not only are they there, but they are also the source of most medium-size and large private firms. If supply-side intervention in favor of small firms is planned, the objective must be to spot potential winners and speed them on their way. Since it is difficult to spot winners at the starting gate, it follows that such intervention is more likely to be successful at a stage when there is already evidence of some success.

• The differences in labor intensity between industries dwarf the differences found to exist between enterprises within an industry. This points to the fact that there is much more mileage, so far as increasing the demand for labor is concerned, in altering the pattern of demands in favor of labor-intensive industries than in supply-side efforts to change the size structure of industries in favor of the more labor-intensive part of the size range, which anyway varies from industry to industry and is hard to find. The general economic policy measures that shift demand toward labor-intensive industries are the promotion of exports, and of agriculture, and income redistribution. Of these, the first is the most certain and well-documented. The East Asian export-oriented economies have been the most labor-demanding of all developing countries, and at the same time they have seen the most rapid relative decline of small firms.

• Where supply-side interventions are made in favor of the small—other than removing policy-induced obstacles—the small should be defined by some measure of capital employed, as in India. Since market value is too difficult to determine for policy purposes, and working capital variable and subject to window dressing, the original constant price value of fixed assets is probably the best measure.

• Lending agencies tend to lend toward the upper end of any limit established as the definition of "small." There is no reason to suppose that this is to be regretted.

• India's policies toward the small are peculiar. In some industries, those to be assisted are effectively defined in terms of technology (handlooms, soap, khandsari), or partially so (shoes, powerlooms). Technology is certainly a more effective discriminator so far as labor intensity is concerned than is the number of workers employed. There is no doubt that India has greatly increased direct employment by protecting traditional processes. However, it is far more doubtful whether the demand for labor has been increased when indirect effects are brought into the account. The indirect discouragement of clothing exports resulting from India's textile policies is the most notable example.

• It is also probable that India's policies toward textiles and sugar, and possibly toward other sectors, have bought even direct employment too dearly. A rough cost-benefit analysis suggests that this is true for powerlooms and khandsari. Intermediate technologies and products are not always socially desirable.

• More generally, the policy of reserving a great many products for the small (despite the fact that small is defined in terms of capital when not in terms of the technology used) has serious disadvantages. It tends to segment product markets and reduce the element of competition not only for the protected sector but also for large firms. It tends also to freeze the structure of the industry and inhibit the organic growth of firms beyond the protected size. It must also inhibit exports in that larger firms have economies of scale in foreign markets, as evidenced by the fact that small firms typically export a much lower portion of their production than larger firms. These effects are unquantifiable but may be serious in the long run.

• India has engineered a large increase in commercial bank loans to the small sector. In terms of loanable funds, a high proportion has gone to the upper half of the sector, and probably as much as half to firms that are of medium size in terms of employment (more than 50 workers). This does not go against our analysis of credit markets, which suggests that it may be a mistake for formal lending institutions to try to penetrate far into the jungle of very small firms. But there is also some pressure on banks to lend to the very small, and the number of small loans is very large. As yet the *overt* record of defaults is quite low and compares favorably with that of other developing countries, but there is a question as to whether undue pressure is being put on bank costs (profits have been falling).

• In India, bank loans to the small are at lower interest rates than those to the large. This normally results in such "favored" sectors getting less credit. This has been avoided by directives to the banks. But it implies, especially when increased banking costs are taken into account, that such loans are quite heavily subsidized. Our analysis of capital markets and of the economies of small enterprises does not support the view that such subsidization is called for, especially as there is no prima facie case that there are normally benefits from the point of view of overall income distribution. In general, banks should be permitted to charge higher rates for loans to the small, since they are both more risky and more costly to process.

Appendix 1. Description of the Survey Instrument, Response Rates, and Sample Selection

This appendix describes the survey instrument used to collect most of our firm-level data on small-scale enterprises in India and analyzes the rates of response to various important questions in the survey. It also presents definitions of the main variables used in the econometric analyses of production contained in chapters 10 and 11 and discusses the criteria employed in selecting observations from the sample for use in the econometric work.

The Survey Instrument and Response Rates

In describing the survey instrument, the response frequency of the various sections is reviewed for the five industries analyzed in chapters 10–12 and is broken into five size categories within each industry to see whether there were any significant patterns of nonresponse.[1] The "Question Code Key" gives only those questions for which response rates were analyzed.

The response rates were very high—over 90 percent—with little variance among industries. Machine tools exhibited the highest response rate, followed closely by soap.[2] The printing and shoe industries showed a slightly lower response rate but were still close to 90 percent for most sections of the survey. Metal casting responded worst, but the rate was still over 80 percent

1. The response rates reported for the various questions are those of enterprises that completed Module VIII, which included the core economic questions concerning the value of inputs and outputs. Unfortunately, no record was retained of the number of enterprises which were rejected from the sample because they were unable or unwilling to respond to these questions. A somewhat different instrument was used for powerlooms. Response rates have not been analyzed for this industry survey.

2. The soap industry does, however, show a poor response rate for three questions: III C3, IV A4.1, and VB 3.2.

for nearly all the questions. The breakdown of the industries into five size groupings does not reveal any general variation of response rate by size. Details are given in the Industry Response Tables.

With regard to response patterns across modules, Module VII, which reviews finances, had the lowest response rate. The remaining modules had very high response rates, the highest being Module IV, Labor and Training. There is only one question which stands out by virtue of its very low response rate: Question III C3, which asks into which income levels the firm's customers fall. A more detailed description and analysis of each module follows.

Module I. Identification and Ownership of Enterprise

Module I gives basic information on the interviewee and his/her establishment including its location and whether or not it is registered. It also describes the legal liability and ownership characteristics of the firm. Since this module had to be answered by each firm included in the sample, the response rate is not reviewed.

Module II. Start-Up and Growth of the Enterprise and Entrepreneurial History

Section A asks for the personal history of the entrepreneur, including his job history and education and also his father's occupation and education. Table 2.1, the entrepreneur's job history, is the only part of this section that displayed any significant pattern of nonresponse, ranging from a 34 percent response rate for metal casting to a 95 percent rate for machine tools. All other questions yielded a very high response rate in all industries.

Section B examines the history of the enterprise and concentrates on historical changes in employment and sales (Table 2.3). All industries produced extremely high response rates for this section.

Section C is concerned with the start-up of the enterprise and focuses on problems in starting the business, initial capital requirements, and details about borrowed capital. Over 90 percent of firms in four industries described problems encountered in starting the business (Question II C1). The soap industry response rate was about 80 percent, with the smallest firms (1–5 employees) accounting for the bulk of the nonresponses. Except for metal casting almost all firms reported how much of the required initial capital was available from savings (Question II C3). Less than half the firms in every industry borrowed a portion of the initial capital requirement, with only one-fifth of firms in the soap industry borrowing to start the business. Of those firms that did borrow a portion of the initial capital requirement, most (85–100 percent) reported details about the borrowings (Table 2.4).

Module III. Products and Markets

Section A asks for information on the enterprise's sales. Table 3.1, which supplies the most recent sales statistics, was answered by 95 percent or more of the firms in all the industries. All firms in the machine tool and soap industries and nearly all the printing firms also answered the question as to whether sales fluctuate (Question III A3). Only 10 percent of firms in the shoe industry and metal casting did not answer this question.

Section B asks for information on the firm's stocks of products. Questions in this section were not reviewed for frequency of response.

Section C is concerned with the market for the firm's product and concentrates mostly on the firm's customers. The first question—categories of the firm's customers—was answered by all firms in the soap and machine tool industries and almost all firms in the shoe and metal casting industries. Fifteen percent of firms in the printing industry did not answer the question.

Very few firms, with the notable exception of the shoe industry, answered the question concerning the income breakdown of the firm's household customers. Even in the shoe industry only 70 percent of firms responded, with half of the responding firms falling in the smallest size category. The relatively high response by firms in the shoe industry results no doubt from the presence of household customers, whereas the other industries supply products to nonhousehold customers.

Section D asks who the firm's main competitors are (by category) and where they are located. Every firm, except a handful in the soap and metal casting industries, ranked the order of importance of the firm's competitors. Other small enterprises were overwhelmingly the main competitors.

Module IV. Labor and Training

Tables 4.1 and 4.2 provide data on the number of workers employed and their monthly earnings in each skill category, for men and women separately. The response rate for these tables is over 95 percent for all industries.

Question IV 4.1 is concerned with training of the skilled and semiskilled workers. Surprisingly, only one firm in the soap industry answered this question, but over 95 percent of the firms in the other industries responded. But the question was not asked in the metal casting survey.

Question IV 8 addresses the length of employment of the firm's employees. Almost 90 percent or more of firms in all industries responded. Nearly all firms in every industry, except metal casting, also responded to Question IV 11 concerning the problem of turnover. In the case of metal casting the question was not asked.

Module V. Production Methods and Fixed Capital

No set of questions is provided for the firm's production methods. Rather, a general description of the production methods was given by the interviewer, based on the general background of the industry, on observations of the individual firm, and on discussions with the owner.

The section on fixed capital asks for details about the land and buildings of the firm but was not reviewed for response frequency. Table 5.1 provides information on the type and purchase price of machinery. The response rate ranges from 73 percent for firms in the metal casting industry to 100 percent for machine tools. Except for soap, for which there was no response, and metal casting, for which the response rate was 84 percent, virtually all firms supplied information on the book and replacement value of machinery and equipment (Questions VB 3.1 and VB 3.2).

Module VI. Materials and Supplies

Section A asks for information on materials and supplies processed in the firm. All firms in each industry supplied information on at least one of the main raw materials. Further, all firms in machine tools and shoes, and 98 percent of firms in the soap industry, gave the total value of the stock of materials and components held by the firm, while 90 percent of printing and 75 percent of metal casting firms answered the question.

Virtually all firms in the soap, printing, machine tool, and metal casting industries, and 93 percent of firms in the shoe industry, answered the question asking how often a lack of materials held up work.

Section B asks the firm about services it uses, such as electricity, telephone, and gas, as well as transport costs. This section was not reviewed for frequency of response.

Module VII. Finance and Technical Assistance

Section A is concerned with borrowings during the last three years. All firms in the soap industry and nearly all firms (99 percent) in the machine tool and printing industries stated whether the firm had borrowed or not. Ninety-three percent of firms in the shoe industry and 75 percent in metal casting answered the question. The number of firms that did borrow varied quite widely among the four industries, from a low of 42 percent in the shoe industry to a high of 73 percent in machine tools. Details of the borrowings were supplied by 93 percent or more of borrowing firms in the printing, machine tool, and soap industries, but were supplied by only 78 percent or more of those in the shoe and metal casting industries. Fewer firms provided information on problems in borrowing money (except firms in the soap industry, all of which provided this information). The range of response rates was greater as well—from 37 percent in printing to 81 percent in machine tools. There was a significant number of firms which did not borrow money but supplied information concerning the problems they encountered in borrowing. Section B on technical assistance was not reviewed for response rates.

Module VIII. Economics of the Unit

Information provided in Module VIII consists of sales and production statistics as well as expenditure items such as depreciation, wages, and interest. The annex asks for the book value of the various types of capital assets. Since one requirement for a firm to be included in the total sample was full completion of this module, the response rate is necessarily 100 percent.

Frequency of Response Tables

Question Code Key

MODULE I. Identification and Ownership of Enterprise

MODULE II. Start-Up and Growth of Enterprises and Entrepreneurial History

A. Personal Background

II A1. Age
II A2. Father's occupation
II A3. Father's education
II A4. Own education

Table 2.1.[3] Job History of Entrepreneur

II A9.1. Years in job (first job)
II A9.2. Occupation (first job)

B. Enterprise History

Table 2.3. Change in Employment and Sales

II B2.1. Number employed in first year
II B2.3. Number employed at present

3. Both subquestions required an answer for a positive response to be recorded for Table 2.1. This applies also to Tables 2.3, 3.1, and 5.1.

C. Start-Up of Enterprise

II C1. Main problems encountered in starting up business
II C3. Proportion of initial capital available from personal savings

Table 2.4.[4] Details about Borrowings

II C4.1. Amount borrowed from friends
II C4.2. Interest rate charged from banks

Module III. Products and Markets

A. Sales

Table 3.1. Sales in Last Year/Month/Week

III A1.1. Product, process, or service
III A1.1(b). Total value sold of III A1.1
III A3. Do sales fluctuate much during the year?

B. Stocks

C. Markets

III C1. Who are the main customers for your products or services?
III C1.1. }
III C1.2. } Various categories of customers
III C1.10. }

III C3. What percentage of your products or services are for "poor," "medium," and "wealthy" households?

III C3.1. Poor—percent sales
III C3.2. Medium—percent sales
III C3.3. Wealthy—percent sales

D. Competitors

III D1. Rank the order of importance of your competitors:
- ☐ Small enterprises
- ☐ Large modern enterprises
- ☐ Imports

4. An answer to either subquestion sufficed for a positive response to be recorded for Table 2.4. This applies also to Tables 4.1 and 4.2.

Module IV. Labor and Training

Table 4.1. Number of Workers

IV 13. Skilled labor (male)
IV 15. Unskilled labor (male)

Table 4.2. Monthly Earnings of Employees

IV 21. Maximum for skilled labor (male)
IV 23. Maximum for unskilled labor (male)
IV 4.1. How long to train skilled and semiskilled workers?
IV 8. How many "regular" workers have been with you:
Since your business started?
For more than 10 years?
For 5 to 10 years?
For 1 to 5 years?
For less than 1 year?
IV 11. Is turnover a problem?

Module V. Production Methods and Fixed Capital

Table 5.1. Data on Major Machines

VB 2(a). Type of machine
VB 2(b). Purchase price of machine
VB 3.1. Total book value of all machines and equipment
VB 3.2. Total replacement value of all machines and equipment

Module VI. Materials and Supplies

A. Materials

VI A1.1. Most important main raw material or component that is processed in the firm
VI A3. Total value of stocks and components at the present time
VI A7. Does lack of materials/supplies hold up your work?

B. Services

Module VII. Finance and Technical Assistance

A. Finance

VII 1. Have you borrowed money in the last three years?

Table 7.1.[5] Details about borrowings

VII 21. Amount
VII 21B. Source
VII 21C. Interest
VII 3.2. What were the difficulties in borrowing?

B. Technical Assistance

The Econometric Data Set

The data set used in the econometric analyses of chapters 10 and 11 is a subset of the survey data set and consists of firms which supplied complete information on the variables listed and defined in appendix 3. Not surprisingly, given the high response rates encountered in the samples, very few observations were excluded from the main survey sample to form the econometric sample. In two cases, the soap and metal casting industries, we were able to supplement our survey sample with data collected on a contemporaneous basis from large-scale firms on the basis of an instrument which was restricted to the variables detailed in appendix 3.[6] For this reason in these two cases the econometric data set is larger than the main sample. Sample sizes for the two data sets are given below:

Industry	*Econometric sample*	*Full survey sample*
Printing	66	71
Machine tools	78	89
Shoes	99	99
Soap	57	50
Metal casting	45	44

For purposes of classification by employment size, in the econometric work we have taken the total labor force of the firm including unpaid family labor, which corresponds to the labor input variables defined for use in the production function estimate. This results in some upward shifting of the size classification of firms, particularly at the lower end of the size range, relative to a classification which counts only paid employees.

5. All three subquestions required an answer for a positive response to be recorded for Table 7.1.

6. This supplementary survey was conducted by one of the authors when it became apparent that the sample survey data set had excluded power-using firms.

Appendix 2. Specification and Estimation of the Translog Production Function

This appendix provides a brief technical introduction to the specification and estimation of the production function used in chapters 10 and 11. The production function adopted in these chapters is the transcendental logarithmic (translog) form developed by Christensen, Jorgensen, and Lau (1971), which has been widely applied in recent empirical studies of the structure of production.

The translog function is the best known of the so-called flexible functional forms. Flexible functional forms of the production function (which are second-order approximations to an arbitrary production structure) impose relatively few a priori restrictions on the properties of the underlying technology. The translog function does not restrict the values of the elasticity of substitution at any point in input space. Homotheticity, separability, and constant returns to scale can be imposed by testable restrictions on the parameters, and the form reduces to the multiple input Cobb-Douglas specification as a special case.

Thus the translog production function provides a suitable empirical framework within which to test a number of hypotheses concerning the structure of production. Of particular interest in the context of the present study are the properties of substitution between labor and capital, the nature of returns to scale, and the presence of homotheticity in the production technology.

The translog function for an industry is

$$\text{(1)} \qquad \ln X(s) = \alpha_0 + \Sigma_m \alpha_m \ln Z_m(s) + \frac{1}{2} \Sigma_m \Sigma_n \beta_{mn} \ln Z_m(s) \ln Z_n(s)$$

where $X(s)$ is an index of output and the $Z_m(s)$ are indexes of input levels at each observation s. As specified, equation 1 constrains each establishment within an industry to lie on the same production function, regardless of size or other characteristics of the firm. However, because the translog form allows the elasticity of substitution to differ at each data point, the size of the establishment will affect the substitution properties of the technology. It is also possible to test for the stability of coefficients across subsets of the sample.

The output elasticity of each variable input is

$$\frac{\partial \ln X(s)}{\partial \ln Z_m(s)} = \alpha_m + \Sigma_n \beta_{mn} \ln Z_n(s). \tag{2}$$

The marginal factor productivities therefore depend upon input levels via the "share elasticities," β_{mn}, although the underlying structure of factor substitution possibilities described by subscript *mn* is invariant across *s*.

Under the assumption of competitive factor and product markets and price efficiency of producers, equation 2 gives the cost share of inputs in total output S_m as a linear function of the logs of each of the inputs. Indeed, the conventional method for estimation of the translog production function has been to work with these side conditions rather than to attempt to estimate the production function directly.[1]

A production function is considered to be "well behaved" if it has positive marginal products for each input (monotonicity) and if it is quasi-concave. Monotonicity requires that $\partial X(s)/\partial Z_m(s) > 0$. Since $X(s)$ and the Z_m are always positive, this is equivalent to requiring that the logarithmic marginal products in equation 2 be positive at each data point. Quasi-concavity, which is equivalent to the requirement that the isoquants of the production function be convex, requires that the bordered Hessian matrix of first and second partial derivatives of the production function be negative semidefinite. The translog function does not fulfill these requirements globally, but if the conditions are met over a sufficient number of the observed levels of inputs and outputs the function is considered to be well behaved. Since the partial derivatives of the production function depend on the values of inputs and outputs as well as on the coefficients of the estimated production function, both monotonicity and quasi-concavity may be verified at each data point.

Because the translog production function is neither monotonic nor concave over any arbitrary (positive) range of inputs, it may be necessary to impose monotonicity and global concavity on the production structure in cases when the estimated function is not locally well behaved. Concavity of the production function requires that the matrixes of second-order partial derivatives be negative semidefinite. Necessary and sufficient conditions for the Hessian matrix to be negative semidefinite are that the matrixes of share elasticities are themselves negative semidefinite. It can be demonstrated that under constant returns to scale the matrixes of share elasticities are negative semidefinite if and only if the diagonal elements β_{mm} are nonpositive. Thus in

1. An exception is Corbo and Meller (1979), who present single-equation estimates of the translog production function.

the estimation of the production function parameters, concavity can be imposed by imposing constant returns to scale and nonpositivity of the own-share elasticities.

To impose global concavity on the translog form we represent the B parameters in terms of their Cholesky factorizations (Lau 1978). Under the restrictions of symmetry and constant returns to scale, there is a one to one correspondence between share elasticities and parameters of the Cholesky factorization. Thus it is possible to estimate the model in terms of the parameters of the Cholesky factorization as well as the remaining parameters $(\alpha_k, \alpha_u, \alpha_s)$. With three factors of production, there are five possible cases for which the convexity restrictions may be fulfilled; one case exists if all of the own-share elasticities are strictly negative, three cases allow for one of the own-share elasticities to equal zero, while the fourth case reduces to the constant returns to scale Cobb-Douglas form. The case for which the nonpositivity constraints are satisfied and which has the highest likelihood function value provides parameter estimates of the restricted globally concave translog production function.

If the production function is well behaved, it can then be employed to explore a number of characteristics of the production technology. Three main characteristics of the production structure are examined in chapter 9: homotheticity, constant returns to scale, and the elasticities of substitution between pairs of factors of production.[2]

A production function is said to be homothetic if marginal rates of substitution between pairs of inputs are constant at constant relative levels of factor intensity. In the context of the translog function this is equivalent to requiring that

$$\Sigma_m \beta_{mn} = 0 \qquad (n = 1, \ldots, N) \tag{3}$$

which may be tested in the usual linear hypothesis framework. If the restriction is accepted, the translog function takes the general form of a multifactor Kmenta (1967) approximation to the CES production function. While this approximation does not have constant elasticities of substitution, it is a homothetic form with elasticities of substitution that depend only on the ratios of the inputs.

Production under constant returns to scale implies an additional restriction on the parameters of equation 1 in combination with equation 3:

$$\begin{aligned} &\Sigma_m \alpha_m = 1, \\ &\Sigma_m \beta_{mn} = 0 \qquad (n = 1, \ldots, N). \end{aligned} \tag{4}$$

2. It is also possible to test for the separability of inputs and hence for the existence of a consistent aggregate among inputs. Since the test for separability involves pairwise restrictions on the partial elasticities of substitution, however, the estimated elasticities provide sufficient information on the existence of a consistent aggregate for present purposes.

Again, this may be tested directly from equation 1 by imposing the restrictions in equation 4.

Allen (1938) has defined the partial elasticity of substitution between inputs *m* and *n* as

$$(5) \qquad \sigma_{mn} = \sum_{h=1}^{m} F_h Z_h / Z_m Z_n (|H_{mn}|/|H|)$$

where $F_h = \partial X/\partial Z_h$ and where $|H_{mn}|$ and $|H|$ are the cofactor of *mn* in the Hessian matrix of equation 1 and the determinant of the Hessian matrix respectively. The parameter estimates of the translog function provide sufficient information to estimate the Allen elasticities of substitution between all pairs of inputs at each data point.

Econometric Specification

The estimates of factor substitution possibilities contained in chapter 9 are based on a conventional stochastic specification of the production function, equation 1, and the share equation, equation 2. With the addition of a stochastic error, ϵ, the parameters of equation 1 may be estimated directly using ordinary least squares (OLS).

A difficulty with the use of OLS, however, is that the regressors are endogenous variables for an establishment within an industry. Thus parameter estimates derived from equation 1 will be biased and inconsistent. In cross-section analysis, the traditional approach to the simultaneous equation bias problem has been to employ a consistent instrumental variables estimator. In the present context, however, there is no logical variable or set of variables in the data set to employ as instruments.

The problem can be overcome in the standard competitive setting by exploiting the properties of the logarithmic marginal product equations in equation 2. Brendt and Christensen (1973a) have shown that under symmetry and constant returns to scale, the $m - 1$ independent share equations, after one of the equations has been dropped, can provide efficient estimates of the parameters. Alternatively, equations 1 and 2 can be estimated jointly after one of the share equations is eliminated in order to obtain a nonsingular system. This is the approach followed in chapter 9.

In a single period cross section, it is least likely that the conditions underlying the capital share equation will hold. Indeed capital may be truly a predetermined variable in a single production period. Thus the appropriate system to estimate consists of equation 1 and the set of labor equations that make up equation 2. The disturbances ϵ are considered stochastically independent, and the disturbance vector on the system of equations is assumed to be joint normally distributed with zero mean and nonsingular covariance matrix. Under the assumed stochastic specification with cross-equation constraints, the technique of multivariate regression will yield maximum likelihood estimates of the parameters on convergence.

Specification of the Frontier Production Function

The estimates of technical efficiency contained in chapter 11 are based on a frontier production function of the translog form. Measures of technical efficiency derived from a production frontier can be directly related to the assumed causes of output variation underlying the specification of the residual and to the method of estimation chosen. The approach adopted in chapter 10, which permits the recovery of a firm-specific index of technical efficiency, is to minimize the sum of the deviations from the frontier, subject to the constraint that all observations lie on or below it. The resulting "deterministic" frontier ascribes all variations from predicted best practice output to technical inefficiency.

The translog frontier form corresponding to equation 1 is

$$\text{(6)} \qquad \ln \hat{X}(s) = \partial_o + \Sigma_m \, \partial_m \ln Z_m(s) + \frac{1}{2} \sum_m \Sigma_n \hat{\beta}_{mn} \ln Z_m(s) \ln Z_n(s)$$

where $X(s)$ is an index of maximum potential output and the $Z_m(s)$ are indexes of input levels at each observation s. Given this specification, the estimation procedure is an application of linear programming (Nishimizu and Page 1982). The objective function to be minimized may be written as follows:

$$\text{(7)} \quad \text{Min: } \sum_{s=1}^{S} \hat{\alpha}_o + \sum_m \hat{\alpha}_m \ln Z_m(s) + \frac{1}{2} \Sigma\Sigma \hat{\beta}_{mn} \ln Z_m(s) \ln Z_n(s) - \ln X(s)$$

which is linear in the unknown parameters.

The constraints of the model are imposed by the S restrictions on the observations securing the observed input-output combinations to lie on or below the frontier:

$$\text{(8)} \qquad \alpha_o + \sum_m \hat{\alpha}_m \ln Z_m(s) + \frac{1}{2} \sum_m \sum_m \hat{\beta}_{mn} \ln Z_m(s) \ln Z_n(s) \geq \ln X(s).$$

The need to impose two important characteristics on the production function, monotonicity (the requirement that marginal products be nonnegative at all observations) and concavity, implies an additional set of constraints on the model. In order to impose monotonicity it is necessary and sufficient to restrict the subscript m and the output elasticity of each factor input to be nonnegative:

$$\text{(9)} \qquad \hat{\alpha}_m + \sum_n \hat{\beta}_{mn} \ln Z_n(s) \geq 0 \qquad (n = 1, \ldots, N).$$

Under monotonicity, the necessary and sufficient conditions for concavity require imposition of constant returns to scale and the nonpositivity of own-share elasticities. Thus, it is possible to impose concavity on the frontier by the restriction

$$\sum_m \hat{\alpha}_m = 1$$

$$(10) \qquad \sum_m \hat{\beta}_{mn} = 0 \qquad (n = 1, \ldots, N)$$

$$\beta_{mm} \leq 0.$$

A firm's actual output $X(s)$, given observed input levels, is equal to predicted output $\hat{X}(s)$ only if the firm operates on the production frontier. Each firm is assigned an efficiency index (Farrell index) equal to the ratio of its actual to predicted output, $X(s)/\hat{X}(s) \leq 1$, which may be found as the antilog of the slack variable in the programming problem. The value of the index provides a measure of relative technical efficiency.

With the properties of constant returns to scale and global concavity imposed, it can be shown that the Farrell indexes provide a set of transitive multilateral total factor productivity indexes (Caves, Christensen, and Diewert 1982; Bateman, Nishimizu, and Page 1983). The translog frontier production function therefore provides an attractive framework within which to study both the relationship between actual and best practice productivity at any one data point and level differences in total factor productivity among pairs of observations.

Appendix 3. Variable Definitions for Production Function Estimates and the Analysis of Sources of Technical Inefficiency

Variable	*Definition*
Output	Gross value of output at 1979–80 ex-factory prices.
Capital services	Annualized capital service flow based on 10 percent return and assumed economic lifetimes by asset category. All capital stock values are in historic cost at constant 1979–80 prices.
Skilled labor	Labor services in person-months of all employee categories enumerated as skilled by respondents. Includes managerial and supervisory labor and working proprietors.
Unskilled labor	Labor services in person-months of all labor categories not classified as skilled. Includes semiskilled and unskilled labor, casual labor, and unpaid family workers.
Material inputs	Value of all purchased material and service inputs at 1979–80 prices.
Literacy	Dummy variable taking on value of 1 for entrepreneurs who were reported to be illiterate.
Father's occupation	Dummy variable taking a value of 1 if entrepreneur's father was engaged in the ownership or management of an enterprise in the same industry.
Turnover problem	Dummy variable taking a value of 1 for firms which reported no difficulty with labor turnover.
Lab < 10	Dummy variable taking a value of 1 for firms with total employment, including family employment of less than 10.

Total employment	Total number of workers, including family workers.
Experience of entrepreneur	Total years of experience of the entrepreneur in his current position and in all related positions in the industry.
Age of firm	Age in years of enterprise under current ownership.
Employee experience	Mean of distribution of regular employees by firm-specific experience in years.
Vintage of plant	Value weighted average age of capital stock in years.
Capacity index	Dummy variable taking a value of 1 for firms which reported higher constant price values of output three years prior to the survey than in the survey year.

Appendix Table 1. *Industry Response: Metal Casting*
(number; percent)

Module, question, and table	*Total*	*Firm size group* 1–5	6–10	11–25	26–50	51 plus
Total response	44;100	0;0	6;13.6	20;45.5	6;13.6	12;27.3
Module II. Start-Up and Growth of Enterprise and Entrepreneurial History						
A1	41;93.1		6;100	20;100	6;100	9;25
A2	39;88.6		6;100	19;95	6;100	8;66.6
A3	37;84.1		6;100	19;95	6;100	6;50
A4	36;81.8		5;83.3	19;95	6;100	6;50
Table 2.1	15;34.1		1;16.7	8;40	5;83.3	1;8.3
Table 2.3	36;81.8		6;100	17;85	3;50	10;83.3
C1	42;95.4		6;100	18;90	6;100	12;100
C3	35;79.6		4;66.7	17;85	6;100	8;66.6
Table 2.4[a]	12;85.7		1;100	6;1—	2;100	3;25
(Out of)[b]	14;31.8[c]		1;16.6	6;30	2;33.3	5;41.7
Module III. Products and Markets[d]						
Table 3.1	43;97.7		6;100	19;95	6;100	12;100
A3	40;90.9		6;100	20;100	6;100	8;66.6
C1	43;97.7		6;100	20;100	6;100	11;91.7
C3						
D1	40;90.9		6;100	20;100	6;100	8;66.6
Module IV. Labor and Training[e]						
Table 4.1	42;95.5		6;100	20;100	6;100	10;83.3
Table 4.2	41;93.2		6;100	20;100	6;100	9;75
IV 4.1	0;0		0;0	0;0	0;0	0;0
IV 4.8	41;93.2		6;100	20;100	6;100	9;75
IV 11						
Module V. Production Methods and Fixed Capital						
Table 5.1	32;72.7		4;66.6	17;85	5;83.3	6;50
V B3.1	39;88.6		6;100	20;100	5;83.3	8;66.6
V B3.2	37;84.1		6;100	18;90	5;83.3	8;66.6
Module VI. Materials and Supplies						
A1.1	43;97.7		6;100	20;100	6;100	11;91.7
A3	33;75		5;83.3	18;90	3;50	7;58.3
A7	43;97.7		6;100	20;100	6;100	11;91.7
Module VII. Finance and Technical Assistance						
1	33;75		6;100	17;85	4;66.7	6;50
Table 7.1[f]	16;84.2		2;100	6;85.7	2;100	6;75
3.2[g]	12;63.2		1;50	7;100	2;100	2;25
(Out of)[h]	19;43.2		2;33.3	7;35	2;33.3	8;66.6

a. Included only if the answer to II C3 is not 100.

b. Number of firms included in Table 2.4.

c. Number of firms included in Table 2.4 as a percentage of total number of firms in size group.

d. Question III C3—What percentage of your products or services are for "poor," "medium," and "wealthy" households?—was not asked in the survey.

e. Question IV 22—Is turnover a problem?—was not asked in the survey.

f. Included only if the answer to VII 1 is yes.

g. Number of firms included in Table 7.1 and Question 3.2.

h. Number of firms included in Table 7.1 and Question 3.2 as a percentage of total number of firms in size group.

Appendix Table 2. *Industry Response: Printing*
(number; percent)

Module, question, and table	*Total*	*Firm size group* 1–5	*6–10*	*11–25*	*26–50*	*51 plus*
Total response	71;100	6;8.5	31;43.7	18;25.4	9;12.7	7;9.9
Module II. Start-Up and Growth of Enterprise and Entrepreneurial History						
A1	71;100	6;100	31;100	18;100	9;100	7;100
A2	71;100	6;100	31;100	18;100	9;100	7;100
A3	67;94.4	6;100	29;93.5	18;100	9;100	5;71.4
A4	71;100	6;100	31;100	18;100	9;100	7;100
Table 2.1	64;90.1	5;83.3	30;96.8	15;83.3	9;100	5;71.4
Table 2.3	70;98.6	5;83.3	31;100	18;100	9;100	7;100
C1	65;91.5	6;100	29;93.5	15;83.3	8;88.9	7;100
C3	67;94.4	6;100	31;100	15;83.3	8;88.9	7;100
Table 2.4[a]	29;93.5	1;100	10;90.9	10;90.9	5;100	3;100
(Out of)[b]	31;43.7[c]	1;16.7[c]	11;35.5[c]	11;61.1[c]	5;55.6[c]	3;42.9[c]
Module III. Products and Markets						
Table 3.1	67;94.4	6;100	28;90.3	18;100	8;88.9	7;100
A3	69;97.2	6;100	30;96.8	17;94.4	9;100	7;100
C1	60;84.5	6;100	24;77.4	16;88.9	7;77.8	7;100
C3	13;18.3	4;66.7	5;16.1	2;11.1	1;11.1	1;14.3
D1	71;100	6;100	31;100	18;100	9;100	7;100
Module IV. Labor and Training						
Table 4.1	70;98.6	5;83.3	31;100	18;100	9;100	7;100
Table 4.2	70;98.6	5;83.3	31;100	18;100	9;100	7;100
IV 4.1	69;97.2	6;100	31;100	17;94.4	8;88.9	7;100
IV 4.8	70;98.6	6;100	30;96.8	18;100	9;100	7;100
IV 11	70;98.6	6;100	30;96.8	18;100	9;100	7;100
Module V. Production Methods and Fixed Capital						
Table 5.1	70;98.6	5;83.3	31;100	18;100	9;100	7;100
V B3.1	71;100	6;100	31;100	18;100	9;100	7;100
V B3.2	71;100	6;100	31;100	18;100	9;100	7;100
Module VI. Materials and Supplies						
A1.1	71;100	6;100	31;100	18;100	9;100	7;100
A3	65;91.5	4;66.7	29;93.5	17;94.4	9;100	6;85.7
A7	71;100	6;100	31;100	18;100	9;100	7;100
Module VII. Finance and Technical Assistance						
1	70;98.6	5;83.3	31;100	18;100	9;100	7;100
Table 7.1[d]	35;92.1	0;0.0	12;92.3	11;91.7	7;100	5;100
3.2[d]	14;36.8	0;0	6;46.2	2;16.7	4;57.1	2;40.0
(Out of)[e]	38;53.5[f]	1;16.7[f]	13;41.9[f]	12;66.7[f]	7;77.8[f]	5;71.4[f]

a. Included only if the answer to II C3 is not 100.

b. Number of firms included in Table 2.4.

c. Number of firms included in Table 2.4 as a percentage of total number of firms in size group.

d. Included only if the answer to VII 1 is yes.

e. Number of firms included in Table 7.1 and Question 3.2.

f. Number of firms included in Table 7.1 and Question 3.2 as a percentage of total number of firms in size group.

Appendix Table 3. *Industry Response: Machine Tools*
(number; percent)

Module, question, and table	*Total*	*Firm size group* *1–5*	*6–10*	*11–25*	*26–50*	*51 plus*
Total response	89;100	6;5.6	27;30.3	26;29.2	11;12.4	20;22.5
Module II. Start-Up and Growth of Enterprise and Entrepreneurial History						
A1	89;100	5;100	27;100	26;100	11;100	20;100
A2	89;100	5;100	27;100	26;100	11;100	20;100
A3	88;98.9	5;100	27;100	26;100	11;100	19;95.0
A4	89;100	5;100	27;100	26;100	11;100	20;100
Table 2.1	85;95.5	5;100	27;100	23;88.5	10;90.9	20;100
Table 2.3	87;97.8	5;100	27;100	25;96.2	11;100	19;95.0
C1	84;94.4	5;100	27;100	26;100	11;100	15;75.0
C3	89;100	5;100	27;100	26;100	11;100	20;100
Table 2.4[a]	35;100	2;100	7;100.0	10;100	5;100	11;100
(Out of)[b]	35;39.3[c]	2;40.0[c]	7;25.9[c]	10;38.5[c]	5;45.5[c]	11;55.0[c]
Module III. Products and Markets						
Table 3.1	88;98.9	5;100	27;100	25;96.2	11;100	20;100
A3	89;100	5;100	27;100	26;100	11;100	20;100
C1	89;100	5;100	27;100	26;100	11;100	7;100
C3	1;1.1	0;0	0;0	0;0	1;5.0	1;5.0
D1	89;100	5;100	27;100	26;100	11;100	20;100
Module IV. Labor and Training						
Table 4.1	89;100	5;100	27;100	26;100	11;100	20;100
Table 4.2	89;100	5;100	27;100	26;100	11;100	20;100
IV 4.1	89;100	5;100	27;100	26;100	11;100	20;100
IV 4.8	89;100	5;100	27;100	26;100	11;100	20;100
IV 11	88;98.9	5;100	27;100	25;96.2	11;100	20;100
Module V. Production Methods and Fixed Capital						
Table 5.1	89;100	5;100	27;100	26;100	11;100	20;100
V B3.1	89;100	5;100	27;100	26;100	11;100	20;100
V B3.2	89;100	5;100	27;100	26;100	11;100	20;100
Module VI. Materials and Supplies						
A1.1	89;100	5;100	27;100	26;100	11;100	20;100
A3	89;100	5;100	27;100	26;100	11;100	20;100
A7	89;100	5;100	27;100	26;100	11;100	20;100
Module VII. Finance and Technical Assistance						
1	88;98.9	5;100	26;96.3	26;100	11;100	20;100
Table 7.1[d]	66;95.7	2;100	19;100	17;85.0	11;100	17;100
3.2[d]	56;81.2	1;50.0	16;0	16;80.0	4;82.7	15;88.2
(Out of)[e]	69;72.6[f]	2;40.0[f]	19;70.4[f]	20;76.9[f]	11;100	17;85.0

a. Included only if the answer to II C3 is not 100.
b. Number of firms included in Table 2.4.
c. Number of firms included in Table 2.4 as a percentage of total number of firms in size group.
d. Included only if the answer to VII 1 is yes.
e. Number of firms included in Table 7.1 and Question 3.2.
f. Number of firms included in Table 7.1 and Question 3.2 as a percentage of total number of firms in size group.

Appendix Table 4. *Industry Response: Shoes*

(number; percent)

Module, question, and table	*Total*	Firm size group *1–5*	*6–10*	*11–25*	*26–50*	*51 plus*
Total response	99;100	41;43.2	26;27.4	23;24.2	2;2.1	7;3.2
Module II. Start-Up and Growth of Enterprise and Entrepreneurial History						
A1	98;99	41;100	26;100	23;100	2;100	6;85.7
A2	93;93.9	40;97.6	24;92.3	21;91.3	2;100	6;85.7
A3	94;95	40;97.6	23;88.5	23;100	2;100	6;85.7
A4	98;99	41;100	26;100	23;100	2;100	6;85.7
Table 2.1	91;91.9	39;95.1	25;96.2	19;82.6	2;100	6;85.7
Table 2.3	97;98	40;97.6	25;96.2	23;100	2;100	7;100
C1	95;96	39;95.1	25;96.2	23;100	2;100	6;85.7
C3	97;98	39;95.1	26;100	23;100	2;100	7;100
Table 2.4[a]	44;86.2	21;95.5	7;77.8	12;85.7	0;0	0;0
(Out of)[b]	51;51.5[c]	22;57.3[c]	9;34.6[c]	14;60.9[c]	0;0[c]	6;85.7
Module III. Products and Markets						
Table 3.1	94;95	41;100	24;92.3	21;91.3	2;100	6;85.7
A3	90;90.9	37;90.2	25;96.2	20;87	2;100	6;85.7
C1	94;95	37;90.2	26;96.2	23;100	2;100	7;100
C3	70;70.7	33;80.5	14;53.8	15;65	1;50	7;100
D1	98;100	41;100	26;100	23;100	2;100	7;100
Module IV. Labor and Training						
Table 4.1	96;97	39;95.1	26;100	23;100	2;100	6;85.7
Table 4.2	95;96	37;90.2	26;100	23;100	2;100	7;100
IV 4.1	95;96	39;95.1	26;100	23;100	2;100	5;71.4
IV 4.8	88;88.9	34;82.9	24;92.3	21;91.3	2;100	7;100
IV 11	98;99	40;97.6	26;100	23;100	2;100	7;100
Module V. Production Methods and Fixed Capital						
Table 5.1	93;93.9	37;90.2	26;100	23;100	2;100	5;71.4
B3.1	97;98	41;100	26;100	23;100	2;100	5;71.4
B3.2	97;98	41;100	26;100	23;100	2;100	5;71.4
Module VI. Materials and Supplies						
A1.1	99;100	41;100	26;100	23;100	2;100	7;100
A3	99;100	41;100	26;100	23;100	2;100	7;100
A7	92;92.9	39;95.1	24;92.3	20;87	2;100	7;100
Module VII. Finance and Technical Assistance						
1	92;92.9	38;92.7	24;92.3	22;95.7	2;100	6;85.7
Table 7.1[d]	35;79.6	16;80	9;75	4;66.7	0;0	6;100
3.2[d]	27;61.4	8;40	9;75	4;66.7	0;0	6;100
(Out of)[e]	44;44.4[f]	20;48.8[f]	12;50[f]	6;26.1[f]	1;50[f]	6;85.7[f]

a. Included only if the answer to II C3 is not 100.

b. Number of firms included in Table 2.4.

c. Number of firms included in Table 2.4 as a percentage of total number of firms in size group.

d. Included only if the answer to VII 1 is yes.

e. Number of firms included in Table 7.1 and Question 3.2.

f. Number of firms included in Table 7.1 and Question 3.2 as a percentage of total number of firms in size group.

Appendix Table 5. *Industry Response: Soap*

(number; percent)

Module, question, and table	Total	Firm size group 1–5	6–10	11–25	26–50	51 plus
Total response	50;100	13;26	18;36	10;20	3;6	6;12
Module II. Start-Up and Growth of Enterprise and Entrepreneurial History						
A1	50;100	13;100	18;100	10;100	3;100	6;100
A2	50;100	13;100	18;100	10;100	3;100	6;100
A3	49;98	12;92.3	18;100	10;100	3;100	6;100
A4	50;100	13;100	18;100	10;100	3;100	6;100
Table 2.1	31;62	10;76.9	12;66.7	6;60	2;66.7	1;16.7
Table 2.3	50;100	13;100	18;100	10;100	3;100	6;100
C1	41;82	9;69.2	16;88.9	9;90	2;66.7	5;83.3
C3	49;98	13;100	18;100	10;100	3;100	5;83.3
Table 2.4[a]	11;100	3;100	4;100	1;100	0;0	3;100
(Out of)[b]	11;22[c]	3;23.1[c]	4;22.2[c]	1;10[c]	0;0[c]	3;50[c]
Module III. Products and Markets						
Table 3.1	50;100	13;100	18;100	10;100	3;100	6;100
A3	50;100	13;100	18;100	10;100	3;100	6;100
C1	50;100	13;100	18;100	10;100	3;100	6;100
C3	3;6	0;0	0;0	0;0	0;0	3;50
D1	48;96	13;100	18;100	8;80	3;100	6;100
Module IV. Labor and Training						
Table 4.1	48;96	11;84.6	18;100	10;100	3;100	6;100
Table 4.2	48;96	11;84.6	18;100	10;100	3;100	6;100
IV 4.1	1;2	0;0	0;0	0;0	1;33.3	0;0
IV 4.8	50;100	13;100	18;100	10;100	3;100	6;100
IV 11	49;98	12;92.3	18;100	10;100	3;100	6;100
Module V. Production Methods and Fixed Capital						
Table 5.1	45;90	12;92.3	18;100	9;90	3;100	3;50
V B3.1	50;100	13;100	18;100	10;100	3;100	6;100
V B3.2	1;2	0;0	1;5.6	0;0	0;0	0;0
Module VI. Materials and Supplies						
A1.1	50;100	13;100	18;100	10;100	3;100	6;100
A3	49;98	12;92.3	18;100	10;100	3;100	6;100
A7	49;98	13;100	18;100	10;100	3;100	5;83.3
Module VII. Finance and Technical Assistance						
1	50;100	13;100	18;100	10;100	3;100	6;100
Table 7.1[d]	29;0	4;0	11;100	6;100	3;100	5;100
3.2[d]	30;100	5;61.5	11;100	6;100	3;100	5;100
(Out of)[e]	30;60[f]	5;38.5[f]	11;61.1[f]	6;60[f]	3;100[f]	5;83.3[f]

a. Included only if the answer to II C3 is not 100.

b. Number of firms included in Table 2.4.

c. Number of firms included in Table 2.4 as a percentage of total number of firms in size group.

d. Included only if the answer to VII 1 is yes.

e. Number of firms included in Table 7.1 and Question 3.2.

f. Number of firms included in Table 7.1 and Question 3.2 as a percentage of total number of firms in size group.

Bibliography

Aigner, D. J., and S. E. Chu. 1968. "On Estimating the Industry Production Function." *American Economic Review* 58.

Allen, R. G. D. 1938. *Mathematical Analysis for Economists*. New York: St. Martin's.

Amjad, Rashid. Ed. 1981. *The Development of Labor Intensive Industry in ASEAN Countries*. Geneva: International Labour Organisation, Asian Regional Team for Employment Promotion.

Anderson, Dennis. 1982a. *Small Industry in Developing Countries—Some Issues*. World Bank Staff Working Paper 518. Washington, D.C.

———. 1982b. "Small Industry in Developing Countries: A Discussion of Issues." *World Development* 10, no. 2.

Anderson, D., and F. Khambata. 1981. *Small Enterprises and Development Policy in the Philippines: A Case Study*. World Bank Staff Working Paper 468. Washington, D.C.

Armington, Catherine, and Marjorie Odle. 1982. "Small Business: How Many Jobs?" *Brookings Review*, no. 2.

Banerjee, B. 1977. "Growth of Industrial Employment in India and the Structure of Manufacturing Industries in Uttar Pradesh 1961–1967." World Bank Development Economics Department. Washington, D.C. Processed.

Banerji, R. 1978. "Average Plant Size in Manufacturing and Capital Intensity: A Cross Country Analysis by Industry." *Journal of Development Economics* 5, no. 2 (June).

Bannock, Graham. 1976. *The Smaller Business in Britain and Germany*. London: Wilton House.

Bateman, D. A., Mieko Nishimizu, and J. M. Page. 1983. "Regional Productivity Differentials in Yugoslavia." World Bank Development Research Department. Washington, D.C. Processed.

Baumol, William J. 1977. *Economic Theory and Operations Analysis*, 4th ed. Englewood Cliffs, N.J.: Prentice-Hall.

Bautista, R. M. 1981. "The Development of Labor Intensive Industry in the Philippines." In Rashid Amjad, ed., *The Development of Labor Intensive Industry in*

ASEAN Countries. Geneva: International Labour Organisation, Asian Regional Team for Employment Promotion.

Berna, J. J. 1960. *Industrial Entrepreneurship in Madras State*. New Delhi: Asia Printing House; Stanford, Calif.: Stanford Research Institute.

Berndt, Ernst R., and L. R. Christensen. 1973a. "The Internal Structure of Functional Relationships: Separability, Substitution, and Aggregation." *Review of Economic Studies* 40(3), no. 123 (July).

———. 1973b. "The Translog Function and the Substitution of Equipment Structures and Labor in U.S. Manufacturing 1929–68." *Journal of Econometrics* 1, no. 1 (March).

———. 1974. "Testing for the Existence of a Consistent Aggregate Index of Labor Inputs." *American Economic Review* 4, no. 3 (June).

Berry, A. 1983. *Essays on Industrialization in Colombia*. Tempe, Ariz.: Arizona State University Center for Latin American Studies.

Bhatt, V. V., and Alan R. Roe. 1979. *Capital Market Imperfections and Economic Development*. World Bank Staff Working Paper 338. Washington, D.C.

Bolton Committee. 1971. *Bolton Committee Report: Small Firms*. Cmnd 4811. London: Her Majesty's Stationery Office.

Bombay Labour Office. 1936. "Report on an Enquiry into Wages, Hours of Work and Condition of Employment in the Retail Trade of Some Towns of Bombay Presidiary." Appendix B.

Bombay Textile Research Association. 1963. "An Investigation into Socio-Economic Life of a *Badli* Worker." *Bombay Textile Research Association Report*, no. 4.

Bose, A. N. 1978. *Calcutta and Rural Bengal: Small Sector Symbiosis*. Calcutta: T. K. Mukherjee.

Boswell, J. 1972. *The Rise and Decline of Small Firms*. London: Allen and Unwin.

Bowden, R. J., and D. Mazumdar. 1982. "Segmentation and Earning Profiles in LDCs." World Bank Development Research Department. Discussion Paper 4. Washington, D.C. Processed.

Bruch, Mathias. 1982. "Financial Sources of Small Scale Manufactures: A Micro-Analysis for Malaysia." *Kieler Arbeitspapiere*, no. 157 (October).

Caves, D. W., L. R. Christensen, and W. E. Diewert. 1982. "Multilateral Comparisons of Output, Input, and Productivity Using Superlative Index Numbers." *Economic Journal* 92, no. 365 (March).

Chapman, B. J., and H. W. Tan. 1980. "Specific Training and Inter-Industry Wage Differentials in U.S. Manufacturing." *Review of Economics and Statistics* 62, no. 3 (August).

Child, F. C. N.d. "An Empirical Study of Small Scale Rural Industry in Kenya." Institute of Development Studies Working Paper 127. University of Nairobi, Kenya.

Child, F. C., and Hiromitsu Kaneda. 1975. "Links to the Green Revolution: A Study of Small Scale, Agriculturally Related Industry in the Pakistan Punjab." *Economic Development and Cultural Change* 23, no. 2 (January).

Christensen, L. R., D. W. Jorgenson, and L. J. Lau. 1971. "Conjugate Duality and the Transcendental Logarithmic Production Function." *Econometrica* 39, no. 4 (July).

———. 1973. "Transcendental Logarithmic Production Frontiers." *Review of Economics and Statistics* 55, no. 1 (February).

Corbo, Vittorio, and Patricio Meller. 1979. "The Translog Production Function: Some Evidence from Establishment Data." *Journal of Econometrics* 10.

Cortes, Mariluz, Albert Berry, and Ashfaq Ishaq. 1983. "What Makes for Success in Small and Medium-Scale Enterprises: The Evidence from Colombia." Washington, D.C.: World Bank Development Research Department. Processed.

———. 1987. *Success in Small and Medium-Scale Enterprises: The Evidence from Colombia.* New York: Oxford University Press.

Deshpande, L. R. 1979. "The Bombay Labor Market." Bombay University. Processed.

Dhar, P. N. 1960. *Small-Scale Industries in Delhi.* Bombay: Asia Publishing House.

Dhar, P. N., and H. F. Lydall. 1961. *The Role of Small Enterprises in Indian Economic Development.* Bombay: Asia Publishing House.

Dorfman, Robert, and Peter O. Steiner. 1964. "Optimal Advertising and Optimal Quality." *American Economic Review.*

Farrell, M. J. 1957. "The Measurement of Productive Efficiency." *Journal of the Royal Statistical Society,* no. A120.

Financial Times Survey, June 12, 1984.

Fisseha, Yacob, and Omar Davies. 1981. *The Small Scale Manufacturing Enterprises in Jamaica: Socio-Economic Characteristics and Constraints.* Michigan State University Rural Development Series Working Paper 16. East Lansing.

Fong, Pang Eng, and Augustine Tan. 1981. "Employment and Export-Led Industrialization: The Experience of Singapore." In Rashid Amjad, ed., *The Development of Labor Intensive Industry in ASEAN Countries.* Geneva: International Labour Organisation, Asian Regional Team for Employment Promotion.

Forsund, F. R., C. A. Knox Lovell, and P. Schmidt. 1980. "A Survey of Frontier Production Functions and of Their Relationship to Efficiency Measurement." *Journal of Econometrics* 13, no. 1.

Freeman, C. 1974. *The Economics of Industrial Innovation.* Harmondsworth, England: Penguin.

———. N.d. "The Role of Small Firms in Innovation in the United Kingdom Since 1945." Bolton Committee Research Report 6. London: Her Majesty's Stationery Office.

Galenson, Walter. Ed. 1979. *Economic Growth and Structural Change in Taiwan: The Post-War Experience of the Republic of China.* Ithaca, N.Y.: Cornell University Press.

Gandhi, M. K. 1940. *An Autobiography,* 2d ed. Ahmedabad, India: Navajivan.

Goldar, B. N. 1982. "Unit Size and Economic Efficiency: A Study of Small Scale Soap Industry in India." Delhi: Delhi Institute of Economic Growth. Processed.

Greene, W. H. 1980. "Maximum Likelihood Estimation of Econometric Frontier Functions." *Journal of Econometrics* 13, no. 1 (May).

Haan, H. H. de. 1980. *The Industrial Distribution of the Labor Force in India, 1961–71*

(Part II). Center for Development Planning Discussion Paper 55B. Rotterdam: Erasmus University.

———. N.d. "The Viability of Small-Scale Industrialization, The Case of Sugarcane Processing." Delhi: Institute of Economic Growth. Processed.

Habbakuk, J. H. 1962. *American and British Technology in the Nineteenth Century: The Search for Labor Saving Inventions*. Cambridge: Cambridge University Press.

Haggblade, Steve, Jacques Defay, and Bob Pitman. 1979. *Small Manufacturing and Repair Enterprises in Haiti: Survey Results*. Michigan State University Working Paper 4. East Lansing.

Harris, John R. 1971. "Nigerian Entrepreneurship in Industry." In Carl Eicher and Carl Liedholm, eds., *Growth and Development of the Nigerian Economy*. East Lansing: Michigan State University Press. Reprinted in P. Kilby, ed., *Entrepreneurship and Economic Development*. New York: Free Press, 1971.

Hay, D. A., and D. J. Morris. 1984. *Unquoted Companies*. London: Macmillan.

Helm, Franz C. 1981. "A View of the Effectiveness of Small Scale Enterprise Lending," World Bank Industry Department. Processed.

Ho, S. 1980. *Small-Scale Enterprises in Korea and Taiwan*. World Bank Staff Working Paper 384. Washington, D.C.

Hugo, Graeme. 1977. "Circular Migration." *Bulletin of Indonesian Economic Studies* 13, no. 3.

India, Government of. 1942. *Report of the Fact Finding Committee on the Textile Industry (Handlooms and Mills)*. Delhi.

———. 1964. *National Sample Survey: Report of the Seventeenth Round, 1961–62*.

———. 1970. Bureau of Costs and Prices. "An Enquiry into Costs and Prices in the Soap Industry." Processed.

———. 1971. *Census of Population, 1971*. New Delhi: Registrar General.

———. 1978. Planning Commission. "The Economic Costs of Alternative Technologies in the Textile Industry." Processed.

———. 1979. *Wages and Productivity in Organized Manufacturing Sector, 1960–61 to 1976–77*. Bulletin 15D/5. New Delhi: Department of Planning.

Indian Cotton Manufacturers' Federation. Annual. *Handbook of Statistics*. Bombay.

Indian Labour Journal. 1981. December.

Ingham, G. K. 1970. *Size of Industrial Organisation and Worker Behavior*. Cambridge: Cambridge University Press.

International Labour Organisation. 1961. *Services for Small Scale Industry*. Geneva.

———. 1974. *Sharing in Development in the Philippines*. Geneva.

Ishikawa, S. 1967. *Economic Development in Asian Perspective*. Tokyo: Kinokinuya.

James, Jeffrey, and Frances Stewart. 1981. "New Products: A Discussion of the Welfare Effects of the Introduction of New Products in Developing Countries." *Oxford Economic Papers*, new series, vol. 33.

Japan Economic Planning Agency. 1959. *Employment Structure and Business Fluctuation*. Economic Research Institute Bulletin 2. Tokyo.

Jardavpur University. 1964. "Survey of Small Engineering Units in Howrah." Bombay: Reserve Bank of India.

Jewkes, J. 1952. "The Size of the Factory." *Economic Journal*, June.

Joshi, Heather, and Vijay Joshi. 1976. *Surplus Labour and the City: A Study of Bombay.* Delhi: Oxford University Press.

Kahnert, F. 1985. "The Small Scale Enterprise Credit Programme under the Second and Third Calcutta Urban Development Projects." World Bank Water Supply and Urban Development Department. Washington, D.C. Processed.

———. N.d. "Case Study: The Small Business Support Component in the Second Urban Project of the World Bank in El Salvador." World Bank Urban Projects Department. Washington, D.C. Processed.

Kaneda, Hiromitsu. 1980. "Development of Small and Medium Enterprises and Policy Response in Japan: An Analytical Survey." Studies in Employment and Rural Development 32. World Bank Development Economics Department. Washington, D.C. Processed.

Kangle R. P. Ed. 1965. *The Kautilya Arthasastra (A Fourth Century B.C. Work on Kingship and Statecraft).* Bombay: University of Bombay.

Kmenta, J. 1967. "On Estimation of the CES Production Function." *International Economic Review*, no. 8, pp. 180–99.

Knight, J., and R. H. Sabot. 1982. "From Migrants to Proletarians: Employment Experience, Mobility, and Wages in Tanzania." *Oxford Bulletin of Economics and Statistics* 44.

Krishnamurty, J. 1973. "Working Force in 1971 Census: Some Exercises on Provisional Results." *Economic and Political Weekly*, special number, August.

Kulkarni, V. B. 1979. *History of the Indian Cotton Textile Industry.* Bombay.

Lakdawala, D. T., and J. C. Sandesara. 1960. "Small Industry in a Big City." Bombay.

Lau, L. J. 1978. "Testing and Imposing Monotonicity, Convexity, and Quasi-Convexity Constraints." In Melvin Fuss and Daniel McFadden, eds., *Production Economics: A Dual Approach to Theory and Applications*, vol. 1. Amsterdam: North-Holland.

Leibenstein, H. 1957. *Economic Backwardness and Economic Growth.* New York: Wiley.

Lester, R. 1967. "Pay Differentials by Size of Establishment." *Industrial Relations* 71.

Levitsky, J. 1983. "Assessment of Bank Small Scale Enterprise Lending." World Bank Industry Department. Washington, D.C. Processed.

Liedholm, Carl, and E. Chuta. 1977. *The Economics of Rural and Urban Small Scale Industries in Sierra Leone.* African Rural Economy Paper 14. East Lansing: Michigan State University.

Lim, Chee Peng, Donald Lee, and Foo Kok Thye. 1981. "The Case for Labor Intensive Industries in Malaysia." In Rashid Amjad, ed., *The Development of Labor Intensive Industry in ASEAN Countries.* Geneva: International Labour Organisation, Asian Regional Team for Employment Promotion.

Little, Ian M. D. 1982a. *Economic Development Theory, Policy, and International Relations.* New York: Basic Books.

———. 1982b. "Indian Industrialization 1857–1947." In M. Gersowitz and others, *The Theory and Experience of Economic Development: Essays in Honour of Sir Arthur Lewis.* London: Allen and Unwin.

Lluch, C., and D. Mazumdar. 1983. "Wages and Employment in Indonesia." World Bank Development Research Department Report 3586-Ind. Washington, D.C. Processed.

Mahalanobis, P. C. 1963. *The Approach of Operational Research to Planning in India.* Written 1955. Bombay: Asia Publishing House.

Malcolmson, James M. 1981. "Unemployment and the Efficiency Wage Hypothesis." *Economic Journal* 91, no. 364 (December).

Manning, Christopher S. 1979. "Wage Differentials and Labor Market Segmentation in Indonesian Manufacturing." Ph.D. thesis, Australian National University.

Marshall, A. 1920. *Principles of Economics,* 8th ed. London: Macmillan.

Martin, J. P., and J. M. Page. 1983. "The Impact of Subsidies on LDC Industry: Theory and Empirical Test." *Review of Economics and Statistics.*

Mazumdar, Dipak. 1958. "Factors in Wage Disparities between Agriculture and Industry in a Developing Country." Ph.D. thesis, Cambridge University.

———. 1973. "Labor Supply in Early Industrialization: The Case of the Bombay Textile Industry." *Economic History Review,* 2d series, vol. 26 (August).

———. 1981. *The Urban Labor Market and Income Distribution: A Study of Malaysia.* New York: Oxford University Press.

———. 1983. "The Role of Small-Scale Enterprises in the Indian Economy." World Bank Development Research Department. Washington, D.C. Processed.

———. 1984a. *The Issue of Small versus Large in the Indian Textile Industry: An Analytical and Historical Survey.* World Bank Staff Working Paper 645. Washington, D.C.

———. 1984b. "The Rural-Urban Wage Gap Migration and the Working of Urban Labor Market: An Interpretation Based on a Study of the Workers of Bombay City." *Indian Economic Review* 18, no. 2.

Mehta, B. V. 1969. "Size and Capital Intensity in Indian Industry." *Oxford Bulletin* 31, no. 3 (August).

Meller, P. 1976. "Efficiency Frontiers for Industrial Establishments of Different Size." *Explorations in Economic Research* 3.

Mohan, Rakesh. 1981. *The Determinants of Labour Earnings in Developing Metropolises.* World Bank Staff Working Paper 498. Washington, D.C.

Mubin, A. K., and David J. C. Forsyth. 1982. *Appropriate Products, Employment and Income Distribution in Bangladesh: A Case Study of the Soap Industry.* International Labour Organisation Technology and Employment Program Working Paper 74. Geneva.

Mukerji, K. 1959. "Trends in Real Wages in Cotton Textile Mills in Bombay City and Island from 1900 to 1951." *Artha Vijnana* 1, no. 1 (March).

Murtz, M. N., and N. H. Stern. 1982. "Price and Tax Policies for Cotton Textiles in India." Delhi: Indian Statistical Institute. Processed.

Nayyar, D. 1976. *India's Exports and Export Policies in the 1960s.* London: Oxford University Press.

Nishimizu, Mieko, and John M. Page, Jr. 1982. "Total Factor Productivity Growth, Technological Progress, and Technical Efficiency Change: Dimensions of Productivity Change in Yugoslavia, 1965–1978." *Economic Journal* 92 (December).

Ohkawa, K., and M. Tajima. 1976. *Small-Medium Scale Manufacturing Industry: A Comparative Study of Japan and Developing Nations*. Tokyo: International Development Center of Japan, March.

Ojha, P. D. 1982. "Finance for Small-Scale Enterprise in India." *Reserve Bank of India Bulletin*, November.

Organisation for Economic Co-operation and Development. 1971. *Problems and Policies Relating to Small and Medium-Sized Business*. Paris.

Padranabhan, M. 1974. "The Sick Mill Problem in the Indian Cotton Textile Industry." Ph.D. thesis, Bombay University.

Page, John M., Jr. 1979. *Small Enterprises in African Development: A Survey*. World Bank Staff Working Paper 363. Washington, D.C.

———. 1980. "Technical Efficiency and Economic Performance: Some Evidence from Ghana." *Oxford Economic Papers* 32 (July).

———. 1984. "Firm Size and Technical Efficiency: Applications of Production Frontiers to Indian Survey Data." *Journal of Development Economics*.

Park, Fun Koo. 1981. "An Analysis of Wage Differentials According to Firm Size." *Korea Development Review*, December.

Park, Yung Chul. 1976. "The Unorganized Financial Sector in Korea 1945–1975." World Bank Development Economics Department Domestic Finance Studies 28. Washington, D.C. Processed.

Pencavel, John H. 1974. "Relative Wages and Trade Unions in the United Kingdom." *Economica* 41 (May).

Pitt, M., and L. F. Lee. 1981. "The Measurement and Sources of Technical Inefficiency in the Indonesian Weaving Industry." *Journal of Development Economics* 9, no. 1.

Poot, H. 1981. "The Development of Labour Intensive Industries in Indonesia." In Rashid Amjad, ed., *The Development of Labor Intensive Industry in ASEAN Countries*. Geneva: International Labour Organisation, Asian Regional Team for Employment Promotion.

Prais, S. J. 1981. *Productivity and Industrial Structure*. Cambridge: Cambridge University Press.

Pryor, F. 1972. "The Size of Production Establishments in Manufacturing." *Economic Journal* 82, no. 326 (June).

Rangarajan, C. 1980. "Innovations in Banking: The Indian Experience, Impact on Deposits and Credit." World Bank Development Economics Department Domestic Finance Studies 63. Washington, D.C. Processed.

"Report of the Royal Commission on Labour." 1892. *Parliamentary Papers* 36.

Reserve Bank of India. 1979. "Survey of Small Industrial Units 1977." Department of Statistics. *Statistical Report*, vols. I and II. Bombay.

Saito, K. A., and D. P. Villanueva. 1980. "Transaction Cost of Credit to the Small-Scale Sector in the Philippines." World Bank Development Economics Department Domestic Finance Studies 53. Washington, D.C.

Sandesara, J. C. 1966. "Scale and Technology in Indian Industry." *Oxford Bulletin of Economics and Statistics*, August.

———. 1969. "Size and Capital Intensity in Indian Industry: Some Comments." *Oxford Bulletin of Economics and Statistics* 31, no. 4 (November).

———. 1981. "Small Industry in India: Evidence and Interpretation." *Indian Planning and Economic Policies*, Gujarat Economic Association, Ahmedabad.

———. 1982. *Efficiency of Incentives for Small Industry.* Bombay: Industrial Development Bank of India.

Schmidt, Peter, and C. A. Knox Lovell, 1980. "Estimating Stochastic Production and Cost Frontiers when Technical and Allocation Efficiency are Correlated." *Journal of Econometrics* 13, no. 1 (May).

Shah, M. N. 1941. "Labor Recruitment and Turnover in the Textile Industry of Bombay." Ph.D. thesis, University of Bombay.

Shapiro, K. H., and J. Muller. 1977. "Sources of Technical Efficiency: The Roles of Modernization and Information." *Economic Development and Cultural Change* 25 (January).

Shetty, S. L. 1982. "Industrial Growth and Structure." *Economic and Political Weekly*, October 2 and 9.

Smith, Norman R. 1967. *The Entrepreneur and His Firm: The Relationship between Type of Man and Type of Company.* East Lansing: Michigan State University Press.

Staley, Eugene, and Richard Morse. 1965. *Modern Small Industry for Developing Countries.* New York: McGraw-Hill.

Statistical Package for Social Sciences, 2d ed. 1975. New York: McGraw-Hill.

Steel, W. 1977. *Small-Scale Employment and Production in Developing Countries: Evidence from Ghana.* New York: Praeger.

Stiglitz, J. E. 1974. "Alternative Theories of Wage Determination and Unemployment in LDCs: The Labor Turnover Model." *Quarterly Journal of Economics* 88, no. 2 (May).

Stiglitz, J. E., and A. E. Weiss. 1981. "Credit Rationing in Markets with Imperfect Information." *American Economic Review* 71, no. 3 (June).

Storey, D. J. 1982. *Entrepreneurship and the New Firm.* London: Croom Helm.

Subramanian, K. N. 1979. *Wages in India.* New Delhi: McGraw-Hill.

Suri, K. B. 1987. "Technology, Firm Size, and Product Quality: A Study of Laundry Soap in India." In K. B. Suri, ed., *Small-Scale Enterprises in India.* New Delhi: Sage.

Tambunlertchai, Somsak, and Chesada Loohawenchit. 1981. "Labor Intensive and Small-Scale Manufacturing in Thailand." In Rashid Amjad, ed., *The Development of Labor Intensive Industry in ASEAN Countries.* Geneva: International Labour Organisation, Asian Regional Team for Employment Promotion.

Timberg, T. A. 1978. "Report of the Survey of Small-Scale Industrial Units in Bombay." World Bank Development Economics Department. Washington, D.C. Processed.

Timberg, T. A., and C. V. Aiyar. 1980. "Informal Credit Markets in India." World Bank Development Economics Department Domestic Finance Studies 62. Washington, D.C. Processed.

Thingalaya, N. K. 1978. "Innovations in Banking: The Syndicate's Experience." World Bank Development Economics Department Domestic Finance Studies 46. Washington, D.C. Processed.

Tulsi, S. K. 1980. *Incentives for Small Scale Industries: An Evaluation*. Delhi: Kunj.

Tyler, W. E. 1979. "Technical Efficiency in Production in a Developing Country: An Empirical Examination of the Brazilian Plastics and Steel Industries." *Oxford Economic Papers* 31, no. 3 (November).

UNIDO (United Nations Industrial Development Organization). 1978. "Alternative Technologies in the Indian Textile Industry." Paper presented at the Conference on Appropriate Technology, Delhi.

U.S. Bureau of the Census. 1977. *Census of Manufactures*. Washington, D.C.: U.S. Government Printing Office.

U.S. Small Business Administration. 1984. *The State of Small Business: A Report to the President*. Washington, D.C.: Government Printing Office. March.

Virmani, Arvind. 1982. *The Nature of Credit Markets in Developing Countries: A Framework for Policy Analysis*. World Bank Staff Working Paper 524. Washington, D.C.

White, L. J. 1978. "The Evidence on Appropriate Factor Proportions for Manufacturing in Less Developed Countries: A Survey." *Economic Development and Cultural Change* 27, no. 1 (October).

———. 1981. *Measuring the Importance of Small Business in the American Economy*. New York: Salomon Brothers Center for the Study of Financial Institutions and New York University Graduate School of Business Administration.

Wilkinson, R. 1973. "Earnings and Size of Plant in the Iron and Steel Industry.' Cambridge University Department of Applied Economics. Processed.

Wilson, M. 1981. *Some Problems in Operating a Loan Program for Craft and Emerging Small-Scale Non-Farm Enterprises in Jamaica*. Michigan State University Working Paper 15. East Lansing.

Wolf, Martin. 1982. *India's Exports*. New York: Oxford University Press.

Wood, Adrian. 1978. *A Theory of Pay*. Cambridge: Cambridge University Press.

World Bank. 1979. *World Report*, vol. 1, no. 2490-IND. Processed.

———. 1984. *Abstracts of Current Studies*. Washington, D.C.

Zachariah, K. C. 1968. *Migrants in Greater Bombay*. Bombay: Asia Publishing House.

Index

The most recent World Bank publications are described in the catalog *New Publications,* which is issued in the spring and fall of each year. The complete backlist of publications is shown in the annual *Index of Publications,* which is of value principally to libraries and institutional purchasers. The latest edition of each is available free of charge from Publications Sales Unit, The World Bank, 1818 H Street, N.W., Washington, D.C. 20433, U.S.A., or from Publications, The World Bank, 66, avenue d'Iéna, 75116 Paris, France.